VALUES OF t

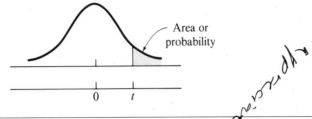

Area or probability

0   t

Degrees of Freedom (df)	Right Tail Areas				
	.10	.05	.025	.01	.005
1	3.078	6.314	12.706	31.821	63.657
2	1.886	2.920	4.303	6.965	9.925
3	1.638	2.353	3.182	4.541	5.841
4	1.533	2.132	2.776	3.747	4.604
5	1.476	2.015	2.571	3.365	4.032
6	1.440	1.943	2.447	3.143	3.707
7	1.415	1.895	2.365	2.998	3.499
8	1.397	1.860	2.306	2.896	3.355
9	1.383	1.833	2.262	2.821	3.250
10	1.372	1.812	2.228	2.764	3.169
11	1.363	1.796	2.201	2.718	3.106
12	1.356	1.782	2.179	2.681	3.055
13	1.350	1.771	2.160	2.650	3.012
14	1.345	1.761	2.145	2.624	2.977
15	1.341	1.753	2.131	2.602	2.947
16	1.337	1.746	2.120	2.583	2.921
17	1.333	1.740	2.110	2.567	2.898
18	1.330	1.734	2.101	2.552	2.878
19	1.328	1.729	2.093	2.539	2.861
20	1.325	1.725	2.086	2.528	2.845
21	1.323	1.721	2.080	2.518	2.831
22	1.321	1.717	2.074	2.508	2.819
23	1.319	1.714	2.069	2.500	2.807
24	1.318	1.711	2.064	2.492	2.797
25	1.316	1.708	2.060	2.485	2.787
26	1.315	1.706	2.056	2.479	2.779
27	1.314	1.703	2.052	2.473	2.771
28	1.313	1.701	2.048	2.467	2.763
29	1.311	1.699	2.045	2.462	2.756
30	1.310	1.697	2.042	2.457	2.750
40	1.303	1.684	2.021	2.423	2.704
60	1.296	1.671	2.000	2.390	2.660
120	1.289	1.658	1.980	2.358	2.617
∞	1.282	1.645	1.960	2.326	2.576

Source: From M. Merrington, "Table of Percentage Points of the *t*-Distribution," *Biometrika*, 1941, *32*, 300. Reproduced by permission of the *Biometrika* Trustees.
Example: The *t*-value for df − 15 and a right-hand tail area of .05 is 1.753.

BASIC
BUSINESS
STATISTICS

BASIC BUSINESS STATISTICS

E. J. FREED
University of Portland

Harcourt Brace Jovanovich, Publishers
San Diego   New York   Chicago   Austin
Washington, D.C.   London   Sydney
Tokyo   Toronto

ISBN: 0-15-504900-3
Library of Congress Catalog Card Number: 90-82260
Printed in the United States of America

To Sally, Amy, and Dan

PREFACE

Basic Business Statistics is intended to provide the basis for a single-semester first course in statistics. It can also serve as an easy-to-read review for students who might want to reacquaint themselves with elementary statistics in preparation for a class requiring the application of basic statistical principles. The informal style and commonsense explanations that characterize this textbook should help to reduce the anxiety that frequently marks a student's first exposure to the field. The intent here is to provide a book that supports rather than threatens, one that will offer students who possess modest-to-good mathematical skills a resource that is interesting, intelligible, and perhaps even a little entertaining.

Unlike many competing books, which are often no more than abridged versions of textbooks intended for use in a two-semester course, *Basic Business Statistics* was conceived and designed specifically for a one-semester course in business statistics. The emphasis is on simple illustrations rather than on formal proofs and theorems. Potentially confounding mathematical notation is kept to a minimum. And a serious effort has been made to help students distinguish between core issues and those of secondary importance, something absolutely critical in promoting student confidence and comprehension. Particular attention has been paid to the development of a clear, comprehensible introduction to *confidence interval estimation* and *statistical hypothesis testing*—two crucial areas of study in beginning statistics. The nature and role of sampling distributions are carefully established and continually reinforced.

Several features have been incorporated into *Basic Business Statistics* that are specifically designed to aid students in their attempts to master statistical concepts and procedures.

- *In-Chapter Exercises* are included in each chapter to provide students with an immediate opportunity to apply the concepts covered in the chapter.
- *Problem Sets* have been designed to augment and extend chapter discussions, rather than simply to provide opportunities for drill and repetition. The end-of-chapter exercises cover a broad range of applications that are both challenging and interesting.

- *Chapter Supplement Sections* are provided for the chapters covering Descriptive Statistics (Chapter 2), Probability Theory (Chapter 3), Statistical Inference (Chapter 5), and Regression Analysis (Chapter 9). These optional sections are included for instructors who wish to cover these topics in greater depth.

- *Comprehensive Solutions*—not just answers—are provided in the book for nearly half of the end-of-chapter problems and for all the in-chapter exercises.

- *Chapter Objectives* give students a clear understanding of what they can expect to encounter in the material that follows.

- *Chapter Summaries* reinforce and clarify the chapter material that has just been covered.

- *Key Terms* for discussion and retention are listed at the end of the chapter in which they first appear.

Basic Business Statistics is not intended to be all things to all people. What it does offer is an effective alternative to the sometimes-imposing tomes that are poorly suited to a one-semester format. Extensive class testing of the manuscript at the University of Portland has been extremely useful in the attempt to achieve this and other ends.

Ancillaries

A *Solutions Manual* containing completely worked-out solutions to all in-text exercises and end-of-chapter problems is available.

A *Testbook* containing true/false, multiple-choice, and other problems is available.

Acknowledgments

Professor William Maurer (Naval Postgraduate School) and Professor Richard Spinetto (University of Colorado), by their example, provided me with a model for effective teaching. Professor George Chou (University of Portland) provided support and encouragement throughout the writing of the book. Professor James Jurinski (University of Portland) was helpful in finding a suitable publisher. And Theresa Vu, Linnea Aguiar, and Dan Freed provided valuable assistance in preparing problem solutions and in checking computational accuracy. My thanks to all of them.

I also thank Mickey Cox and Scott Isenberg of Harcourt Brace Jovanovich for their efforts in transforming a rough outline into a finished product. Finally, I thank the reviewers for their many constructive comments and criticisms: Dan Albrechtson, Milwaukee Area Technical College; Les Dlaby, Lake Forest College; Darrell Christie, University of Wisconsin–Stevens Point; Gerald Evans, University of Montana; W. F. Mackara, East Tennessee State University; and Norman Rittgers, Pasadena City College.

CONTENTS

8 HYPOTHESIS TESTING FOR PROPORTIONS, MEAN DIFFERENCES, AND PROPORTION DIFFERENCES   235

9 REGRESSION ANALYSIS   272

STATISTICAL TABLES   317

SOLUTIONS TO STARRED EXERCISES   331

INDEX   393

AN INTRODUCTION TO STATISTICS

CHAPTER OBJECTIVES

Chapter 1 should enable you to:

1. Define the term *statistics*.
2. Identify the two main branches of statistics: descriptive statistics and inferential statistics.
3. Appreciate the role of probability theory in statistical analysis.
4. Understand the nature of data and the concept of levels of measurement.
5. Anticipate the topics to be discussed later in the text.

1.1 STATISTICS DEFINED

In common usage, the term *statistics* can take on a variety of meanings. For example, it's frequently used to describe numerical data of almost any sort—height, weight, batting average, GPA, age, flash point, and the like. Some people might connect the term to the results of surveys, polls, and questionnaires. We'll use *statistics* in this text primarily to designate a specific academic discipline focused on methods of data collection, analysis, and presentation. In nearly every case,

> **Statistics** involves the transformation of *data* into *information*.

That is, statistical analysis—in whatever form—generally entails translating raw numbers into meaningful information in order to establish the basis for sound decision making.

1.2 BRANCHES OF STATISTICS

It's possible to divide the discipline of statistics into two principal branches: descriptive statistics and inferential statistics.

Descriptive Statistics

Descriptive statistics focuses directly on the summarization and presentation of data. Here, we're typically looking for effective ways to refine, distill, and describe data so that, as we've discussed above, the numbers involved become more than simply numbers. While descriptive statistics can potentially involve some rather complex measures, we will, at the introductory level, deal in measures not much more complicated than simple averages.

Inferential Statistics

When dealing with especially large data sets, it's often necessary, for reasons of time, cost, or convenience, to draw conclusions (or to make *inferences*) about an entire data set by using only a subset of the values involved. **Inferential statistics** (sometimes called **inference statistics** or **inductive statistics**) deals with the selection and use of *sample* data to produce information about the larger *population* from which the sample was selected.

Note: The term **statistic** is sometimes used to refer specifically to characteristics of a sample. For example, the average value in a sample of 10 measurements could be considered a sample "statistic." In contrast, population characteristics are frequently referred to as **parameters**. In inferential statistics, we'll routinely use sampling theory to relate sample "statistics" to population "parameters."

1.3   PROBABILITY

Probability theory—the study of likelihood or chance—can be seen as the link between descriptive and inferential statistics. A knowledge of basic probability will enable us to connect what we see in a sample to what we would *likely* see in the population being represented by the sample. Figure 1.1 details this simple classification scheme.

FIGURE 1.1   **BRANCHES OF STATISTICS**

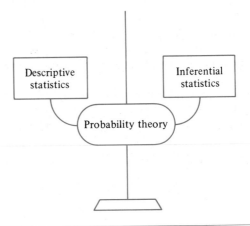

1.4   A BROADER PERSPECTIVE

To put statistical analysis into a somewhat broader context, we can identify two distinct analytic views of the world—two sets of beliefs about "how things are." In a *deterministic* view of the world, the world we see is essentially a world of certainty. It's a world that consists primarily (if not exclusively) of one-to-one, cause-and-effect relationships: given some cause, there's a unique and identifiable effect; given some effect, there's a clearly discernible cause. The world here, while not necessarily simple, is at least well defined.

In contrast, under a *probabilistic* view of the world, things don't appear quite as straightforward. Here, given some cause, there may well be a number of possible effects; given some effect, there may be a number of possible causes. In this conception of reality, the world is dominated by a real, and complicating, *un*certainty.

It turns out that for each of these two contrasting world views, an appropriate set of analytic tools is available. The tools of probability and statistics (especially inferential statistics) are designed to deal with problems arising in a distinctly uncertain world. (We can, in fact, think of probability theory as the means to measure—and thus control for—uncertainty.) In contrast, optimization tools such as linear programming and differential calculus are better suited to deal with problems encountered in a "certain" (i.e., deterministic) world.

1.5   THE STRUCTURE OF THE TEXT

The chapters that follow are intended to introduce the student or practitioner to the principles of basic statistical analysis without the rigid formality of many of the books written in the area. In general, we've tried to use instinctive, intuitive arguments to replace formal proofs and theorems. For the most part, key concepts and procedures are developed around simple illustrations. Excessive mathematical notation is avoided. At the same time, we've tried not to exclude complex issues from the discussion.

Importantly, substantial effort has been made to separate "wheat" from "chaff"—key and fundamental issues versus issues of only secondary or supporting significance. (In my own view, failure to differentiate clearly for the beginning student those ideas that are absolutely fundamental from those that are peripheral to the central theme is the main deficiency of most introductory texts in the field.) We've attempted to keep each chapter brief and to the point. Problem sets have been designed to augment and extend chapter discussions rather than solely to provide opportunities for drill and repetition. Solutions to nearly half of the end-of-chapter exercises are provided in the back of the text.

For a class proceeding at a moderate pace, the material presented fits readily into a single-semester, introductory-level course. What follows is a capsule summary of the remaining chapters.

Chapter 2: Descriptive Statistics

Chapter 2 introduces basic descriptive measures useful in converting data into information. Included are measures of center (or central tendency)—the mean, the median, and the mode, as well as measures of dispersion—the range, the mean absolute deviation, the variance, and the standard deviation. The chapter supplement adds a measure of relative dispersion (the coefficient of variation) and two measures of position—percentiles and quartiles. Data presentation possibilities are also described, with a primary focus on simple bar charts (histograms).

Chapter 3: Probability Theory

Chapter 3 outlines a number of basic probability concepts: simple, joint, and conditional probabilities; statistically independent and mutually exclusive events;

and the two principal rules of basic probability—the additive and the multiplicative rules. Visual formats—probability trees, Venn diagrams, and joint probability tables—are shown to be useful in organizing and solving a variety of problems.

Chapter 4: Probability Distributions

Probability distributions, which involve the comprehensive listing of probabilities assigned to a full range of possible events, are viewed as a logical extension of Chapter 3 principles. Special-purpose distributions (the binomial, the normal, and the Poisson), which can be shown as compact mathematical functions, are introduced in order to efficiently produce probabilities under certain clearly defined conditions. The chapter supplement adds the exponential distribution to the special-purpose list.

Chapter 5: Statistical Inference: Estimating a Population Mean

Chapter 5 initiates a discussion of basic sampling procedures. Here statistical theory is introduced in order to link sample results to the population from which the sample was selected. The focus is on the construction of **confidence interval estimates** of a population mean (or average). (For example, estimating from a *sample* of items in inventory the average age of *all* the items in inventory, or estimating from a *sample* of components produced by a certain firm the average useful life of *all* the components produced by that firm.)

Chapter 6: Interval Estimation for Proportions, Mean Differences, and Proportion Differences

Here the methods of Chapter 5 are extended to address three additional estimation cases: building confidence interval estimates of a *population proportion* (e.g., the proportion of items in a population that are large, or over 30, or flawed, etc.); building confidence interval estimates of the *difference between the means* of two populations (e.g., the difference between the average recovery time for a test group of patients and the average recovery time for a control group of patients); and building confidence interval estimates of the *difference between two population proportions* (e.g., the difference between the proportion of Democrats who support a temporary budget freeze and the proportion of Republicans who favor such a freeze).

Chapter 7: Statistical Hypothesis Testing: Hypothesis Tests for a Population Mean

Chapter 7 introduces the second side of statistical inference, statistical **hypothesis testing**. In hypothesis testing, a statement is first made about a population of values. (For example, "The average life of all tires produced by company XYZ is 40,000 miles," or, "The average years of experience for all employees of the firm is 6.5 years.") A sample is then selected to help decide whether or not the statement is true (or "believable"). The trick, as we'll see, is to balance the chances of reaching wrong conclusions: the chance of believing that the statement is true when it's actually false versus the chance of believing the statement is false when it's actually true.

Chapter 8: Hypothesis Testing for Proportions, Mean Differences, and Proportion Differences

Here hypothesis testing principles are applied to the three cases listed in the chapter title.

Chapter 9: Regression Analysis

Chapter 9 introduces regression analysis, a statistical procedure commonly used to explore the relationship that may exist between certain key variables in a broad range of business and nonbusiness situations. For example, regression analysis might be used to investigate the possible relationship between the amount a firm spends on advertising and the firm's annual sales revenue; or the possible relationship between the number of inspections performed on units assembled by a company's production people and the number of units returned by customers as unsatisfactory; or the possible relationship between a baseball team's average player salary and its win-loss record. The application possibilities are almost unlimited.

1.6   THE NATURE OF DATA: LEVELS OF MEASUREMENT

One final note. Whenever data is collected, whether from historical records, casual observation, or controlled experimentation, the process of measurement—assigning proper values to observed phenomena—is involved. And whenever the process of measurement is involved, the issue of **levels of measurement** presents itself. Since most statistical techniques are suited to data measured only at certain levels, any user of statistical analysis should be aware of the level of measurement involved in his or her data. We can, in fact, describe four distinct measurement levels.

Nominal Data

Nominal data represents the lowest level of measurement. With nominal data, each value serves strictly as a label or a name. For example, a country of origin data set might include the possible "values" France (which might be designated country 1), the United States (country 2), Japan (country 3), and so on. With this sort of data, there's no natural ordering of values; the "value" of France (1) is no larger or smaller than the "value" of Japan (3). Each value is just an identifier. 2 . F

Ordinal Data

Ordinal data represents a step up on the measurement scale. Here, in contrast to the nominal data case, values can be meaningfully rank-ordered. For example, we might create an ordinal data set by asking an audience to rank-order five previewed television shows. Or we might want to rank all-time great baseball players. In most such cases, a rank (or value) of 1 would be considered "higher" than (or superior to) a rank of 2, a rank of 2 would be considered higher than a rank of 3, and so on. It's important to note, however, that with ordinal data, even though rank-ordering is possible, measuring the precise distance between successive ranks is normally difficult or impossible. (A rank of 3 is higher than a rank of 4, but HOW MUCH higher?) Furthermore, there's no reason to believe that the distance between a number 1 ranking and a number 2 ranking is the same as the distance between a number 3 and a number 4 ranking. Nor is it clear that a rank of 4 is twice as "bad" as a rank of 2.

Interval Data

Interval data allows not only a rank-ordering of values, but also shows standardized, well-defined distances between successive values on the measurement scale. Temperatures measured on the Fahrenheit scale are frequently cited as an example of interval data. A 40° day is exactly 10° warmer than a 30° day, the same distance that separates a 70° day from a 60° day. (However, it would be incorrect to say that a 60° day is precisely twice as warm as a 30° day. This point brings us to the next category.)

Ratio Data

At the highest level of measurement, **ratio** data has all the properties of interval data, plus a so-called natural "zero" point, allowing for ratio comparisons. For example, a data set consisting of the heights of family members would show data measured on a ratio scale. We could say Bob (72 inches) is twice as tall as Jim (36 inches) but only two-thirds as tall as Peter (108 inches). Not only are there measurable and meaningful differences between the heights, but ratio comparisons also have meaning. (The natural "zero" point on the height scale indicates a complete absence of height. It's this fact that allows for effective ratio comparisons. Compare this to the case of Fahrenheit temperatures, where zero degrees clearly doesn't indicate a complete absence of heat.)

As a general rule, any statistical procedure that is appropriate to data at one level of measurement will be appropriate at any higher level of measurement, but not necessarily at lower levels. For example, computing the median value to describe the center of a data set—something we'll discuss in Chapter 2—makes sense for data at the ordinal level of measurement and above (i.e., for interval and ratio data), but not for nominal data. The inferential procedures described in Chapters 5 through 9 require at least interval data.

██████████ **CHAPTER SUMMARY**

Chapter 1 provided an introduction to the study of probability and statistics. Statistics was seen generally as the means to convert data into information.

Descriptive statistics offers ways of summarizing, describing, and presenting data effectively. Inferential statistics (sampling theory) provides the means to draw conclusions (inferences) about large data sets (populations) based on the information acquired from smaller, representative subsets (samples). Probability theory, the study of risk or uncertainty, was seen as a link between descriptive and inferential statistics.

To put statistical methods in a broader perspective, we discussed two analytic views of the world and asserted that the tools of probability and sta-

tistics were suited to a probabilistic, rather than a deterministic, world view.

Finally, the levels of measurement involved in the collection and analysis of data were discussed, since various statistical procedures may require specific levels of measurement. Nominal (name), ordinal (rank), interval (showing standardized distances between values), and ratio (with a natural zero point) data were described and differentiated.

KEY TERMS

Descriptive statistics
Hypothesis testing
Inferential (or inductive) statistics
Interval estimation
Levels of measurement
   Nominal
   Ordinal

   Interval
   Ratio
Parameter
Probability theory
Statistic
Statistics

DESCRIPTIVE STATISTICS

CHAPTER OBJECTIVES

Chapter 2 should enable you to:

1. Effectively summarize and present numerical data.

2. Compute and interpret measures of central location (mean, median, and mode).

3. Describe dispersion in data with a variety of measures (range, mean absolute deviation, variance, and standard deviation).

4. Use frequency and relative frequency tables to display data.

5. Create histograms to visually summarize data.

6. Deal with situations in which grouped data is presented.

7. Recognize and apply supplementary descriptive measures such as the coefficient of variation, the geometric mean, weighted means, and percentiles.

As mentioned in Chapter 1, descriptive statistics involves finding ways to summarize, condense, *describe*, and represent data effectively—in sum, doing what's necessary to transform *data* into *information*. With the almost unlimited access to huge data banks that modern computer technology provides, it would seem critical that we have a basic capacity to summarize and distill data, if only to avoid being completely overwhelmed by an avalanche of numbers. And while data summary and description are not necessarily the most exciting aspects of statistical analysis, they do provide a necessary foundation for our discussion of statistical inference later in the text.

In this chapter, we'll consider some of the most common descriptive measures, beginning with basic measures of central tendency (averages) and dispersion (range, mean deviation, variance and standard deviation). In the Chapter Supplement, we'll discuss three or four additional measures including the coefficient of variation and the geometric mean.

2.1 MEASURES OF CENTRAL LOCATION OR CENTRAL TENDENCY

Consider the following "raw" data set,

$$1, 3, 2, 2, 5, 4, 3, 3, 4, 3$$

To give the data some meaning, assume that what we have here is actually a list of responses to a question put to 10 students before the start of their most recent statistics class: "How many previous statistics classes have you taken?" The first student responded "one," the second "three," the third "two," and so forth. (You can see that one poor soul answered "five.")

The Arithmetic Mean

Suppose you were asked to produce a single value to summarize all 10 responses—one number to characterize the entire data set. What number would you select? Most of us, I think, would respond pretty quickly. We'd simply add the 10 values together and divide by 10 to produce an *average*. (Here, $30 \div 10 = 3$.) This simple average, which might more accurately be labeled the **mean** (or, even more accurately, the **arithmetic mean**), is easily the most common of all descriptive measures. We tend to use the mean almost instinctively to identify, for summary purposes, the "typical" value in any given data set. In its way, it identifies a center point in the data and uses that center point to represent all of the values.

To formalize things, if we use x_i (read "x-sub-i") to designate each of the numbers in our 10-student data set—with i serving here as an identifying subscript ranging in value from 1 to 10—we can show the *mean* computation as

$$\frac{\sum_{i=1}^{10} x_i}{10} = \frac{x_1 + x_2 + \cdots + x_{10}}{10} = \frac{1 + 2 + 2 + \cdots + 5}{10}$$

$$= \frac{30}{10} = 3$$

(The uppercase Greek letter Σ (sigma) shown here is frequently used to represent the summation operation. The full numerator of the mean expression can be read as the "sum of the x-sub-i values as i takes on values from 1 to 10.")

Using the common $\bar{x}$ (read "x-bar") notation for the arithmetic mean, and generalizing to a data set involving a total of n values, we can show the mean computation as

$$\bar{x} = \frac{\sum_{i=1}^{n} x_i}{n}$$

Note: For brevity, we'll often show the $\bar{x}$ expression without explicitly identifying the i subscript or the $i = 1$ to n limits. That is, we'll show

$$\bar{x} = \frac{\sum x}{n}$$

In our student response example, we can now succinctly report results: We spoke with 10 students and discovered that, *on average*, these students had enrolled in three previous statistics classes. (It should be mentioned that the mean need *not* be one of the actual values in the original data set. For example, the mean of the numbers 4 and 8 is 6.)

The Median

The arithmetic mean isn't the only central tendency measure that might be used to represent the "typical" value in a data set. At least two other possibilities are frequently cited.

The **median** offers a slightly different, and sometimes superior, measure of center. By definition, 3.7

> the median value marks the data set midpoint by establishing a value such that (at least) half the numbers in the data set are at or above the designated point, and (at least) half the numbers are at or below.

More concisely, the median is simply the middle value in an ordered list of the data. For our data set of 10 student responses, this means that by rearranging the list to produce the ordered array

$$1, 2, 2, 3, 3, 3, 3, 4, 4, 5$$

$$\uparrow$$
$$\text{MED}$$

we can establish the median at 3. (Notice here that the data set consists of an even number of values. Counting left to right to the halfway point actually puts us between a pair of center values, in this case between the second and the third 3, or between the fifth and sixth values overall. The median is consequently set midway between the second 3 and the third 3. Had the two middle values in our data set been 3 and 4, the median would have been 3.5. As in the case of the mean, the median value need not be one of the values in the original data set.)

> **Note:** Putting things a bit more formally, we might describe the median as
>
> $$\text{MED} = x_{(n+1)/2}$$
>
> For example, if $n = 15$, then the median would be $x_{(15+1)/2} = x_8$, the eighth entry in the ordered list. If $n = 10$, the median would be $x_{(10+1)/2} = x_{5.5}$, the 5.5th entry—halfway between entry 5 and entry 6.

The Mode

As an alternative to the mean or the median, the **mode** (or modal value) can also be used to represent a "typical" data set member. The mode is simply the *most frequently occurring* value. (This would mean that for our 10-student data set, the mode—or modal response—was 3.)

It should be pointed out that not all data sets have a mode. For example, the data set 1, 2, 3 has no mode. On the other hand, some data sets may have more than one mode: 1, 1, 3, 4, 3, 5 is actually *bimodal*, that is, it has two modes (1 and 3).

Summing Up

We can now report results from the student survey by citing any one of three measures of central location: the mean (the "average" response was three previous classes); the median (at least half the responses were three classes or less, at least half the responses were three classes or more); or the mode (the most frequent response was three classes).

> **Note:** For this particular data set, it doesn't seem to matter much which of the three measures we choose. No matter how we slice it, 3 seems to be the magic number. This won't always be the case, however. As we'll see later in the chapter, there will be occasions when one measure may be preferred over another.

2.2   MEASURES OF DISPERSION

In summarizing data, it's frequently useful, and sometimes critical, to augment any measure of central tendency with a corresponding measure of data *dispersion* (or scatter, or spread, or variability). In data description, simply reporting the mean (or the median or the mode) provides only part of the overall picture. For example, suppose I tell you that I have two numbers in mind, and that the mean of the two numbers is 100. Without some measure of dispersion—some indicator of how the numbers are spread out around the mean—you really don't have much of a clue as to what the two numbers might be. They might both be 100, or one could be 0 and the other 200, or -1004 and . . . , well, you see the point. Something is missing from our summary description. (The old story comes to mind of the man who was asked to put one foot in a bucket of boiling water and one foot in a bucket of ice. He was assured that on average he would be perfectly comfortable. Without some measure of dispersion or variability, however, he couldn't fully evaluate his situation.)

The Range

Back to our two-numbers example. Suppose now I tell you that for the two numbers I have in mind (the two with a mean of 100), the distance between the smaller value and the larger value is 10. Combining this added piece of information with what you already know about the mean should enable you to identify the full data set almost immediately—95 and 105. By reporting the **range** of the data—the difference between the smallest value and the largest value in the data set—we've painted a much clearer picture of the values involved.

For our student survey data, we can now report a range of $5 - 1 = 4$ courses to usefully support any of the measures of central tendency that we had earlier computed.

Unfortunately, although the range is obviously a simple measure to describe and compute, its capacity to measure data dispersion effectively is somewhat limited. The problem is that only two values in the data set (the smallest and the largest) are actively involved in the calculation. None of the values in between has any influence at all. (Consider the data set 3, 7, 5, 2, 4, 5, 1000. The range (998) gives a rather misleading view of the degree of dispersion involved here, since all the values but one are clustered fairly closely together.) The measures described next are intended to correct for this apparent deficiency.

Mean Absolute Deviation

In contrast to the range, the *mean deviation* (or the *mean absolute deviation*) provides a much more comprehensive measure of dispersion. Here, *every* value in the data set—not just the two extremes—plays an influencing role. Specifically, the mean deviation measures the average distance (or deviation) of all the values in the data set from the data set mean. To compute it, we'll need to calculate the distance of each value from the mean (i.e., the difference between each value and the mean), sum the distances, and determine the "average" distance by dividing the total distance by the number of values involved.

For the data in our 10-student survey, this would seem to translate easily to

$$\frac{(1-3)+(2-3)+(2-3)+(3-3)+(3-3)+(3-3)+(3-3)+(4-3)+(4-3)+(5-3)}{10}$$

$$= \frac{(-2)+(-1)+(-1)+0+0+0+0+1+1+2}{10} = \frac{0}{10} = 0$$

Can you see the problem? Our computation has produced a pretty strange result. It suggests that the average distance of the 10 data points from the center (i.e., from the mean) is 0. What's happened here is that the positive distances (deviations) in the numerator have canceled the negative distances to create a net 0. In fact, if we were to follow this same procedure for any data set, the results would always be identical. The basic nature of the arithmetic mean (it serves as a kind of "balance point" for the data) guarantees that the sum of deviations computed around the mean will always be 0.

This being the case, it appears that the mean deviation provides a generally ineffective (read "useless") measure of data dispersion. Fortunately, however, a small adjustment in our procedure will quickly resolve the problem. If we simply insert into the "sum of distances" numerator the *absolute value* operator (usually shown as two vertical bars around the quantity involved) to indicate that it's the *magnitude* of the distances that's important, not the *sign* (which indicates direction), the cancellation-to-0 problem disappears. The positive distances stay positive, the negative distances become positive, and the result is a far more useful indicator of data dispersion.

To emphasize the adjustment, we'll call this modified measure the **mean absolute deviation** (MAD) and describe it simply as the average *absolute* distance of data points from the center (i.e., from the mean) of the data set. Symbolically,

$$\text{MAD} = \frac{\sum |x_i - \bar{x}|}{n}$$

where $\bar{x}$ = the mean
n = number of data points
x_i = individual x values

For our example,

$$\text{MAD} = \frac{2 + 1 + 1 + 0 + 0 + 0 + 0 + 0 + 1 + 1 + 2}{10} = \frac{8}{10} = .8$$

What we have here, then, is a set of 10 values with a mean of 3 and a mean absolute deviation of .8, indicating that, on average, responses were .8 units (here, courses) from the mean. Had we reported a set of 10 values with a mean of 3 and a MAD of, say, 106, our image of the data would obviously be very different.

Variance

Although the MAD appears to be a straightforward, easily interpreted value, it's not, it turns out, the most frequently used measure of data dispersion in statistical analysis. Much more common are the variance and the standard deviation—two very closely related descriptors of dispersion that possess more desirable statistical properties.

The calculation of **variance** begins much as the MAD calculation began—by focusing on the distance of each data point from the mean. Here, however, to avoid the cancellation-to-0 problem, the variance computation involves a *squaring* of individual data point distances (rather than applying the absolute value operator). The *squared* deviations are summed and then averaged. Symbolically,

$$\text{VAR}(x) = \frac{\sum(x_i - \bar{x})^2}{n}$$

where $\bar{x}$ = the mean
n = number of data points
x_i = individual x values

Applied to our student response example,

$$\text{VAR}(x) = \frac{(1 - 3)^2 + (2 - 3)^2 + (2 - 3)^2 + \cdots + (4 - 3)^2 + (5 - 3)^2}{10}$$

$$= \frac{-2^2 + -1^2 + -1^2 + \cdots + 1^2 + 2^2}{10}$$

$$= \frac{12}{10} = 1.2$$

We have, then, a set of 10 responses in which the mean response is 3 courses, and the variance—the average squared distance of responses from the mean—is 1.2 *courses squared*.

> **Note:** If it seems to you that our description of variance is a little awkward, and that the "courses squared" measurement is difficult to interpret, you're not alone. The range and MAD are clearly easier dispersion measures to compute and describe. Yet variance is a more frequently used statistical measure. This isn't (solely) because statisticians are unusually perverse and only interested in making simple things difficult. Rather, the variance measure of data dispersion possesses, as we've mentioned, useful mathematical properties not shown by the others.
>
> For example, the arithmetic mean, our primary measure of central tendency, can be described as the *variance-minimizing* value for any data set. That is, if we were to use any value other than the mean as the point around which we computed the average squared distance, the result would be larger than the result produced by choosing, as we did, the arithmetic mean as the center value.
>
> You might test this proposition by making another variance-type computation for the student response data we've been using. Rather than computing the squared distances around the mean value 3, compute the squared distances around some other value, say 2. Average the sum of the squared distances produced (i.e., divide by 10), then compare your result to the 1.2 variance that we've already calculated.

Standard Deviation

Largely because the squared units produced in the variance computation are so unwieldy, **standard deviation**—the square root of the variance—is often used instead of the variance to report data dispersion. In general,

$$\text{STD DEV}(x) = \sqrt{\text{VAR}(x)}$$

$$= \sqrt{\frac{\sum(x_i - \bar{x})^2}{n}}$$

In our student response example,

$$\text{STD DEV}(x) = \sqrt{\text{VAR}(x)} = \sqrt{1.2} = 1.1 \text{ courses}$$

Taking the square root of the "courses squared" units produced in the variance computation gets us back to the more natural "courses" units of the original data. We can thus report our set of 10 responses as having a mean of 3 courses and a standard deviation of 1.1 courses. By changing from variance to standard deviation, we can measure both center and dispersion in the same basic units.

> **Note:** In various settings, the symbols used to represent the variance and standard deviation formulas may change, but the essential nature of the computations remains. For example, we'll later draw a distinction between the variance and standard deviation of a population, and the variance and standard deviation of a sample. For now, though, we'll stick with our generic variance and standard deviation expressions.

EXERCISE 1   For the following set of values, compute the mean, median, mode, range, MAD, variance, and standard deviation.

$$10, 30, 15, 35, 30, 40, 45, 10, 30, 5$$

Ans: mean = 25   med = 30   mode = 30

range = 40   MAD = 12   VAR = 175

STD DEV = 13.23

(For a more detailed solution, see the end of the chapter.)

2.3   DATA PRESENTATION

To this point, we've examined ways of representing a full data set with one or two simple summary measures (for example, a mean and a standard deviation). In the process of such summarization, we inevitably sacrifice detail for ease of description. It may often be useful, however, to present the data in a partially summarized form, thus sacrificing some, but not all, of the data set detail.

Frequency Distributions

One possibility is to present the data in a so-called **frequency distribution** format. Here we'll simply identify the unique value possibilities shown in the data and report the number of data set members that correspond to each value. Think back to our student response data:

$$1, 3, 2, 2, 5, 4, 3, 3, 4, 3$$

What are the unique value possibilities here? It seems clear that we can identify five distinct responses: 1, 2, 3, 4, and 5 previous courses. We have, in fact, one student who responded "1," two students who responded "2," four who responded "3," two who responded "4," and one who responded "5." Reporting these counts in a simple frequency distribution table produces

Value	Frequency
1	1
2	2
3	4
4	2
5	1

If we use x to represent the five possible values and $f(x)$ to represent frequencies, the table can be shown as

x	$f(x)$
1	1
2	2
3	4
4	2
5	1

Frequency Histograms

Translating the table to a graphical equivalent (Figure 2.1) provides a useful visual representation of the data. This kind of bar chart (or line chart) shows frequency as the length (height) of the vertical bar (or line segment) above each value of x.

If you prefer, you can choose to use wider bars in your picture, as shown in Figure 2.2. We'll routinely refer to any of these bar-chart representations as a frequency **histogram**.

Connecting the tops of the bars (or lines) on a frequency histogram produces a *frequency polygon*; smoothing the sketch and creating a continuous contour produces a *frequency curve*. (See Figure 2.3.) 5. F

The Shape of the Data Set

Histograms, frequency polygons, and frequency curves show the basic shape of the data. For example, the data set in Figure 2.1 is perfectly *symmetric*. We could, in this case, establish a center point along the x-axis that would split the distribution into two identical halves. If we were to "fold" the distribution at this center mark, the right-hand side of the fold would match up perfectly with the left-hand side. (It's precisely this symmetry that, in our student survey data, accounted for the equivalence we saw between the mean, the median, and the mode—all of which turned out to have a value of 3.)

Clearly, not all data sets are symmetric. In fact, any number of shapes are possible. Figure 2.4 shows a few of the possibilities.

Figure 2.4(a) and (b) display what is commonly called **skewness** in data—an asymmetry in which an elongated "tail" extends in either the right-hand direction (*positive* skewness) or the left-hand direction (*negative* skewness). (It might be worth noting here that for positively skewed data, the mean is larger than both the median and the mode; for negatively skewed data, the mean is smaller. In general, the

FIGURE 2.1   **FREQUENCY BAR CHART FOR THE STUDENT RESPONSE DATA**

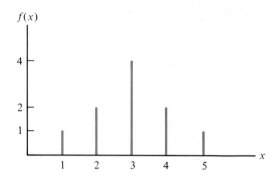

FIGURE 2.2   **HISTOGRAM ALTERNATIVES**

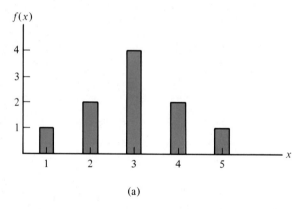

(a)

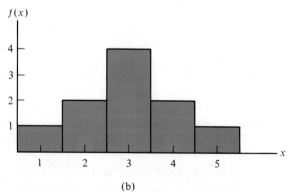

(b)

FIGURE 2.3   **FREQUENCY DIAGRAMS**

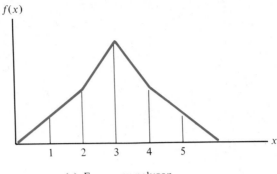

(a) Frequency polygon

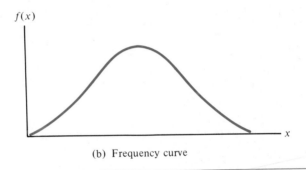

(b) Frequency curve

mean tends to be more severely influenced by extreme values in the data, making it a less reliable measure of center in highly skewed cases. We'll say more on this later in the chapter.)

Figure 2.4(c) shows a *bimodal* distribution—describing a case in which the data reaches two separate "peaks" (i.e., shows two distinct modes). Figures 2.4(d) and (e) show contrasting degrees of "peakedness" or steepness, often referred to as *kurtosis*. (Descriptive statistics are available to measure kurtosis, but we won't take the time to develop them here.)

Relative Frequency Distributions

The **relative frequency distribution** format offers an alternative to the frequency table as an effective means for presenting data in partially summarized form. Rather than reporting the number of data set members having value 1, 2, 3, . . . , we'll report the percentage (or proportion) of the members associated with each of the values. A relative count (we'll label it $p(x)$) is substituted for the absolute count ($f(x)$) that was shown in the frequency table.

FIGURE 2.4   **DISTRIBUTION SHAPES**

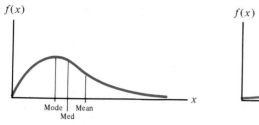

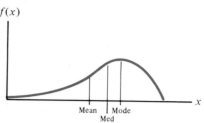

(a) Positive skewness   (b) Negative skewness

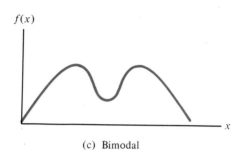

(c) Bimodal

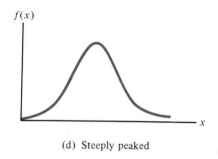

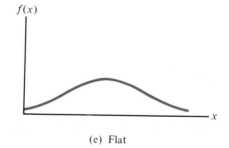

(d) Steeply peaked   (e) Flat

The relative frequency table for the 10-student data appears below:

Value	Relative frequency	or	x	p(x)
1	1/10 = .1		1	.1
2	2/10 = .2		2	.2
3	4/10 = .4		3	.4
4	2/10 = .2		4	.2
5	1/10 = .1		5	.1

(You should note that the $p(x)$ values shown here are simply $f(x)/n$, where n, as before, represents the total number of values in the data set.)

FIGURE 2.5   RELATIVE FREQUENCY BAR CHART FOR THE STUDENT RESPONSE DATA

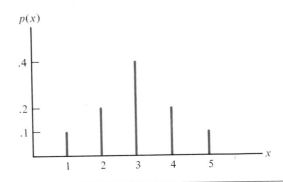

Relative Frequency Histograms

Figure 2.5 shows our relative frequency results graphically. As you can see, this bar chart (histogram) reveals the same symmetry that characterized our frequency histogram. All that's really changed is the label on the vertical scale.

2.4 MODIFYING THE COMPUTATION OF DESCRIPTIVE MEASURES TO SUIT THE FREQUENCY DISTRIBUTION FORMAT

Earlier we established expressions to compute descriptive measures such as the mean and the variance for data presented in raw form (i.e., when each value in the data set is individually reported). We'll need to modify these basic expressions in order to accommodate those cases in which the data is presented in the frequency or relative frequency formats. To demonstrate, we've reproduced the student survey frequency table

x	$f(x)$
1	1
2	2
3	4
4	2
5	1
	10

To compute the mean here, we'll need to change our approach only slightly. Instead of using

$$\bar{x} = \frac{\sum x}{n}$$

we'll use

$$\bar{x} = \frac{\sum (x \cdot f(x))}{n}$$

to produce

$$\bar{x} = \frac{1(1) + 2(2) + 3(4) + 4(2) + 5(1)}{10} = \frac{30}{10} = 3$$

As you can see, we've (1) weighted (multiplied) each unique value for x by the frequency with which it occurs in the data, (2) summed these product terms, and (3) divided by n, the total count of the values involved (i.e., $n = \sum f(x)$). Not surprisingly, we've produced precisely the same mean (3) that we computed for the original raw data.

Computing the variance in this case follows the same pattern:

$$\text{VAR}(x) = \frac{\sum (x - \bar{x})^2 f(x)}{n}$$

For the student response data, this translates to

$$\text{VAR}(x) = \frac{(1 - 3)^2(1) + (2 - 3)^2(2) + (3 - 3)^2(4) + (4 - 3)^2(2) + (5 - 3)^2(1)}{10}$$

$$= \frac{12}{10} = 1.2$$

which is precisely the variance we calculated for the original data set. As usual, the standard deviation is simply the square root of the variance:

$$\text{STD DEV}(x) = \sqrt{\frac{\sum (x - \bar{x})^2 f(x)}{n}}$$

2.5 MODIFYING THE COMPUTATION OF DESCRIPTIVE MEASURES TO SUIT THE RELATIVE FREQUENCY DISTRIBUTION FORMAT

When the data are presented in a relative frequency format, the mean, variance, and standard deviation expressions can be similarly modified. Given our student response data in relative frequency form

x	$p(x)$
1	.1
2	.2
3	.4
4	.2
5	.1

we can produce the mean simply by weighting (multiplying) each unique x value by its relative frequency and summing the results. That is,

$$\bar{x} = \sum x \cdot p(x)$$

For the 10-student data, then,

$$\bar{x} = 1(.1) + 2(.2) + 3(.4) + 4(.2) + 5(.1) = 3$$

Notice we've shown no division by n here. In actuality, the division by n is done implicitly. Remember, each $p(x)$ is defined as $f(x)/n$. The division by n is thus carried implicitly in each of the $p(x)$ terms.

The adjustment for variance follows a similar pattern:

$$VAR(x) = \sum (x - \bar{x})^2 p(x)$$
$$= (1 - 3)^2(.1) + (2 - 3)^2(.2) + (3 - 3)^2(.4) + (4 - 3)^2(.2) + (5 - 3)^2(.1)$$
$$= 1.2$$

As before, standard deviation is just the square root of the variance.

$$STD\ DEV(x) = \sqrt{\sum (x - \bar{x})^2 p(x)}$$

2.6 THE EFFECT OF DISTRIBUTION SHAPE ON MEASURES OF CENTRAL TENDENCY

We noted earlier that for the sort of perfectly symmetrical case that our student response data represents, the mean, the median, and the mode will all be equal (3 in the case of our 10 student responses). When the distribution of values is skewed, however, this equivalence disappears. In severely skewed cases, the commonly used arithmetic mean may actually become the least effective measure of center and a potentially misleading indicator of the "typical" data set member.

To demonstrate, suppose we were to collect data on the hourly wage of the people who work for Firm ABC. The first person reports a wage of $4; the second says $6, and so does the third. The fourth person reports making $8 an hour, and the

FIGURE 2.6   **HOURLY WAGE DISTRIBUTION**

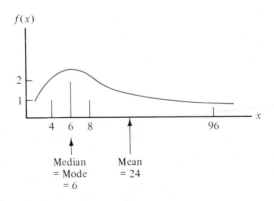

fifth (the nephew of the CEO?) says $96. As Figure 2.6 illustrates, the data set here is clearly not symmetric.

What single value would you choose to summarize results? the mean? The mean turns out to be $24, not a very representative value. The one extreme response ($96) has, in a sense, distorted the mean—it's pulled it well to right of most of the wages reported. With the distribution so markedly skewed, either the mode or the median (both $6 here) would appear to provide a much more representative indicator of the "typical" wage. Unlike the mean, neither of these measures has been seriously influenced by the single $96 extreme.

2.7   GROUPED DATA

When a data set involves a large number of distinct values, effective data presentation may require putting data points together in manageable groups. Suppose, for example, we've collected data on housing prices in the local area. The listing below shows the raw data results of a 2000-home survey:

Home	Selling price	Home	Selling price
1	$78,500	1001	$102,300
2	56,700	1002	68,450
3	145,600	1003	206,100
4	92,100	1004	45,250
5	57,900	1005	32,150
6	121,800	1006	87,300
7	76,400	1007	167,800
8	154,300	1008	103,900
9	82,000	1009	94,200
10	112,700	1010	43,600
11	34,900	.	.
12	87,300	.	.
.	.	.	.
.	.	.	.
.	.	.	.
1000	47,800	2000	74,200

By establishing intervals (or "classes") within which to group the raw data, we can effectively create the kind of histogram representation that we saw earlier. We might, for example, group selling price results within the following 10 intervals:

$10,000 to under $30,000
$30,000 to under $50,000
$50,000 to under $70,000
$70,000 to under $90,000
$90,000 to under $110,000
$110,000 to under $130,000
$130,000 to under $150,000
$150,000 to under $170,000
$170,000 to under $190,000
$190,000 to under $210,000

FIGURE 2.7   **GROUPED-DATA HISTOGRAM FOR HOUSING PRICES**

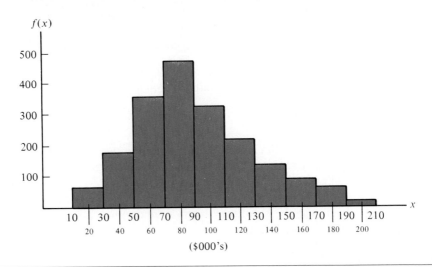

($000's)

We could then count the number of data points (i.e., selling prices) in each interval to produce a table like the one shown below:

Interval	Interval midpoint	Frequency count	Relative frequency
$10,000 to under $30,000	$20,000	60	.03
$30,000 to under $50,000	$40,000	182	.091
$50,000 to under $70,000	$60,000	358	.179
$70,000 to under $90,000	$80,000	491	.2455
$90,000 to under $110,000	$100,000	319	.1595
$110,000 to under $130,000	$120,000	230	.115
$130,000 to under $150,000	$140,000	168	.084
$150,000 to under $170,000	$160,000	102	.051
$170,000 to under $190,000	$180,000	70	.035
$190,000 to under $210,000	$200,000	20	.01
		2000	1.0000

The corresponding frequency histogram is given in Figure 2.7.

Approximating Summary Measures for Grouped Data

We can easily adapt some of our earlier computational expressions to produce *approximate* summary measures of central tendency and dispersion for the grouped-data case. We'll simply treat the *midpoint* of each interval as a representative value for all the group (or class) members involved.

To demonstrate, suppose we want to approximate the mean of the selling price data from the frequency histogram of Figure 2.7 (or from the corresponding frequency table). We need only draw on the "frequency" form for the arithmetic mean computation given earlier in the chapter:

$$\bar{x} = \frac{\sum x \cdot f(x)}{n}$$

Substituting interval midpoints for x in this case produces

$$\bar{x} = \frac{\begin{array}{l}\$20{,}000(60) + 40{,}000(182) + 60{,}000(358) + 80{,}000(491) + 100{,}000(319) \\ + 120{,}000(230) + 140{,}000(168) + 160{,}000(102) + 180{,}000(70) + 200{,}000(20)\end{array}}{2000}$$

$$= \$92{,}590$$

(Suggestion: In checking the computation here, you may want to work in units of $1000s so that, for example, $20,000 becomes simply 20. This should expedite your computational effort.)

The variance and standard deviation calculations follow this same pattern. We'll simply substitute interval midpoints for x in the standard frequency distribution formulas.

EXERCISE 2   Use the frequency information from the table above to approximate the variance and standard deviation of the selling price data. (Suggestion: Work in units of $1000. The mean that we've computed above would then be treated as 92.59. If you follow this form, the variance will be reported in units of 1000^2. Taking the square root to produce the standard deviation will return you to $1000 units.)

Ans: VAR$(x) = 1{,}537{,}292{,}000$ (squared dollars)

STD DEV$(x) = \$39{,}208$

(For a more detailed solution, see the end of the chapter.)

For a relative frequency display in the grouped-data case, summary measures can be similarly produced by substituting interval midpoints in the corresponding relative frequency expressions.

EXERCISE 3   Use the bar chart of Figure 2.8 to approximate the mean, variance, and standard deviation for the data involved.

Ans: Mean $= 28$ VAR$(x) = 201$ STD DEV$(x) = 14.18$

(For a more detailed solution, see the end of the chapter.)

A Final Note on Grouped Data

We have admittedly sidestepped discussion of some of the finer points of grouped-data description. Although some guidelines exist to help determine such things as

FIGURE 2.8

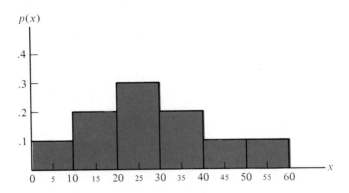

the proper number of intervals to use and how best to set interval widths, much of the methodology involved is art as much as science. Given the availability of computer routines to quickly examine any number of options, experimentation may be the best way to establish the most effective way to summarize and present the data in grouped-data form. The trick is to balance the loss of data detail with the need to provide a simple, comprehensible summary. Our only concrete suggestion in grouping the data is to maintain equal interval widths, but even here there may be some exceptions.

CHAPTER SUPPLEMENT: ADDITIONAL DESCRIPTIVE MEASURES

Although we've covered most of the primary measures of descriptive statistics, there are other measures that may occasionally prove useful.

Coefficient of Variation

Having established the standard deviation as a commonly used measure of data dispersion, we'll now consider a derivative measure that uses the ratio of the standard deviation to the mean in order to describe a kind of "relative" dispersion. To see how it works, take a look at the two data sets below:

A	B
$1, $2, $3	$100,001, $100,002, $100,003

Assume data set A consists of the price of the same brand of pretzels at three different stores, and that data set B shows the value of three sealed bids on a construction project. Which data set seems to show the greater degree of dispersion?

It turns out that in standard deviation terms, the dispersion of the two data sets is exactly the same (about $.82). Yet somehow this doesn't seem right. Given the chance to compare, we would likely conclude that in the case of data set A the price of the pretzels varies significantly. On the other hand, in data set B we would most likely report that the three project bids are virtually the same.

A *relative* standard deviation measure—which we'll label the **coefficient of variation**—has the potential to resolve this apparent inconsistency. We'll simply show the standard deviation as a proportion (or percentage) of the mean:

$$\text{CV} = \text{Coefficient of Variation} = \frac{\text{STD DEV}(x)}{\bar{x}}$$

Using the values in data set A produces

$$\text{CV}_A = \frac{.8165}{2} = .4083$$

For data set B, the result is markedly different:

$$\text{CV}_B = \frac{.8165}{100,002} = .0000082$$

As measured by the coefficient of variation, the variability reported for data set A appears to be much greater than that for data set B, a result more consistent with our instinctive response to the original data.

The coefficient of variation can also be used to compare variability in cases where the data sets being compared involve unlike units. For example, which of the two data sets below shows greater variability?

A (income in dollars)	B (weight in ounces)
20,000	39
42,000	22
35,000	25
33,000	14
Mean $32,500	25 oz
STD DEV $7953	9.03 oz

Because the units involved (dollars and ounces) are dissimilar, comparing standard deviations is not especially meaningful. Using the coefficients of variation, how-ever, clearly shows that there is greater relative variability in the list of weights (9.03/25 = .36) than in the list of incomes (7953/32,500 = .24).

The Geometric Mean

The arithmetic mean for a data set of size n can be thought of as that unique value which, if used to replace each of the n data set values, would produce precisely the same sum as the original numbers. (Example: For the data set 5, 10, 15 the sum is 30. Replacing each of the three values with the mean value 10 produces the same total, 30.)

On occasion, we may need to use a slightly different averaging process. The **geometric mean** for a data set of size n is that unique value which, when used to replace each of the n data set values, will produce precisely the same *product* as the original numbers. Simply defined,

$$\text{Geometric mean} = \sqrt[n]{(x_1)(x_2)\cdots(x_n)}$$

It's simply the nth root of the product of the original n values.

The geometric mean is especially appropriate when averaging ratios, particularly when growth rates or compound rates of return are involved. Suppose, for example, that over the past 3 years a $1000 investment of yours returned 4% the first year, 12% the second year, and 8% the third year, producing at the end of the third year an accumulated total of $1257.98:

$$\$1000(1.04)(1.12)(1.08) = \$1257.98$$

The table below shows in detail the growth pattern involved.

Year	Beginning amount ($)	Ending amount ($)	Ratio (end/begin)	Rate of return
1	1000	1040	1.04	.04
2	1040	1164.80	1.12	.12
3	1164.80	1257.98	1.08	.08

How would you report the average rate of return?

The arithmetic mean of the three rates is 8% ($(.04 + .12 + .08)/3 = .08$). But is this really the measure we want? Applying the average 8% rate in each of the three years would produce an accumulated total of $1259.71, slightly more than the actual amount ($1257.98):

$$\$1000(1.08)(1.08)(1.08) = \$1259.71$$

Now compute the geometric mean of the ratios shown in the table:

$$\sqrt[3]{(1.04)(1.12)(1.08)} = 1.0795$$

The geometric mean of these ratios suggests an "average" rate of return of 7.95% ($1.0795 - 1.0 = .0795$). Applying this measure of average to our 3-year investment produces the correct total accumulation of $1257.98:

$$\$1000(1.0795)(1.0795)(1.0795) = \$1257.98$$

Although the difference between arithmetic and geometric mean results is minor here, it can often be significant.

EXERCISE 4 Suppose over the past 4 years the local economy has grown at a somewhat erratic rate: 8% the first year, then 14%, 22%, and 6% in succeeding years.

(a) Compute the "average" growth rate using (1) the arithmetic mean and (2) the geometric mean.
(b) Suppose at the beginning of the 4-year period, the local economy was valued at $1,000,000. Trace the growth of the economy and report its value today (i.e., at the end of the fourth year).
(c) Use the arithmetic mean and the geometric mean to compute the current value, and compare the results to the total in part (b).

Note: To compute the fourth root on a hand-held calculator, you might just use the square root key twice, or if you have a Y^x key, use it with an exponent of $1/4$ (i.e., $Y^{1/4}$ or $Y^{.25}$).

Ans: (a) arith. mean $= 12.5\%$; geom. mean $= 12.33\%$
 (b) accumulated value today $= \$1,592,187$
 (c) using the arithmetic mean $= \$1,601,807$
  using the geometric mean $= \$1,592,187$

(For a more detailed solution, see the end of the chapter.)

In cases where we have only a beginning value and an ending value n time periods later, the geometric mean can be just as easily computed. For example, suppose you invested \$100 four years ago and now have \$180. The average annual rate of return can be computed by taking the fourth root of the 1.8 ratio (end/begin $= 180/100 = 1.8$), to produce an average ratio of 1.1583. The resulting annual average rate is 15.83%. Compare this to the arithmetic mean rate of return of $.80/4 = 20\%$.

Weighted Averages

Situation: A Freshman student at Acme Tech has completed five courses to date:

Course	Credit hrs	Grade	Grade point
History	3	C	2.0
Chemistry	5	B+	3.33
Calculus	4	B	3.0
Philosophy	3	B	3.0
Statistics	3	A	4.0

Your job is to compute the student's GPA (grade point average).
 Simply adding the five individual grade points and dividing the total by 5 is inappropriate here. What you'll need to do is (1) weight (i.e., multiply) the five individual grade points by the corresponding credit hour value, (2) sum these weighted results, and (3) divide by the total credit hours involved. (We're suggesting a procedure similar to the averaging we did earlier in the frequency distribution section of the chapter.) The result is

$$\text{GPA} = \frac{3(2.0) + 5(3.33) + 4(3.0) + 3(3.0) + 3(4.0)}{3 + 5 + 4 + 3 + 3} = \frac{55.65}{18} = 3.09$$

This sort of **weighted average** (or weighted mean) procedure can be generally represented as

$$\bar{x}_w = \frac{\sum(w_i x_i)}{\sum w_i}$$

where $\bar{x}_w$ = weighted average
 x_i = individual values
 w_i = corresponding weights

EXERCISE 5 An investor has three stocks: 300 shares of A, 400 shares of B, and 100 shares of C. Stock A is currently worth \$2 per share; stock B is currently worth \$10 per share; and stock C is currently worth \$20 per share. Compute the average value per share for the stocks in the investor's portfolio.

Ans: \$6600/800 = \$8.25

Percentiles and Quartiles

The relative position of any value within a data set is sometimes defined in terms of its **percentile**. To illustrate, if 15% of the values in a given data set are 106 or less, 106 is said to be the *15th percentile*. If 77% of the values in that same data set are 189 or less, 189 is said to be the *77th percentile*. In general, for any data set, the pth percentile corresponds to the value that bounds the lower p% of all the values in the data set.

Note: Defined in these terms, the 50th percentile is the median value.

Quartiles can also be used to identify relative position within a data set. The 1st quartile is simply the boundary marking the lower 25% of the values. The second quartile marks the boundary between the lower 50% and the upper 50% (i.e., it's the median). The third quartile bounds the lower 75%. The *interquartile range*—the distance between the 1st quartile and the 3rd quartile—is occasionally used to describe dispersion in data.

▉▉▉▉ CHAPTER SUMMARY

The ability to summarize and present data is critical to the process of translating data into information. We became acquainted with a variety of measures used to essentialize data. Measures of central tendency or central location, including the mean, the median, and the mode, offer an opportunity to represent with a single descriptor an entire set of raw data values. Augmented by measures of data dispersion, these descriptors provide an effective capsule summary of data results.

Variance and standard deviation are the preferred statistical measures of data dispersion (variability). Variance is (not so simply) "the average squared distance of data set members from the overall mean of the data set values." The standard deviation is produced by taking the square root of the variance, thereby removing the awkwardness of the squared units associated with the variance.

To find ways to present data in a visually effective way, we examined frequency and relative frequency table alternatives, together with corre-

sponding histogram (or bar chart) displays. These forms allow us to quickly communicate key data set traits: shape and dispersion, skewness and center are immediately apparent. We also noted necessary adjustments in our calculation of appropriate descriptive measures when the data is presented in these frequency or relative frequency formats.

We established that the shape of the data will affect which of the measures of central tendency might be most appropriate in showing a "typical" (i.e., characteristic) data set value. In cases where the data is fairly symmetric, the familiar arithmetic mean is most commonly used. However, when the data set is noticeably skewed, the median (the data set midpoint) or the mode (the most frequently occurring value) may provide a more accurate measure.

If the data set contains a large number of values, we suggested that a grouping procedure is often in order. Forming clusters of data points within uniform intervals (or classes) allows for an effective histogram-type presentation of the data. Trading detail for conciseness is part of the bargain in trying to find the proper number and width of the intervals to be formed.

In the chapter supplement, we introduced some special descriptive measures: the coefficient of variation, to measure relative dispersion; the geometric mean, to find average growth rates or rates of return; the weighted average, a special-case mean; and percentiles and quartiles, to identify relative position within a data set.

KEY TERMS

Arithmetic mean
Coefficient of variation
Frequency distribution
Geometric mean
Grouped data
Histogram
Mean absolute deviation
Median

Mode
Percentiles
Range
Relative frequency distribution
Skewness
Standard deviation
Variance
Weighted average (or weighted mean)

IN-CHAPTER EXERCISE SOLUTIONS

EXERCISE 1

$$\text{Mean} = \bar{x} = \frac{10 + 30 + 15 + 35 + \cdots + 5}{10} = \frac{250}{10} = 25$$

Putting the values in ascending order, we have

$$5, 10, 10, 15, 30, 30, 30, 35, 40, 45$$
$$\uparrow$$
$$\text{Med}$$

The median is the middle value (here 30, the midway point between the 1st 30 and the second 30 in the data set).
The mode is 30, the most frequently occurring value.
The range is the difference between the largest and the smallest values: $45 - 5 = 40$.
The mean absolute deviation (MAD) is the average absolute distance of the data points from the mean:

$$\text{MAD} = \frac{|10 - 25| + |30 - 25| + |15 - 25| + |35 - 25| + \cdots + |5 - 25|}{10}$$

$$= \frac{120}{10} = 12$$

The variance is the average squared distance of the data points from the mean:

$$\text{VAR}(x) = \frac{(10 - 25)^2 + (30 - 25)^2 + (15 - 25)^2 + (35 - 25)^2 + \cdots + (5 - 25)^2}{10}$$

$$= \frac{1750}{10} = 175$$

Standard deviation is the square root of the variance:

$$\text{STD DEV}(x) = \sqrt{175} = 13.23$$

EXERCISE 2

$$\begin{aligned}
\text{VAR}(x) = & (20 - 92.59)^2(60) + (40 - 92.59)^2(182) + (60 - 92.59)^2(358) \\
& + (80 - 92.59)^2(491) + (100 - 92.59)^2(319) + (120 - 92.59)^2(230) \\
& + (140 - 92.59)^2(168) + (160 - 92.59)^2(102) + (180 - 92.59)^2(70) \\
& \underline{+ (200 - 92.59)^2(20)} \\
& \qquad\qquad\qquad\qquad 2000
\end{aligned}$$

$$= \frac{\begin{aligned}& 316{,}158.48 + 503{,}358.87 + 380{,}234.69 + 77{,}827.477 \\ & + 17{,}515.683 + 172{,}800.86 + 377{,}614.96 + 463{,}499.02 \\ & + 534{,}835.56 + 230{,}738.16\end{aligned}}{2000}$$

$$= \frac{3{,}074{,}583.3}{2000} = 1{,}537.292$$

Remember, however, we were working in units of $1000. Squaring these units produces units of 1,000,000 dollars squared. Our variance result, then, is actually

$$1{,}537{,}292{,}000 \text{ dollars squared}$$

$$\text{STD DEV}(x) = \sqrt{\text{VAR}(x)} = \sqrt{1{,}537{,}292{,}000} = \$39{,}208$$

(The huge variance result here may give further support to our contention that standard deviation is a much more convenient measure of dispersion to report.)

EXERCISE 3

$$\text{Mean} = \bar{x} = 5(.1) + 15(.2) + 25(.3) + 35(.2) + 45(.1) + 55(.1)$$

$$= 28$$

$$\begin{aligned}
\text{VAR}(x) = & (5 - 28)^2(.1) + (15 - 28)^2(.2) + (25 - 28)^2(.3) + (35 - 28)^2(.2) \\
& + (45 - 28)^2(.1) + (55 - 28)^2(.1)
\end{aligned}$$

$$= 52.9 + 33.8 + 2.7 + 9.8 + 28.9 + 72.9$$

$$= 201$$

$$\text{STD DEV}(x) = \sqrt{201} = 14.18$$

EXERCISE 4

(a) Arithmetic mean growth rate $= \dfrac{.08 + .14 + .22 + .06}{4}$

$$= .125 \text{ or } 12.5\%$$

Computing the geometric mean of the ratios

$$\sqrt[4]{(1.08)(1.14)(1.22)(1.06)} = 1.1233$$

suggests a mean growth rate of $1.1233 - 1 = .1233$ or 12.33%.

(b) $1,000,000(1.08)(1.14)(1.22)(1.06) = $1,592,187

Year	Beginning value	Ending value	Ratio (end/begin)	Growth rate
1	$1,000,000	$1,080,000	1.08	8%
2	1,080,000	1,231,200	1.14	14%
3	1,231,200	1,502,064	1.22	22%
4	1,502,064	1,592,187	1.06	6%

(c) With the arithmetic mean, the current value would be overstated:

$$\$1,000,000(1.125)(1.125)(1.125)(1.125)$$

or

$$\$1,000,000(1.125)^4 = \$1,601,807$$

With the geometric mean of the ratios, the current value would be accurately reproduced:

$$\$1,000,000(1.1233)(1.1233)(1.1233)(1.1233)$$

or

$$\$1,000,000(1.1233)^4 = \$1,592,187$$

EXERCISES

★ 1. For the following data set showing eight of your most recent golf scores,

$$100, 90, 110, 80, 120, 140, 100, 60$$

compute the
(a) mean   (b) median   (c) mode
(d) range   (e) MAD   (f) variance
(g) standard deviation

2. For the following data set of demand for your company's product over the last 10 months,

$$36, 24, 24, 30, 28, 32, 26, 24, 26, 200$$

compute the
(a) mean   (b) median   (c) mode
(d) range   (e) MAD   (f) variance
(g) standard deviation

3. For the following data set, which reports the number of hours you spent studying for each of the seven exams you took during the previous semester,

$$5, 6, 3, 5, 7, 10, 13$$

compute the
(a) mean   (b) median   (c) mode
(d) range   (e) MAD   (f) variance
(g) standard deviation

★ 4. In observing a production worker's performance, you record the following completion times (in minutes) for a certain repetitive activity:

$$98, 100, 100, 97, 280$$

Choose one descriptive measure to summarize (characterize) the results. Explain your choice.

5. You have an extremely important job to assign to one of two special project teams. For five somewhat similar jobs, the two teams were rated on a scale of 1 to 100 (100 being a perfect score). Each team's ratings are reported below:

	Ratings				
Team A	80	75	75	70	75
Team B	50	100	60	90	75

To which team would you assign this extremely important job? Explain your answer. (Use descriptive measures as appropriate.)

★ 6. With a little algebraic manipulation, the basic variance expression

$$\text{VAR}(x) = \frac{\sum (x - \bar{x})^2}{n}$$

can be rewritten as

$$VAR(x) = \frac{(\sum x^2) - n\bar{x}^2}{n}$$

This latter expression is often called the *computational form* of the variance expression, and can be a real work-saver if you are computing the variance for a large data set. Once the mean ($\bar{x}$) is calculated, you need only sum the squares of the data point values ($\sum x^2$) and substitute the results into the expression as shown. The variance produced will be perfectly consistent with the variance that would result from use of the original "definitional" expression.

Compute the variance and standard deviation of the data from Exercise 1, using this computational form of the variance expression.

7. Use the computational form of the variance expression to compute the variance and standard deviation of the data in Exercise 2. Check your results with your earlier computations.

★ 8. The following frequency distribution table reports results from a survey of 130 assembly workers who were asked to indicate how many overtime hours (x) they had worked during the past month:

Overtime hrs	No. of workers
x	f(x)
10	31
15	46
20	28
25	15
30	10

(a) Show the histogram (bar chart) representation for the data.
(b) Compute the mean, the variance, and the standard deviation for the data.

9. The following frequency distribution table reports the hourly output rate (x) for 40 workers at a metal fabrication plant operated by Helm Manufacturing:

Units of output	No. of workers
x	f(x)
40	12
42	8
44	6
46	12
48	2

(a) Show the histogram (bar chart) representation for the data.

(b) Compute the mean, the variance, and the standard deviation for the data.

★ 10. The following relative frequency distribution table reports results from a recent study of prices charged by retailers who carry your company's product:

Price ($)	Proportion of retailers
x	p(x)
100	.1
110	.5
120	.3
130	.1

(a) Show the histogram (bar chart) representation for the data.
(b) Compute the mean, the variance, and the standard deviation for the data.

11. The following relative frequency distribution table lists closing finished-goods inventory levels for the last 30 months of your department's operations:

Units of inventory	Proportion of months
x	p(x)
32	.10
40	.20
44	.30
56	.30
100	.10

(a) Show the histogram (bar chart) representation for the data.
(b) Compute the mean, the variance, and the standard deviation for the data.

★ 12. It is sometimes useful to construct *cumulative* frequency or cumulative relative frequency tables to display data. In a cumulative frequency table, we can show explicitly the "number of data set members at or below any specified value." Or we can show "the number of data set members at or above any specified value." We've taken the overtime data from Exercise 8 to demonstrate the idea:

(1)	(2)	(3)	(4)
x	f(x)	f(overtime ≤ x)	f(overtime ≥ x)
10	31	31	130
15	46	77	99
20	28	105	53
25	15	120	25
30	10	130	10

In column 1, x is being used to designate the possible overtime values; column 2 shows the number of workers whose overtime hours were exactly equal to x; column 3 shows the number of workers whose overtime hours were less than or equal to x; column 4 shows the number of workers whose overtime hours were greater than or equal to x.

(a) Following this same pattern, show the output rate data from Exercise 9 in a "less than or equal to" cumulative frequency table.

(b) Draw a bar chart (histogram) for the cumulative results in part (a).

(c) Now construct a "greater than or equal to" cumulative frequency table for the output rate data in Exercise 9.

(d) Draw a bar chart (histogram) for the cumulative results in part (c).

13. Building on your work in Exercise 12, show for the output rate data in Exercise 9

(a) a "less than or equal to" cumulative relative frequency table and the associated histogram.

(b) a "greater than or equal to" cumulative relative frequency table and the associated histogram.

14. Show the price data from Exercise 10

(a) in a "greater than or equal to" cumulative relative frequency table.

(b) in a "less than or equal to" cumulative relative frequency table.

(c) For your results in parts (a) and (b), draw the corresponding bar chart (histogram).

★ 15. The following table of grouped-data values summarizes results from a recent study of 150 firms in the electronics industry. Each firm reported the number of full-time workers it employs:

Interval	Interval midpoint	Frequency count
0 –under 100 employees	50	20
100–under 200	150	50
200–under 300	250	30
300–under 400	350	20
400–under 500	450	15
500–under 600	550	10
600–under 700	650	5
		150

(a) Draw the histogram.

(b) Estimate the mean, the variance, and the standard deviation.

16. The following table of grouped data summarizes recent production cost increases associated with various models of product that your company produces:

Cost increase	Interval midpoint	Proportion of models experiencing the increase
0–under $1	.5	.05
1–under $2	1.5	.10
2–under $3	2.5	.20
3–under $4	3.5	.25
4–under $5	4.5	.20
5–under $6	5.5	.10
6–under $7	6.5	.10

(a) Draw the histogram.

(b) Estimate the mean, the variance, and the standard deviation.

★ 17. Below is a list of dividends paid recently by 60 of the largest firms in the telecommunication industry:

1.23	2.13	4.02	7.92
.66	3.21	5.17	7.14
.21	2.09	4.33	6.77
1.09	3.65	5.78	6.51
1.45	2.54	4.21	7.94
1.87	3.50	5.67	6.56
.56	2.98	5.06	6.41
.12	4.78	5.45	6.83
1.31	5.43	4.80	7.62
.56	5.16	5.67	8.45
1.43	4.12	4.39	9.66
1.22	4.89	6.12	9.07
.97	4.36	6.83	9.54
.80	5.66	6.02	8.22
.43	4.91	7.35	8.49

(a) Show the values in a grouped-data frequency table, using the intervals 0 to under 2, 2 to under 4, 4 to under 6, and so on.

(b) Draw the histogram for the table in part (a).

(c) Using the grouped-data table, estimate the mean, the variance, and the standard deviation.

18. For the data in Exercise 17, set up a grouped-data frequency table, using the intervals 0 to under 1, 1 to under 2, 2 to under 3, and so on.

(a) Draw the corresponding histogram.

(b) Using the grouped-data table, estimate the mean, the variance, and the standard deviation.

19. For the data in Exercise 17, set up a grouped-data frequency table, using the intervals 0 to under 5 and 5 to under 10.
    (a) Draw the corresponding histogram.
    (b) Using the grouped-data table, estimate the mean, the variance, and the standard deviation.

The exercises that follow deal with topics from the chapter supplement.

★ 20. Below are the closing stock prices (over the past 10 weeks) for two of the stocks that comprise your portfolio. For each stock, compute the standard deviation and the coefficient of variation. Which stock showed more variability? Explain.

Stock A	20, 30, 22, 18, 35, 25, 10, 12, 28, 30
Stock B	130, 128, 120, 135, 125, 110, 130, 122, 118, 112

21. As part of your quality control procedures, you monitor the diameter and the weight of each unit you produce. Below are the results of the last eight measurements:

Weight (in ounces)	3.2, 3.5, 3.0, 3.1, 3.2, 3.3, 3.4, 3.3
Diameter (millimeters)	109, 115, 99, 112, 125, 101, 91, 112

Which characteristic—weight or diameter—shows the greater variability? Explain, using an appropriate descriptive measure.

★ 22. You have $10,000 to invest. The anticipated annual returns are provided below:

Year	Begin amount	End amount
1	10,000	10,600
2	10,600	12,720
3	12,720	14,882
4	14,882	20,091

(a) Compute the average annual rate of return, using (1) the arithmetic mean and (2) the geometric mean.

(b) Use each of the average rates that you produced in part (a) to compute your accumulated amount at the end of the 4 years and determine which of the two averages produces the more accurate result.

23. The population of Baker, Oregon, was 10,000 in 1980. Eight years later it was 15,000. Compute the average annual growth rate over this 8-year period.

★ 24. You are checking closing inventory for the week. Your company carries three primary items: $20 ratchets, $30 widgets, and $50 boron synthesizers. You determine that you currently have 150 ratchets, 95 widgets, and 10 boron synthesizers in stock. Compute the average value for the items in your inventory.

25. J. Goveya, a long-distance bike racer, rode (24 minutes) up Milkrun Hill at 3 mph and (6 minutes) down the other side at 50 mph. He computed his average speed over that stretch of road as $(3 + 50)/2 = 26.5$ mph. Use a weighted average approach to produce the real average speed appropriate here.

★ 26. By mixing available fuel, you want to produce 3500 gallons of gasoline for your fleet of trucks this month. You plan to mix 800 gallons of fuel A (80 octane), 1500 gallons of fuel B (92 octane), and 1200 gallons of fuel C (78 octane). Use the weighted average approach to compute the octane rating for the resulting mix.

★ 27. R. D. Climber finally received his test scores for the GSB (Graduate Study in Business) exam. They are

Verbal: 86th percentile   Math: 35th percentile

Interpret the scores for him.

28. You have just learned that the starting salary you have been offered at firm WXY would put you at the first quartile of starting salaries being offered by all firms in the same industry. Are you pleased? Explain.

PROBABILITY THEORY

CHAPTER OBJECTIVES

Chapter 3 should enable you to:

1. Identify the general nature of probability.

2. Describe three distinct approaches to assessing probability (classical, relative frequency, and subjective).

3. Define statistical independence and recognize mutually exclusive events.

4. Apply the multiplicative and additive rules to create additional probabilities from known probabilities.

5. Recognize situations in which a *Venn diagram*, *probability tree*, or *joint probability table* might provide a useful visual framework for problem solving.

6. Count combinations and permutations. (Chapter Supplement)

In dealing with decision-making situations in which uncertainty plays a significant role, basic probability theory provides the means by which uncertainty can be effectively managed, or at least better understood. We plan in this chapter to introduce some of the basic tools and rationale of probability theory, and to demonstrate the potential of these tools to address the difficulties of problem solving in an uncertain world.

3.1 DEFINING PROBABILITY

Let's begin at the beginning. Just what is a probability? For example, when you look out the window and consider the probability that it's going to rain today, precisely what's involved?

Simply stated, a **probability** is intended to somehow represent the chance or likelihood that some event or set of events will occur. Typically taking the form of a number between 0 and 1, probabilities communicate an appropriate degree of uncertainty or level of risk. At the extremes, a probability of 0 means that the event (or events) in question cannot or will not occur; a probability of 1 indicates that the event (or events) will or must occur. We'll tackle the question of where these sorts of numbers come from in the next section.

3.2 PROBABILITY ASSESSMENT STRATEGIES

The Classical Approach

Depending on the situation, several different approaches are available for assessing basic probabilities. To illustrate, consider a simple experiment: I'll toss a coin—a quarter—and you call the result, "heads" or "tails." Before we even get started, though, tell me exactly how likely it is that the coin turns up "heads."

Your immediate response, I suspect—an almost instinctive "50%." The reason? Two outcomes are possible here, one of which is a "head." Conclusion? The probability of a head is 1 out of 2, or 50%! It's a simple counting problem: Count the number of possible outcomes in the experiment (here, 2), count the outcomes favorable to the "heads" event (here, 1), and take the ratio of the two numbers. This basic counting procedure is often labeled the **classical** (or **a priori**) approach to probability assessment and might be summarized as follows: $6.T$

$$P(E) = \frac{F}{T}$$

where $P(E)$ = probability of event E
F = number of outcomes favorable to event E
T = total number of outcomes possible

But is this the only probability assessment approach possible? Even in this simple coin-tossing situation, we may find the F/T strategy inappropriate. Suppose, for example, you suspected that the coin being tossed was not your standard run-of-the-mill coin, but rather a weighted or "loaded" coin that had a significantly greater propensity to land heads up rather than tails up? A critical assumption in the F/T approach is that the T value in the denominator represents a count of "equally likely" outcomes. If we violate this assumption (as we would in the case of a weighted coin), we invalidate the procedure.

Long-Run Relative Frequency

One possible alternative to the classical assessment strategy would involve a little experimentation. In the coin toss situation, you might demand to see me toss the coin repeatedly before you're ready to assign a probability to the event "toss a head." I toss the coin over and over again, you keep track of the outcomes, and when you've seen enough you give me your "heads" probability on the basis of a "number of heads" to "number of tosses" ratio. (For example, if I tossed the coin 1000 times and 520 heads turned up, you would assign a 52% probability to the event "tossing a head.")

This sort of procedure might be characterized as a *relative frequency* or *long-run relative frequency* approach to probability assessment, in which

$$P(E) = \frac{n}{N}$$

where N = total number of observations (trials)
n = number of times the desired outcome (event) occurs

Again, as in the classical case, we have a counting procedure to assess probability, but the counting being done is clearly quite different. In this relative frequency approach, the counting procedure is a little more elaborate and (at least potentially) a lot more time-consuming. In fact, a key question here frequently involves just how many observations (or trials) are needed before a truly valid, defensible probability will emerge. In the weighted coin situation, two or three tosses clearly wouldn't be enough. But how about 100, 1000, 10,000? How much is enough? (Not an easily resolvable question, and one I'll let you think about as we proceed.)

A variation of the long-run relative frequency approach we've described would eliminate the need to "start from scratch" in devising an experiment to assess probabilities. Instead, some form of historical evidence might be used. To demonstrate, suppose I parked my car across the street this morning and I'm now interested in assessing the likelihood of safely recrossing the street to get back to the car. Not only would the classical approach break down (two things can happen when I cross the street: I'll make it safely, or I won't; therefore, the chances of making it safely across are 1 out of 2), but the experimental long-run relative frequency approach also loses much of its appeal. (At least for me it does!)

In such a case, using historical evidence to produce the necessary counts for a relative frequency assessment of probability might make sense. Traffic records might give me a count of the number of accidents involving pedestrians crossing the street in this particular block over, say, the last 5 years; they might also give me some idea as to the amount of pedestrian traffic over that same time span. Number of accidents over number of attempted crossings could then perhaps be used as a defensible measure of likelihood. This is a long-run relative frequency approach, but one that uses available data rather than additional experimentation.

Subjective Probability

In a number of situations, it's possible that neither of the probability assessment strategies we've described so far will fit especially well. How, for example, would you produce the probability that it's going to rain later today? Or the probability that the Seattle Mariners will win their division this year? Or the probability that there will be an innovative new show in the fall TV lineup? (OK, forget the last two. But what about the weather question?) In such circumstances, a well-structured, counting-based approach seems unworkable.

In these kinds of situations, we may find it necessary to fall back on a less structured, less formal approach to probability assessment—one that might best be described as a **subjective** (or "degree of belief") approach. Here we would simply (sometimes not so simply) bring together any number of relevant factors—sifting and sorting and weighing them in an attempt to create a composite number that somehow accurately reflects our "degree of belief" about an event's likelihood of occurring. In deciding whether to bring an umbrella to work, I might, for example, look out the window, notice the cloud formations, listen to a radio weather report, make note of the time of the year, think back to yesterday's weather, feel the twinge of an arthritic knee and eventually settle on a number that somehow subjectively represents my best guess of the chances for rain.

> **Note:** Not everyone is comfortable with subjective probabilities since they involve judgments that can vary markedly from person to person. Some schools of statistical thought, in fact, pointedly reject a major role for subjective probabilities in the decision-making process. Yet it seems inescapable that such probabilities drive many of the choices we make in our everyday activities.

3.3   THE RULES OF PROBABILITY

Fortunately, it turns out that no matter how basic probabilities are established—whether we use a classical approach, a relative frequency approach, or a subjective approach—the same standard operational rules apply, rules that will enable us to manipulate known probabilities to produce additional (and hopefully useful) "new" probability values. We'll spend some time here discussing these rules, along with the necessary terminology, in order to provide the basis for solving a broad range of elementary probability problems.

A simple example will provide the context for our discussion:

Situation: Suppose I'm standing in front of you, holding an ordinary deck of

52 playing cards. In a second, I'll give you an opportunity to draw cards from the deck, but before I do I'm going to ask you to produce some probabilities.

> **Note:** Throughout the chapter, we'll find ourselves involved with decks of cards, rolls of dice, tosses of coins, and so on. It turns out that these sorts of experiments provide clean, unambiguous settings in which most of the concepts (and rationale) of basic probability theory can be developed. You'll have to trust me that at some point in our discussion these kinds of examples will give way to more "real-world" applications.

Simple Probabilities

We'll begin the card selection experiment by having you draw one card from the deck. Tell me how likely it is that the card you select will turn out to be an ace. Easy. Count the number of equally likely outcomes. Count the number of outcomes favorable to the event "drawing an ace," and take the ratio of these two values. The result? 4/52 (1 in 13), or about 8%. Applying the classical approach to probability assessment here has produced a **simple probability**—the likelihood of the single event "drawing an ace."

Introducing some basic symbols, with A representing the simple event "drawing an ace on one draw from the deck of 52 cards" and P representing probability, we can write

$$P(A) = \frac{4}{52} = \frac{1}{13} = .077 \text{ or } 7.7\%$$

> **Note:** Probabilities can be reported as fractions, decimals, or percentages.

> **Second Note:** Having decided to use A to designate the simple event "draw an ace," we'll use A' (A prime) to designate the *complementary* event "draw a card that's not an ace." Correspondingly, if $P(A)$ represents the probability of A, then $P(A')$ represents the probability of not A. Not surprisingly, $P(A') = 1 - P(A)$.

Conditional Probability

Suppose now we extend the experiment to two draws from the deck. Tell me how likely it is that you select an ace on the second draw *given* that you draw an ace on the first draw. What we're looking for here is a **conditional probability**—the probability of one event occurring conditioned on the occurrence of another. Defining A as the event "drawing an ace on the first draw" and B as the event "drawing an ace on the second draw," we'll represent this card selection conditional probability as

$$P(B/A)$$

with the "/" read "given."

And the answer to our "ace given ace" question? Not surprisingly, the answer is "it depends." Clearly, it all depends on what we do with the first card selected. Do I simply take a look at the first card, note that it's an ace, and then replace it before I let you select the second card, or do I hold out the first card and let you draw from a reduced deck?

For the moment, suppose the plan is to hold out the first card. In this case, the probability of drawing an ace on the second draw given that you draw an ace on the first is exactly $3/51$—that is, with 51 equally likely outcomes possible on the second draw, 3 of which are favorable to the event "draw an ace,"

$$P(B/A) = \frac{3}{51} \quad \text{(or approximately 5.9\%)}$$

If the first card selected is *replaced,* the problem again reduces to a simple count. Now, however, the count involves 52 equally likely outcomes on the second draw, 4 of which are favorable to the event "draw an ace." Consequently,

$$P(B/A) = \frac{4}{52}$$

precisely (and not unexpectedly) the same probability we computed for the simple event "draw an ace," which began our discussion.

Statistically Independent Events

This kind of contrast in conditional probability calculations (i.e., in the "with re-placement" versus the "without replacement" cases) raises an important statistical issue: the issue of statistical independence.

Definition: *Two events are said to be **statistically independent** if the occur-rence of one event has no influence on the likelihood of occurrence of the other.*

Stated symbolically, events A and B are statistically independent if

$$P(A/B) = P(A) \quad \text{or} \quad P(B/A) = P(B)$$

In either case, the implication is the same: for statistically independent events, *con-ditional probability equals simple probability.*

In our card selection experiment, the act of replacing the first card and shuf-fling the deck makes the events "ace on first draw" (A) and "ace on second draw" (B) statistically independent. Under this assumption,

$$P(B/A) = P(B)$$

since drawing an ace on the first draw has no bearing on the likelihood that an ace will be drawn on the second.

On the other hand, if the experiment proceeds *without* replacement, the events "ace on first draw" and "ace on second draw" are obviously connected. The chance of drawing an ace on the second draw is clearly influenced by whatever card is selected on the first. The events here are statistically *dependent.*

Joint Probability

Suppose we now want to assess the likelihood of drawing aces on both draws from the deck. In other words, we want to determine the probability that you draw an ace on the first draw *and* on the second draw. The basic rule here is **multiplicative**: to find the **joint probability** of two events occurring together, we need only multiply the (simple) probability of the first event by the (conditional) probability of the second event given the occurrence of the first. That is,

$$P(A \text{ and } B) = P(A)P(B/A)$$

Borrowing the set theory symbol $\cap$ (for "intersection"), we can write the general multiplicative rule as

$$P(A \cap B) = P(A)P(B/A)$$

For the card selection experiment, without replacement, this means

$$P(\text{two aces}) = P(A \cap B) = P(A)P(B/A)$$

$$= \left(\frac{4}{52}\right)\left(\frac{3}{51}\right) = \frac{12}{2652} = .0045$$

Note: The multiplicative rule should seem reasonable. By multiplying two values (probabilities) no greater than 1, the rule serves as a probability reducer. The resulting joint probability (i.e., the probability of both events occurring together) is logically less than, or, at worst, no greater than, the probability of either event occurring separately.

Simplification of the multiplicative rule is possible when the events involved are statistically independent. Here we can produce a "special-case" rule

$$P(A \cap B) = P(A)P(B)$$

which computes the joint probability for two *independent* events by multiplying their respective *simple* probabilities. The "conditional" second term in the general multiplicative expression is replaced here by an "unconditional" (i.e., a simple) probability. For our "with replacement" example, then, the probability of two consecutive aces is

$$P(A \cap B) = P(A)P(B) = \left(\frac{4}{52}\right)\left(\frac{4}{52}\right) = \frac{16}{2704} = .0059$$

Usefully, the simple multiplicative rule extends easily to cases involving more than two independent events:

$$P(A \cap B \cap C \cdots) = P(A) \cdot P(B) \cdot P(C) \cdots$$

The "Conditional Equals Joint over Simple" Rule

An extremely useful corollary to the multiplicative rule is a rule that we'll refer to as the "conditional equals joint over simple" rule. To see how it works, we'll simply divide both sides of the general multiplicative rule

$$P(A \cap B) = P(A)P(B/A)$$

by $P(A)$ to produce

$$P(B/A) = \frac{P(A \cap B)}{P(A)}$$

The rule we've created shows that the conditional probability of event B occurring, given that event A has occurred, can be computed by dividing the *joint* probability of both A and B by the *simple* probability of A—that is, conditional equals joint over simple.

Not surprisingly, it will also be true that

$$P(A/B) = \frac{P(A \cap B)}{P(B)}$$

Notice the condition event (the "given" event) in each case is the one that appears in the denominator of the corresponding joint-over-simple ratio.

We'll see shortly the applicability of this useful relationship.

EXERCISE 1   The probability of event A occurring is .6. The probability of event B occurring is .8. The probability of both events occurring is .5.

(a) Determine the probability of event A occurring *given* that event B occurs.
(b) Determine the probability of event B occurring *given* that event A occurs.

Ans: (a) $P(A/B) = \dfrac{P(A \cap B)}{P(B)} = .5/.8 = .625$

      (b) $P(B/A) = \dfrac{P(A \cap B)}{P(A)} = .5/.6 = .833$

Mutually Exclusive Events

Two events A and B are said to be **mutually exclusive** if the occurrence of one precludes the occurrence of the other; that is, if one event occurs, the other event cannot occur.

For mutually exclusive events, then,

$$P(A/B) = 0 \quad \text{or} \quad P(B/A) = 0$$

or, equivalently,

$$P(A \cap B) = 0$$

Illustration: How likely is it that on a single draw from our card deck you select an ace *and* a king. Easy enough. Since "drawing an ace" and "drawing a king" (on a single draw) are mutually exclusive events, the probability of both events occurring is 0.

The Additive Rule

Suppose now we want to assess the likelihood of drawing *either* an ace *or* a king on a single draw from our standard deck. Letting A represent the event "drawing an ace" and B the event "drawing a king," we need only *add* the two corresponding simple probabilities. That is,

$$P(A \text{ or } B) = P(A) + P(B) = \frac{4}{52} + \frac{4}{52} = \frac{8}{52}$$

It's important to note, however, that this simple **additive rule** works *only* for mutually exclusive events. If the events are not mutually exclusive, the rule becomes a little more complicated:

$$P(A \text{ or } B) = P(A) + P(B) - P(A \cap B)$$

Here the probability of A or B occurring is equal to the probability of A plus the probability of B minus the (joint) probability of A and B. The tail-end joint probability term is needed here to eliminate a "double count" problem that would otherwise cause us to overstate the targeted "either/or" probability.

Illustration: Suppose we want to find the probability of selecting either an ace or a diamond on a single draw from our standard deck. Letting A represent the event "drawing an ace" and B the event "drawing a diamond," we can apply the general additive rule:

$$P(A \text{ or } B) = P(A) + P(B) - P(A \cap B)$$
$$= \frac{4}{52} + \frac{13}{52} - \frac{1}{52} = \frac{16}{52}$$

Since the ace of diamonds was counted as an ace *and* as a diamond, we've appropriately subtracted *one* of the (1/52) ace of diamonds probabilities.

Using the $\cup$ (for "union") notation of set theory, we can write the general additive rule as

$$P(A \cup B) = P(A) + P(B) - P(A \cap B)$$

and show

$$P(A \cup B) = P(A) + P(B)$$

as the special-case rule for mutually exclusive events. Usefully, the special-case additive rule extends readily to cases involving more than two mutually exclusive

FIGURE 3.1   **THE VENN DIAGRAM**

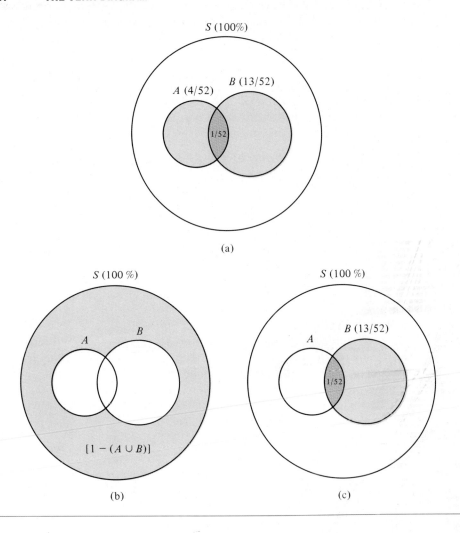

(a)

(b)

(c)

events:

$$P(A \cup B \cup C \cdots) = P(A) + P(B) + P(C) + \cdots$$

A Visual Display: The Venn Diagram

The **Venn diagram**, a visual device borrowed from set theory, is sometimes used to display the elements and rationale of basic probability theory. In Figure 3.1 we've constructed a Venn diagram to represent a simple one-draw card selection experiment. As shown, the larger circle, labeled S for "sample space," includes all of the possible outcomes—i.e., 100% of the 52 possible results associated with the one-draw experiment. (In set theory, this larger circle might be labeled U, for universal set.) We've drawn smaller interlocking circles A and B to represent, respectively, the events "drawing an ace" (4 chances in 52) and "drawing a diamond" (13 chances in 52). The overlap (or intersection) of the two smaller circles corresponds to the joint occurrence of both A and B, in this case drawing the ace of diamonds (1 chance in 52).

To illustrate its usefulness, the diagram might now be used to reaffirm the general additive rule. We can readily identify, for example, the probability of drawing an ace *or* a diamond, $P(A \cup B)$, simply by focusing on the total area bounded by the A and B circles: the area in A plus the area in B minus the (double-counted) area in the overlap. Consequently,

$$P(A \cup B) = P(A) + P(B) - P(A \cap B)$$
$$= \frac{4}{52} + \frac{13}{52} - \frac{1}{52} = \frac{16}{52} = .307$$

See Figure 3.1(a).

Taking things a bit further, we can show the probability of neither A *nor* B occurring—the probability of selecting neither an ace nor a diamond—as the area outside the two event circles. Thus,

$$P(A \cup B)' = 1.0 - P(A \cup B) = 1.0 - \frac{16}{52} = .693$$

See Figure 3.1(b).

We can even see the probability of A *given* B—the conditional probability that the card selected is an ace given that it's a diamond. We'll focus first on the area in B (diamond). Once inside circle B, our interest shifts to the likelihood of being in circle A (ace). (Simply stated, we're interested here in what proportion of B is also in A.) Taking the area in the intersection ($P(A \cap B)$) and dividing it by the area in B ($P(B)$) gives the probability we need. Result:

$$P(A/B) = \frac{P(A \cap B)}{P(B)} = \frac{1/52}{13/52} = \frac{1}{13} = .077$$

See Figure 3.1(c).

You might try testing your understanding with the exercise below.

EXERCISE 2   The probability of event A occurring is .7. The probability of event B occurring is .5. The probability of both A and B occurring is .3. Construct a Venn diagram to represent the situation and use it to determine the following probabilities:

(a) either A or B
(b) neither A nor B
(c) A given that B occurs
(d) B given that A occurs
(e) A occurs but B does not. (*Hint:* Find what's in A, then subtract what's in the intersection of A and B.)

(f) Are the events A and B statistically independent?

Ans: (a) $P(A \cup B) = .7 + .5 - .3 = .9$

   (b) $P(A \cup B)' = 1 - P(A \cup B) = 1 - .9 = .1$

   (c) $P(A/B) = \dfrac{P(A \cap B)}{P(B)} = \dfrac{.3}{.5} = .6$

   (d) $P(B/A) = \dfrac{P(A \cap B)}{P(A)} = \dfrac{.3}{.7} = .429$

   (e) $P(A \cap B') = P(A) - P(A \cap B) = .7 - .3 = .4$

   (f) No, since $P(A/B) \neq P(A)$ and $P(B/A) \neq P(B)$

(This visual analysis should help you solve some of the problems at the end of the chapter. For a more extensive discussion of Venn diagrams in probability analysis, see the chapter supplement.)

Summary

You should by now (1) feel comfortable with some basic terminology (simple, conditional, and joint probabilities, statistical independence, and mutually exclusive events) and (2) understand the core operational rules of probability:

The Multiplicative Rule

$$P(A \cap B) = P(A)P(B/A)$$

or, for statistically independent events,

$$P(A \cap B) = P(A)P(B)$$

The Additive Rule

$$P(A \cup B) = P(A) + P(B) - P(A \cap B)$$

or, for mutually exclusive events,

$$P(A \cup B) = P(A) + P(B)$$

The "Conditional Equals Joint over Simple" Rules

$$P(A/B) = \frac{P(A \cap B)}{P(B)}$$

$$P(B/A) = \frac{P(A \cap B)}{P(A)}$$

Looking Ahead

Believe it or not, you've now seen about all there is to basic probability theory: some standard terminology and a few elementary rules. In a sense, then, solving probability problems should be an easy matter: Either multiply probabilities, or divide, or add them together. Unfortunately, we'll find that things aren't always as simple as they seem. Knowing just when and what to add, and when and what to multiply or divide, very often requires a careful focus and a rather tightly structured analytic approach. We'll spend most of the remainder of the chapter trying to develop an effective problem-solving strategy that should work well in a variety of situations.

3.4 A BASIC PROBLEM-SOLVING STRATEGY

Principle: To determine the probability of any given event or set of events, carefully and completely

> Lay out the way(s) in which the event(s) in question can occur.

In one way or another, this seemingly obvious principle is a recurring theme in virtually every successful probability problem-solving approach.

A Dice-Rolling Example

Situation: Suppose I plan to roll one die (one of two dice) one time. Tell me the probability that I'll roll a 6. Simple enough. There's only one way it can happen. Of the six (equally likely) sides to the die, one corresponds to the event "roll a 6." The probability of rolling a 6 is then 1 in 6, about 16% (a basic classical counting problem).

Situation: Suppose now I plan to roll two dice one time. Determine the probability that I'll roll a 7. Key question: exactly *how* can this happen? Answer? Consider the possibilities:

Die 1	Die 2	Die 1	Die 2
1	1	4	1
1	2	4	2
1	3	4	3**
1	4	4	4
1	5	4	5
1	6**	4	6
2	1	5	1
2	2	5	2**
2	3	5	3
2	4	5	4
2	5**	5	5
2	6	5	6
3	1	6	1**
3	2	6	2
3	3	6	3
3	4**	6	4
3	5	6	5
3	6	6	6

According to our list, there are exactly six ways of producing a "seven." Six ways out of a total of 36 possible outcomes. (As you can see from our list, six possible outcomes on the first die, together with six on the second, produce a total of $6 \times 6 = 36$ possible and equally likely outcomes.) The answer to our question, then, seems clear:

$$P(\text{roll a } 7) = \frac{6}{36} = 16.7\%$$

Situation: Now things get a bit more complicated. Suppose we plan to roll the two dice twice. How likely is it that we produce exactly one 7 on the two rolls? (We're being given two opportunities to make 7. Our interest is in the chances of rolling 7 exactly once.)

Our strategy? Once again, we'll try to lay out the ways in which the event in question can occur. This time, however, we'll do it in a slightly different manner. We'd like to avoid listing all 36×36 possible outcomes and condense our approach to take advantage of work already done and probabilities already known.

In simplest terms, there are really only two ways to produce exactly one 7 in two rolls of the dice:

Way 1: Make 7 on the first roll, then fail to make 7 on the second roll.

Way 2: Fail to make 7 on the first roll, then make 7 on the second roll.

Or showing things a bit more concisely:

	Roll 1	Roll 2
Way 1	7	7′
Way 2	7′	7

Note: Recall that the prime (′) represents the complement of an event.

Clearly, if one or the other of the two ways occurs, we have our "exactly one 7" result. Now if we can just assign a probability to each way, we could simply sum the two values to produce the required "one 7" probability.

Take another look at the two ways we've listed for making 7. Each way actually involves the joint occurrence of two component events. Way 1 is the joint occurrence of 7 on the first roll and 7′ on the second. Way 2 is the joint occurrence of 7′ on the first roll and 7 on the second. Consequently, we can show

	Roll 1		Roll 2
Way 1	7	∩	7′
Way 2	7′	∩	7

or equivalently,

$$\text{Way 1: } 7_1 \cap 7'_2$$
$$\text{Way 2: } 7'_1 \cap 7_2$$

and use the multiplicative rule to determine the probability for each way. For example,

$$P(\text{Way 1}) = P(7_1 \cap 7'_2) = P(7_1) \cdot P(7'_2/7_1)$$

Since each roll is independent of the other, we can actually use the special-case rule in our computation:

$$P(\text{Way 1}) = P(7_1 \cap 7'_2) = P(7_1)P(7'_2)$$
$$= \left(\frac{1}{6}\right)\left(\frac{5}{6}\right) = \frac{5}{36}$$

Similarly,

$$P(\text{Way 2}) = P(7'_1 \cap 7_2) = P(7'_1)P(7_2)$$
$$= \left(\frac{5}{6}\right)\left(\frac{1}{6}\right) = \frac{5}{36}$$

As we've already observed, if Way 1 or Way 2 occurs, we have our "one 7" result. All that's left to do is to apply the basic additive rule to determine the composite either/or probability. Since both ways *cannot* occur together, we have

$$P(\text{exactly one 7}) = P(\text{Way 1} \cup \text{Way 2})$$
$$= P(\text{Way 1}) + P(\text{Way 2})$$
$$= \frac{5}{36} + \frac{5}{36} = \frac{\mathbf{10}}{\mathbf{36}}$$

Summarizing the Procedure

Having carefully and completely laid out the ways in which the "one 7" event could occur, we've used a fairly straightforward application of the multiplicative rule and the additive rule to obtain the required probability (i.e., 10/36). We can use this same pattern to solve any number of similar problems.

EXERCISE 3   Compute the probability of rolling *at least* one 7 on two rolls of two dice. Lay out completely the (three) ways in which this event can occur. Apply the multiplicative rule and the additive rule as appropriate.

Ans: 11/36. (For a more detailed solution, see the end of the chapter.)

3.5   PROBABILITY TREES: A VISUAL AID

For the kind of sequential problem we've seen in the two-roll dice-rolling experiment, sketching a **probability tree** may provide a helpful visual approach to implementing our general "lay out the ways in which the event can occur" strategy.

FIGURE 3.2   **THE FIRST SECTION OF A TREE DIAGRAM FOR THE DICE-ROLLING EXPERIMENT**

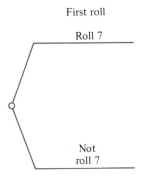

To illustrate, consider the first roll of the two dice in the experiment. We can succinctly describe possible outcomes here as "roll 7" and "not roll 7" and, as shown in Figure 3.2, start to build a tree by creating two corresponding "branches" extending from a starting "node." (A "node" is nothing more than a circle on a tree diagram from which outcome "branches" are "grown.") By creating a node at the end of each of the first two branches, we can then grow a second section of the tree to describe possibilities in the second phase of the experiment ("roll 7" and "not roll 7"). (See Figure 3.3.) Easily constructed, the tree diagram now effectively

FIGURE 3.3   **THE FULL TREE DIAGRAM FOR THE DICE-ROLLING EXAMPLE**

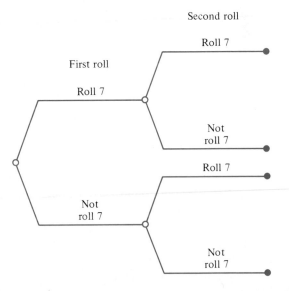

FIGURE 3.4   **ASSIGNING PROBABILITIES TO BRANCHES IN THE FIRST SECTION OF THE TREE**

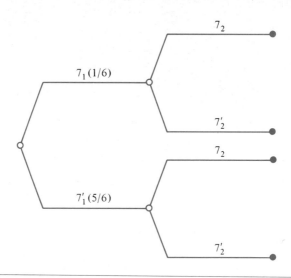

lays out *all* the things that can occur (in "7's" and "not 7's") in our sequential, two-roll experiment. (If the experiment had involved three rolls, extending the tree would be a simple matter.)

    The usual next step is to assign appropriate probabilities to the various outcome branches of the tree. (See Figure 3.4.) For example, the probability of being pushed along the upper branch corresponding to 7 on the first roll is 1/6. We'll show the value in parentheses. The probability of being pushed down the lower "not 7" branch is 5/6, as shown.

    Assigning probabilities to branches in the second section of the tree is just as straightforward. (See Figure 3.5.) (In the general case, once you get beyond the first section of the probability tree, conditional probabilities must be assigned to each branch: for example, the probability of rolling a 7 on the second roll *given* a 7 on the first roll. But since in our current experiment each roll is independent, the simple 1/6, 5/6 probabilities are perfectly acceptable.)

    Once the tree is constructed and probabilities are assigned, we can get down to business. To answer the "one 7 on two rolls" question, we need only identify those *end-nodes* on the tree that correspond to the "one 7" result, compute the probability of reaching each, and sum the associated end node probabilities (see Figure 3.6). To make things easy to reference, we've numbered the end-nodes 1 through 4 and checked-off the appropriate "one 7" nodes ((2) and (3)). Clearly, if we reach either node (2) or node (3), we produce the desired "one 7" result. And how likely is it we'd reach node (2)? Easy. Just multiply the probabilities along the branches that lead there: 1/6 × 5/6 = 5/36. For node (3), 5/6 × 1/6 = 5/36. The chance of arriving at either node (2) or node (3), then, is 5/36 + 5/36 = 10/36. See Figure 3.7.

FIGURE 3.5   **SHOWING THE FULL SET OF PROBABILITIES ON THE TREE**

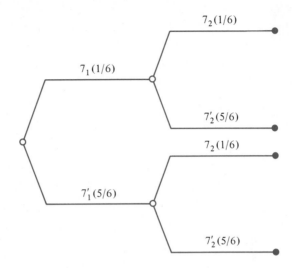

FIGURE 3.6   **IDENTIFYING THE RELEVANT END NODES ON THE TREE**

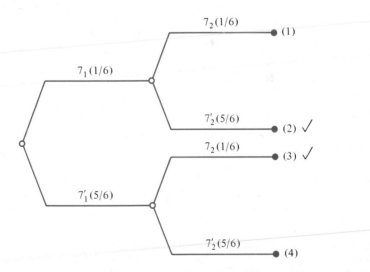

FIGURE 3.7   **PRODUCING END NODE PROBABILITIES**

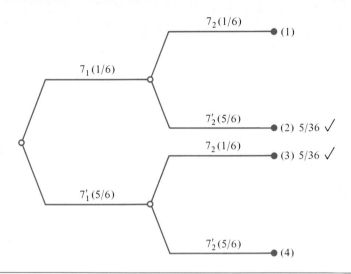

Summing up, we've successfully laid out the ways in which the event in question could occur, with the help of a rudimentary tree diagram, and employed the two basic rules of probability: first the multiplicative rule to produce joint probabilities, then the additive rule to produce an either/or probability. Although the tree approach may not have been essential here, it will become increasingly useful as the problems become more complex.

EXERCISE 4   Use the tree diagram we've developed for the two-dice, two-roll experiment to compute (a) the probability of no 7's in two rolls, and (b) the probability of no more than one 7 on two rolls of the two dice.

Ans: (a) 25/36   (b) 35/36

(For a more detailed solution, see the end of the chapter.)

To convince yourself of the usefulness of the tree diagram in more complex applications, you might want to try problems 21, 22, and 23 at the end of the chapter before proceeding.

A Coin Selection Experiment

Situation: Suppose now I place two similar-sized boxes on the desk beside me, each of which contains a number of coins. Box 1 is known to contain 2 gold coins and

FIGURE 3.8   **A TREE DIAGRAM FOR THE COIN-AND-BOX EXAMPLE**

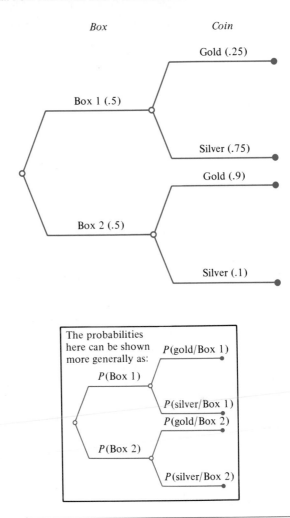

6 silver coins. Box 2 contains 9 gold coins and 1 silver coin:

Box 1	**Box 2**
2 gold	9 gold
6 silver	1 silver

I plan to let you randomly select a coin from one of the two boxes. To be sure things are on the up-and-up, I'll put a blindfold on you and move the boxes around so that you won't know which box you're selecting. Under these circumstances, how likely is it that you would draw a gold coin?

Our solution strategy? Once again, we'll carefully and completely lay out the ways in which the event in question can occur. Exactly how could a gold coin be

FIGURE 3.9   **USING THE TREE TO PRODUCE PROBABILITIES**

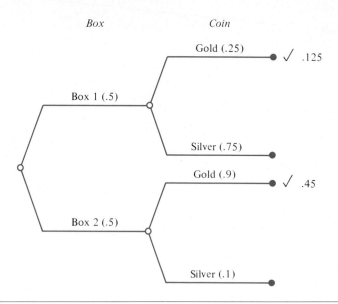

produced? If we portray the exercise as a sequential experiment involving two component steps (phases), we can use a tree diagram to sketch the possibilities (see
Figure 3.8). Notice we're showing box selection as the first phase of our experiment.
The two obvious possibilities, box 1 and box 2, are shown as two branches extending from the initial tree node. Since the boxes are of similar size (and assuming
that selection is truly random) we can assign probabilities of .5 to "selecting box 1"
and .5 to "selecting box 2."

 The second section of the tree focuses on coin selection. Should you find
yourself in box 1, two types of coins are possible: gold or silver. The same is true in
box 2. It's only the probabilities that differ. In box 1, the chances of a gold coin are
clearly 2 in 8, or 25%; the chances of silver are 6 in 8, or 75%. In box 2, the chances
for a gold coin increase significantly, to 9 in 10, while the probability of silver drops
to 10%.

 Now we'll simply check the end nodes on the tree that correspond to the
targeted gold coin event, establish the probability of reaching each of the designated
end nodes by applying the appropriate multiplicative rule, and sum the end node
probabilities to produce the overall gold coin likelihood. (See Figure 3.9.) Result?
The probability of selecting a gold coin equals the probability of reaching into
box 1 and selecting a gold coin (.5 × .25) plus the probability of reaching into
box 2 and selecting a gold coin (.5 × .9). That is,

$$P(\text{gold coin}) = P(\text{box 1}) \times P(\text{gold coin/box 1})$$
$$+ P(\text{box 2}) \times P(\text{gold coin/box 2})$$
$$= (.5)(.25) + (.5)(.9) = .575$$

Our basic approach simply repeats the standard procedure of the preceding section. With the help of the tree diagram, we've been able to identify all the ways in which the gold coin event can occur, and then effectively assigned probabilities to each.

EXERCISE 5   Use the tree developed above to produce the probability of selecting a silver coin. (You should easily confirm that the probability of selecting a silver coin here is 1-P(gold coin).)

Ans: .425. (For a more detailed solution, see the end of the chapter.)

Using the Tree to Produce Conditional Probabilities

Pushing ahead with the coin selection experiment, suppose now you select a coin and find that it is, in fact, gold. How likely is it that you had, in your single random draw, reached into box 1?

(Before we plunge into a formal solution procedure, check your instincts. Think about what's likely to happen. Before the coin was selected, we had settled on a 50-50 chance of being in box 1 or box 2. Now, after the coin is selected, and you find out it's gold, should your box probabilities change? It seems like they should. And if they do change, would you expect the box 1 probability to increase or to decrease? Think which box is more likely to yield a gold coin. It's no contest. Box 2! So, if a gold coin *is* selected, it seems more likely that it came from box 2 and *less* likely that it came from box 1. With a gold coin in hand, the probability of having selected from box 1 should go down, while the probability of having selected from box 2 should go up.)

We can start to devise a solution strategy by focusing precisely on the question being asked (always a good idea in any probability analysis). Are we talking here about a *simple* probability, a *joint* probability, or a *conditional* probability? It should be apparent that we're actually involved with a conditional probability question: *given* that a gold coin is selected, how likely is it that the selection was made from box 1? In short, we need to establish

$$P(\text{box 1/gold coin})$$

Any ideas on how we might proceed? Think back to our earlier discussion of conditional and joint probabilities. There we established a basic rule: *conditional equals joint over simple.* According to this, the conditional probability of box 1 given gold coin should be equal to the joint probability of box 1 *and* gold coin, divided by the simple probability of the condition event, gold coin. Symbolically, we'll find.

$$P(\text{box 1/gold coin}) = \frac{P(\text{box 1} \cap \text{gold coin})}{P(\text{gold coin})}$$

Look again at the tree we devised for our initial pass at the gold coin problem.

The joint probability of "box 1 and gold coin" is shown to the right of the upper-most end node (0.125). And the simple probability of gold? It's the aggregate $P(\text{gold coin}) = .575$ value we found in our earlier analysis. Consequently,

$$P(\text{box 1/gold coin}) = \frac{.125}{.575} = .217$$

Just as we anticipated, the probability of reaching into box 1 has fallen sharply, from 50% before any coin was selected, to only 21.7% after the gold coin is ran-domly drawn. Correspondingly, the probability of box 2 increases to 78.3%. (Use the tree to verify this result.)

3.6   JOINT PROBABILITY TABLES

We might at this point suggest one final "visual aid." For those situations in which joint probabilities are known and conditional probabilities are needed, constructing a **joint probability table** can often provide an extremely effective problem-solving framework.

Recall our coin-and-box example from the previous section. We used a tree diagram there to produce and display a variety of basic probabilities (simple, joint, and conditional). Suppose now we set up a simple 2 × 2 table to summarize results. Across the top of the table we'll reference the box possibilities (box 1 and box 2). Down the left side of the table we'll list the coin possibilities (gold and silver). (See Figure 3.10.) In the four cells we've created, we can place appropriate joint probability values. For example, we'll enter in the upper left-hand cell the prob-ability of reaching into box 1 and selecting a gold coin. In the lower left-hand cell, we'll show the probability of reaching into box 1 and selecting a silver coin. And so on. The completed table appears in Figure 3.11. (Return to the tree to confirm each value.)

Summing Joint Probabilities to Produce Marginals

Having assembled the information in table form, we can now identify (and visually reinforce) some useful relationships. For example, summing across each row, or down each column, produces the so-called **marginal probabilities** for the table—simple probabilities for the table's core events. The simple (or marginal) probability of selecting a gold coin, for example, is the horizontal sum $.125 + .45 = .575$, shown in the right-hand margin of the gold coin row. The simple (or marginal) probability of reaching into box 2 is the vertical sum $0.450 + .05 = .50$, shown in the lower margin of the table. The complete set of marginal probabilities appears in Fig-ure 3.12.

> **Note:** It's only because such values are typically recorded in the right-hand and lower margins of the table that the term *marginal* is applied. As mentioned, these values are actually nothing more than the kinds of values we've identified as simple probabilities.

FIGURE 3.10   **OUTLINING A JOINT PROBABILITY TABLE FOR THE COIN-AND-BOX EXPERIMENT**

Box

	Box 1	Box 2
Gold		
Silver		

Coin

FIGURE 3.11   **ENTERING JOINT PROBABILITIES IN THE JOINT PROBABILITY TABLE**

Box

	Box 1	Box 2
Gold	.125	.450
Silver	.375	.050

Coin

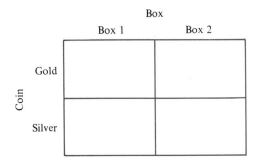

Or, more generally,

	Box 1	Box 2
G	$P(G \cap 1)$	$P(G \cap 2)$
S	$P(S \cap 1)$	$P(S \cap 2)$

Using the Table to Produce Conditional Probabilities

One of the real strengths of the joint probability table format is the ease with which conditional probabilities can be produced. Suppose, for example, we were to repeat the conditional question of the previous section: if a gold coin is selected, how likely

FIGURE 3.12   **SHOWING MARGINAL PROBABILITIES ON THE JOINT PROBABILITY TABLE**

Box

	Box 1	Box 2	
Gold	.125	.450	.575
Silver	.375	.050	.425
	.50	.50	1.00

Coin

Or, more generally,

	Box 1	Box 2	
Gold	$P(G \cap 1)$	$P(G \cap 2)$	$P(G)$
Silver	$P(S \cap 1)$	$P(S \cap 2)$	$P(S)$
	$P(1)$	$P(2)$	

is it that you had selected it from box 1? Simply stated, we need to find

$$P(\text{box 1/gold coin})$$

Guided by the basic rule, conditional equals joint over simple, we can here effectively trace a solution procedure using the joint table as a clear visual reference. The "condition" event, gold coin, immediately puts us in the gold coin row of the table, focusing on only this 57.5% subset of possible results (see Figure 3.13(a)). Once in the gold coin row, the box 1 event puts us in the corresponding first (i.e., box 1) column, pinpointing in the process the .125 joint probability value at the row and column intersection (see Figure 3.13(b)). The required conditional probability is computed by using the ratio

$$\frac{\text{joint}}{\text{simple}} \quad \text{or} \quad \frac{\text{joint}}{\text{marginal}}$$

to produce

$$\frac{.125}{.575} = .217$$

FIGURE 3.13(a)   **LOCATING THE CONDITION ROW ON THE JOINT TABLE**

FIGURE 3.13(b)   **LOCATING THE INTERSECTING COLUMN ON THE JOINT TABLE**

Try following the same pattern to calculate

$$P(\text{box 2/silver coin})$$

Here, the condition event, silver coin, puts us in the second row of the table, keying specifically on this 42.5% segment of experimental results. Once in the silver coin row, the box 2 event puts us in the box 2 column, singling out the .05 entry in the intersection of the second column and second row. Following the standard script,

$$P(\text{box 2/silver coin}) = \frac{\text{joint}}{\text{simple}} = \frac{.05}{.425} = .118$$

The lesson? Once the joint probability table is constructed, *any* conditional is easily accessible. And construction of the joint probability table is often surprisingly simple. (We'll shortly see, for example, that not every joint probability table needs to be derived from a probability tree.)

To test your understanding, consider the following exercise. (You should find it a rather dramatic change from most of the examples so far.)

EXERCISE 6   In a survey of local high school seniors, 30% of the students questioned reported having experimented with drugs, 60% reported that they used alcohol on a fairly regular basis, and 25% reported that they had done both. Using the two simple probabilities and the one joint probability provided, construct a joint probability table to display survey results, and use the table to produce

(a) the probability of drug use given alcohol use.
(b) the probability of a student using alcohol if he or she has not experimented with drugs.
(c) the probability of the student not experimenting with drugs if he or she does not use alcohol.
(d) Based on survey results, are drug use and alcohol use statistically independent?

(*Suggestion:* Use D to represent drug use, D' to represent no drug use, and show these possibilities across the top of your table. Use A to represent alcohol use, A' to represent no alcohol use, and show these possibilities down the left side of the table. Insert the marginal probabilities as appropriate. Now insert the known joint probability $P(D \cap A) = .25$ into the appropriate cell, and use what you know about the relationship between joint and marginal probabilities to unlock the entire table. Remember, the row or column sums of the joint probabilities are always equal to the marginal values.)

Ans: (a) $P(D/A) = .417$ (b) $P(A/D') = .5$
         (c) $P(D'/A') = .875$ (d) The events are statistically dependent.

(For a more detailed solution, see the end of the chapter.)

A Final Illustration

If you're comfortable with things so far, consider the following exercise and use the joint probability table format to answer each question. Since no joint probability value is given directly, you'll need to create one from the simple and conditional probabilities provided. Try using the general multiplicative rule.

EXERCISE 7   In a recent study, 60% of all respondents indicated support for a proposed change in the country's Social Security system. Forty percent of the respondents opposed the measure. Thirty percent of the respondents were 65 years of age or older. You learn that 80% of those respondents in the "65 and older" group supported the proposed change.

(a) Compute the probability of choosing a respondent who is both a supporter of the legislation and under 65 years of age.
(b) Suppose you choose someone randomly from the survey group and find that this

FIGURE 3.14   THE VENN DIAGRAM–JOINT TABLE EQUIVALENCE

Sample space (100%)

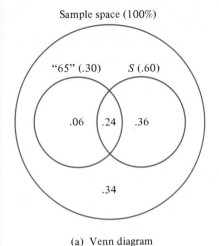

"65" (.30)   S (.60)

.06 .24 .36

.34

(a) Venn diagram

	S	S'	
65	.24	.06	.30
65'	.36	.34	.70
	.60	.40	1.00

(b) Joint probability table

person doesn't support the proposed change. How likely is it that this person is 65 or older?

(c) Suppose you randomly choose someone from the group and find that this person is under 65. How likely is it that this person doesn't support the proposed change?

Ans: Given $P(S) = .60$, $P(65) = .30$, and $P(S/65) = .80$, (a) $P(S \cap 65') = .36$

(b) $P(65/S') = \dfrac{.06}{.40} = .15$ (c) $P(S'/65') = \dfrac{.34}{.70} = .486$

(For a more detailed solution, see the end of the chapter.)

Note: As we mentioned earlier, before trying to solve any probability problem be sure you identify precisely what each of the given probabilities represents and establish appropriate symbols. (In the exercise above, you needed to be especially careful to give proper interpretation to the 80% probability provided as part of the basic problem information. It's a conditional probability rather than a joint probability.) Similarly, make sure you identify the precise nature of the probability questions being asked. Are you looking for a joint probability? a simple probability? a conditional probability? Establish symbols accordingly.

The Relationship between Joint Probability Tables and Venn Diagrams

The joint probability table format can actually be viewed as an extended Venn diagram. Figure 3.14, for example, shows a Venn diagram to match the joint probability table for Exercise 7. Although there is a general equivalence between the joint probability table and the Venn diagram, you may often find it easier to work from the table to produce all the conditionals that might be required in a problem.

EXERCISE 8   Show a Venn diagram representation for Exercise 6. Match each entry in the Venn diagram to a counterpart in the joint probability table. Try to answer each of the questions in Exercise 6 using the Venn diagram you've constructed.

Ans: See the end of the chapter.

3.7   A CLOSING COMMENT AND SOME GENERAL GUIDELINES

It would be unrealistic to argue that you're now prepared to deal easily with any probability problem that might come your way. Yet you do have all the basic tools: a multiplicative rule to produce joint probabilities, an additive rule to produce either/or probabilities, and a joint over simple rule to compute conditional probabilities. You also have a general problem-solving strategy: Carefully lay out all the ways in which the event in question can occur. Lastly, you have at least three visual formats to use in organizing information and to lend structure to your analysis.

Deciding for any given problem which of the visual formats is appropriate can sometimes be difficult. Although there really are no hard-and-fast rules, some general guidelines might be helpful:

1. If you're given two simple probabilities and a joint probability, a Venn diagram or a joint probability table can often be constructed and usefully applied. The joint table provides an especially effective framework for producing all the conditional probabilities that might be required.
2. If you're given a set of simple probabilities or a set of simple and conditional probabilities together, and if the problem can be seen as one that involves a *sequence* of events, a probability tree may be very helpful.
3. If necessary, results from a "tree" analysis can be used to create a joint probability table in order to produce desired conditional probabilities.

All that's needed now is practice, practice, practice.

CHAPTER SUPPLEMENT I: THE VENN DIAGRAM AND PROBABILITY ANALYSIS

This section provides a slightly different look at the elements of basic probability theory, using a Venn diagram and a simple experiment to focus the discussion.

Situation: Suppose I'm now standing by the desk in front of you. Beside me is a large urn containing 10 rubber balls. Five of the balls are large, five are small. Six of the

FIGURE 3.15   **A VENN DIAGRAM FOR THE URN EXAMPLE**

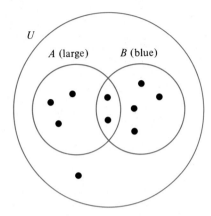

balls are blue, the other four are red. In addition, two of the large balls are blue. I'm going to let you randomly select rubber balls from the urn, but before you do we'll construct a Venn diagram to use in producing a variety of related probabilities (see Figure 3.15).

We've labeled the larger circle U (U for urn or for "universal set"). Circle U contains all 10 of the rubber balls—all possible outcomes once we start selecting from the urn. (As we noted in the chapter, in probability theory what we're calling the universal set would more likely be labeled S for "sample space"—a representation of all possible outcomes in a given experiment.) Within U, we've identified subset A, containing the large rubber balls, and subset B, containing the blue rubber balls. In the overlap, of course, are the large blue balls.

Working from the diagram, how likely is it that given a random draw from the urn you would end up with, say, a large rubber ball? Easy. Count the number of balls in subset A, count the total number of balls in the urn, and take the ratio of the two values: $5/10 = 50\%$.

To formalize things just a bit, we'll use the expression $N(\ )$ to represent a counting function, and show the procedure as

$$P(\text{large ball}) = P(A) = \frac{N(A)}{N(U)} = \frac{5}{10} = .50$$

See Figure 3.16(a).

And what is the probability of selecting a blue rubber ball? We can write

$$P(\text{blue ball}) = P(B) = \frac{N(B)}{N(U)} = \frac{6}{10} = .60$$

That is, we'll count what's in B, count what's in U, and take the ratio of the two

FIGURE 3.16

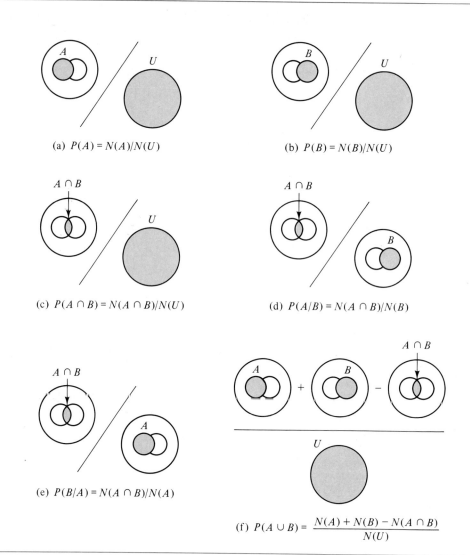

(a) $P(A) = N(A)/N(U)$

(b) $P(B) = N(B)/N(U)$

(c) $P(A \cap B) = N(A \cap B)/N(U)$

(d) $P(A/B) = N(A \cap B)/N(B)$

(e) $P(B/A) = N(A \cap B)/N(A)$

(f) $P(A \cup B) = \dfrac{N(A) + N(B) - N(A \cap B)}{N(U)}$

values. See Figure 3.16(b). In each case we've produced a "simple" probability using an elementary counting procedure and a simple visual display.

Computing a Joint Probability

Suppose now the question is: Given one random draw, what's the probability that you will select a large blue rubber ball? We'll count what's in the intersection of subsets A and B (i.e., the large blue balls), count what's in U (the total count), and take the ratio: $2/10 = 20\%$. Symbolically,

$$P(\text{large blue ball}) = P(A \cap B) = \frac{N(A \cap B)}{N(U)} = \frac{2}{10} = .20$$

See Figure 3.16(c).

Computing Conditional Probabilities

Following the same sort of script, how likely is it that you've selected a large rubber ball given that you know the ball you've selected is blue? Once we acknowledge the "conditional" nature of the question, a routine count to determine the proportion of blue balls that are large is all we'll need. We'll need a count of the balls in B (the "condition" event focuses us immediately on the six blue balls). Once in B, only the large balls (the ones in the intersection of A and B) are of interest to us. So we'll count the number of balls in the intersection of A and B (the large blue balls), count the number of balls in B (all blue balls), and divide: $2/6 = 33\%$. Symbolically,

$$P(\text{large}/\text{blue}) = P(A/B) = \frac{N(A \cap B)}{N(B)} = \frac{2}{6} = .33$$

See Figure 3.16(d).

Now what's the probability of blue given large? Count what's in the intersection of A and B, count what's in the "condition" event A, and take the ratio of the two values.

$$P(\text{blue}/\text{large}) = P(B/A) = \frac{N(A \cap B)}{N(A)} = \frac{2}{5} = .40$$

See Figure 3.16(e).

A Bit of Juggling

Before pushing on much further, we can do a little algebraic manipulation of the terms we've just produced. Consider the expression

$$P(B/A) = \frac{N(A \cap B)}{N(A)}$$

If we divide the numerator and denominator by $N(U)$ (remember, $N(U)$ is just the count of universal set members), we get

$$P(B/A) = \frac{N(A \cap B)/N(U)}{N(A)/N(U)}$$

In this slightly modified form, the conditional computation can be reported as simply "conditional equals joint over simple." Notice that the numerator $(N(A \cap B)/N(U))$ is precisely the expression we used to produce the joint probability of A and B. The denominator $(N(A)/N(U))$ is clearly the simple probability of the condition event A. We have, then,

$$P(B/A) = \frac{P(A \cap B)}{P(A)}$$

As we saw in the chapter, this basic pattern—conditional equals joint over simple—serves as the means to solve a number of common probability problems involving the computation of conditional probabilities.

EXERCISE S1   Use the "conditional equals joint over simple" rule to produce the probability of large ball given blue ball (i.e., $P(A/B)$) in the current example. (In the process, you'll need to convince yourself that $P(A/B) = P(A \cap B)/P(B)$.)

Ans: $P(A/B) = \dfrac{P(A \cap B)}{P(B)} = \dfrac{.2}{.6} = .33$

The Multiplicative Rule Revisited

An easy translation of the conditional relationship shown above reproduces the multiplicative rule. Starting with

$$P(B/A) = \frac{P(A \cap B)}{P(A)},$$

we can multiply both sides of the expression by $P(A)$ to get

$$P(A)P(B/A) = P(A \cap B)$$

or,

$$P(A \cap B) = P(A)P(B/A)$$

This is exactly the multiplicative rule (for joint probabilities) that we earlier stated without proof.

> **Note:** You need to convince yourself that, by similar reasoning, $P(A \cap B)$ is also equal to $P(B)P(A/B)$. Just start with the conditional relationship $P(A/B) = P(A \cap B)/P(B)$.

Establishing Statistical Dependence and Independence

EXERCISE S2   Use the Venn diagram for the urn example to determine whether the events "drawing a large ball" and "drawing a blue ball" are statistically independent. (*Hint:* Is the conditional probability of A/B equal to the simple probability of A? If the answer is yes, then we have statistical independence.)

Ans: The events are not statistically independent. (For a more detailed solution, see the end of the chapter.)

EXERCISE S3   Redraw the Venn diagram to demonstrate a case in which the events "drawing a large ball" and "drawing a blue ball" are statistically independent events.

Ans: See the end of the chapter.

The Additive Rule

How likely is it that you would, on a single selection from the urn, draw either a large rubber ball or a blue rubber ball? Using the same Venn diagram, it seems reasonable that we might simply add the count of what's in A to the count of what's in B, and divide the sum by what's in U. Or can we? We're suggesting

$$P(\text{large or blue}) = P(A \cup B) = \frac{N(A) + N(B)}{N(U)} = \frac{5 + 6}{10} = 1.1$$

Do you see the problem? It's the old double-count difficulty that we talked about earlier. We need to subtract the number of outcomes (rubber balls) in the intersection of A and B (i.e., the large blue rubber balls) once in order to eliminate the double-count overstatement that would otherwise result. Consequently,

$$P(A \cup B) = \frac{N(A) + N(B) - N(A \cap B)}{N(U)} = \frac{5 + 6 - 2}{10} = .90$$

See Figure 3.16(f). By distributing the $N(U)$ denominator to each term in the numerator, we can reproduce the general additive rule of the previous section:

$$P(A \cup B) = \frac{N(A)}{N(U)} + \frac{N(B)}{N(U)} - \frac{N(A \cap B)}{N(U)}$$
$$= P(A) + P(B) - P(A \cap B)$$

Suppose the events A and B were mutually exclusive. Then (as we've seen before) the tail-end joint probability term can be omitted.

EXERCISE S4   Draw a Venn diagram in which the events A and B would clearly be mutually exclusive.

Ans: See the end of the chapter.

Try testing your dexterity with Venn diagrams by completing the next exercise.

FIGURE 3.17

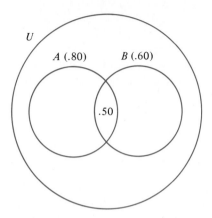

EXERCISE S5 Below is a Venn diagram describing the contents of a slightly different urn, one in which 80% of the balls are large (as shown in circle *A*), 60% of the balls are blue (as shown in circle *B*), and 50% of the balls are both large and blue (as shown in the intersection of *A* and *B*). Use Figure 3.17 to find

(a) the probability of "not *A*" ($P(A')$).
(b) the probability of *A* given *B* ($P(A/B)$).
(c) the probability of either *A* or *B* or both ($P(A \cup B)$).
(d) the probability of *B* given *A* ($P(B/A)$).
(e) Are *A* and *B* statistically independent?
(f) Are events *A* and *B* mutually exclusive?

Ans: (a) .2   (b) .833   (c) .90   (d) .625   (e) No   (f) No

(For a more detailed solution, see the end of the chapter.)

EXERCISE S6 Burns and Allen are co-workers at Acme Shoe, Inc. Thirty percent of the time Allen calls in sick, 20% of the time Burns calls in sick, and 15% of the time they both call in sick. Below is a Venn diagram that might be used to represent the situation. Subset (circle) *A* represents the event "Allen calls in sick," subset (circle) *B* represents the event "Burns calls in sick." Use Figure 3.18 to help answer the questions that follow.

(a) What is the probability that on a given day at least one of the two call in sick?
(b) What is the probability that neither calls in sick?
(c) How likely is it that Allen calls in sick but Burns does not?

FIGURE 3.18

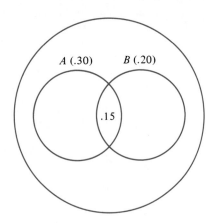

$A\ (.30)$ $B\ (.20)$

.15

(d) If Burns calls in sick, how likely is it that Allen also calls in sick?
(e) Are the events "Burns calls in sick" and "Allen calls in sick" statistically independent? Explain.

Ans: (a) .35   (b) .65   (c) .15   (d) .75
        (e) The events are not statistically independent.

(For a more complete solution, see the end of the chapter.)

CHAPTER SUPPLEMENT II: COMBINATIONS AND PERMUTATIONS

As suggested in the chapter, some probability assessment situations may involve little more than a basic counting procedure: count the number of (equally likely) ways an event can occur, count the total number of possible outcomes that may occur, and take the ratio of the two counts. In some instances, the counting required can be facilitated by two common computational expressions.

Combinations

Suppose you stop your car in the main parking lot to pick up some friends. You have room for four people, but five are waiting. How many different groups of four passengers could you assemble from the larger group of five? Answer: We can use the **combinations** formula

$$_nC_x = \frac{n!}{(n-x)!x!}$$

where ! = factorial operator $n \cdot (n-1) \cdot (n-2) \cdots (1)$
n = size of the larger group
x = size of the smaller "subgroups"

to count

$$_5C_4 = \frac{5!}{(5-4)!4!} = \frac{5 \cdot 4 \cdot 3 \cdot 2 \cdot 1}{(1)(4 \cdot 3 \cdot 2 \cdot 1)} = \frac{120}{(1)(24)} = 5 \text{ distinct groups}$$

which seems to make sense. If we designate group members as friend A, friend B, friend C, friend D, and friend E, the five distinct subgroups are

A,B,C,D   A,B,C,E,   A,B,D,E   A,C,D,E   B,C,D,E

EXERCISE S7   How many different samples (subgroups) of size 6 can be selected from a "population" (i.e., a larger group) of size 10?

Ans: $_{10}C_6 = \dfrac{10!}{4!6!} = 210$

EXERCISE S8   How many distinct poker hands (i.e., subgroups of size 5) can be selected from a standard deck of 52 playing cards?

Ans: $_{52}C_5 = \dfrac{52!}{(52-5)!5!} = 2,598,960$

Note: In making the computation, make sure you do some of the appropriate numerator and denominator "cancellations" so you can avoid having to compute 52! For example, $52!/47! = 52 \times 51 \times 50 \times 49 \times 48 = 311,875,200$.

EXERCISE S9   Refer to exercise S7, where we had determined that there are 210 different samples of size 6 that could be produced from a population of 10 items. Suppose the population here consists of 10 electronic units received by your company in a recent shipment. Suppose further that the shipment contains 3 defective items and 7 nondefective items.

(a) How many of the 210 samples of size 6 would contain exactly 2 defective units and 4 nondefective units? (*Hint:* Compute the number of ways you could end up with exactly 2 out of the 3 defective items together with 4 out of the 7 nondefectives. To accomplish this, you'll need to multiply $_3C_2$ by $_7C_4$.)

(b) Compute the *probability* of randomly choosing a sample of 6 that contains exactly 2 defectives and 4 nondefectives. (*Hint:* Using our F/T approach to

FIGURE 3.19

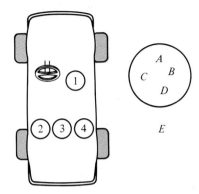

probability assessment, divide your answer to (a) by the total number of samples of size 6 that could be selected.)

Ans: (a) (3)(35) = 105   (b) $\dfrac{105}{210} = .5$

Permutations

Back to your car and your friends. Suppose now you've selected your "in group" of 4, consisting of friends A, B, C, and D. Suppose further you designate each of the four available positions in the car by number. The front passenger seat is position 1; the rear seat, left side is position 2; the rear seat, middle is position 3; and the rear seat, right side is position 4 (see Figure 3.19). How many different seating arrangements could you create with this particular group of four? As you start to make assignments, there are four candidates for position 1. Once someone is assigned to position 1, there are three remaining candidates for position 2, then 2 for position 3, and one for position 4. The result is therefore

$$4 \times 3 \times 2 \times 1 = 24 \text{ different arrangements}$$

If we use x to represent the subgroup size, we can generalize the counting procedure here to simply $x!$. For *any* of the possible subgroups of four, there are $x! = 4! = 24$ possible arrangements.

Now suppose you want to count—before you actually select your chosen few—all possible seating arrangements that could be created from the total group of five friends selected four at a time. With five distinct subgroups (i.e., combinations) possible $(5!/(4!1!) = 5)$ and 24 different arrangements available for each subgroup $(4! = 24)$, a total of $5 \times 24 = 120$ different arrangements are possible. In general, this sort of count can be made simply by multiplying the combinations expression

$$_nC_x = \frac{n!}{(n - x)! \, x!}$$

by the "arrangements per subgroup" expression $x!$ to produce the **permutations** expression.

$$_nP_x = \frac{n!}{(n - x)! \, x!} \cdot x! = \frac{n!}{(n - x)!}$$

As you can see, the number of permutations of n distinct items taken x at a time will always exceed the number of combinations, since permutations involve not only the number of distinct subgroups that can be formed but the number of possible *arrangements* for each subgroup as well.

EXERCISE S10 Determine the number of different possibilities (permutations) that exist if the Senate is to choose, from a list of eight candidates, three senators to head, respectively, the Armed Services Committee, the Commerce Committee, and the Ethics Committee.

Ans: $_8P_3 = \dfrac{8!}{(8 - 3)!} = 336$ possibilities

■ CHAPTER SUMMARY

We described the nature of basic probability establishing that it essentially involves a measurement of uncertainty (or chance or risk). To produce probabilities, we suggested several alternative probability assessment strategies: the classical approach (favorable/total), a relative frequency approach (occurrences/opportunities), and a subjective ("degree of belief") approach.

In the process of developing the major rules of probability manipulation (principally the multiplicative and additive rules), we introduced some of the standard terminology and rudimentary building blocks of probability analysis:

- **simple probability**—the probability of a single event occurring (sometimes called **marginal probability**)
- **conditional probability**—the probability of one event occurring given the occurrence of another event

- **statistical independence**—where the probability of one event occurring is unaffected by the occurrence of another event
- **joint probability**—the probability of two or more events occurring together
- **mutually exclusive events**—events that cannot occur together

Although it appears that virtually all probability problems ultimately reduce to a proper application of the multiplicative rule (for joint probabilities) and/or the additive rule (for either/or probabilities), it's not always obvious just how or when each of the rules should be applied. Consequently, we introduced a set of visual aids to help organize and solve common problem types: Venn diagrams to produce conditional results from simple and joint probabilities, probability trees to effectively address problems of a "sequential" nature, and joint probability tables to extend and adapt Venn diagrams to more complex situations.

Throughout our discussion of problem-solving procedures, there was a single common thread, one generalizable principle: in attempting to assess the probability of any given event, carefully and completely lay out the way(s) in which the event can occur. The visual aids can provide a framework for implementing this general problem-solving strategy.

Given the surprisingly few rules of basic probability theory, solving probability problems becomes mostly a matter of practice and repetition—developing the capacity and the confidence to correctly interpret each problem and to react creatively to problem variations.

KEY TERMS

Additive rule
Classical probability assessment
Combinations
Conditional probability
Joint probability
Joint probability table
Marginal probability
Multiplicative rule

Mutually exclusive events
Permutations
Probability
Probability tree
Simple probability
Statistically independent events
Subjective probability
Venn diagram

IN-CHAPTER EXERCISE SOLUTIONS

EXERCISE 2

Your Venn diagram should look like

$P(A) = .7$
$P(B) = .5$
$P(A \cap B) = .3$

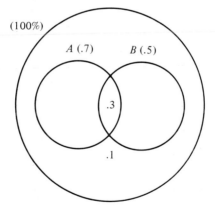

EXERCISE 3

Not surprisingly, we need to lay out the ways in which the event can occur:

Way 1:   "7 on roll 1" and "not 7 on roll 2"

or

Way 2:   "not 7 on roll 1" and "7 on roll 2"

or

Way 3:   "7 on roll 1" and "7 on roll 2"

(Exactly one 7 will do the trick, as will the "two 7's" result.)
For probabilities,

$$P(\text{Way 1}) = \left(\frac{1}{6}\right)\left(\frac{5}{6}\right) = \frac{5}{36}$$

$$P(\text{Way 2}) = \left(\frac{5}{6}\right)\left(\frac{1}{6}\right) = \frac{5}{36}$$

$$P(\text{Way 3}) = \left(\frac{1}{6}\right)\left(\frac{1}{6}\right) = \frac{1}{36}$$

Using the additive rule (for mutually exclusive events) to put it all together, we have

$$P(\text{Way 1 or Way 2 or Way 3}) = P(\text{Way 1}) + P(\text{Way 2}) + P(\text{Way 3})$$

$$= \frac{5}{36} + \frac{5}{36} + \frac{1}{36} = \frac{11}{36}$$

EXERCISE 4

(a) Referring to the tree below, we need only trace out the bottom path, the one involving
not 7 on the first roll and not 7 on the second, leading into end node 4, and multiply
probabilities along the way:

$$P(7' \text{ and } 7') = \left(\frac{5}{6}\right)\left(\frac{5}{6}\right) = \frac{25}{36}$$

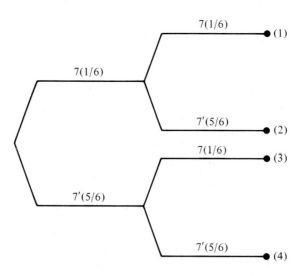

(b) Check those end nodes that correspond to no 7's or exactly one 7 (i.e., end nodes 2, 3,
and 4).

As we saw in part (a), the no 7's path leading into end node 4 involves not 7 on the first and not 7 on the second, producing

$$P(\text{no sevens}) = \left(\frac{5}{6}\right)\left(\frac{5}{6}\right) = \frac{25}{36}$$

We've dealt with the "exactly one 7" paths before (i.e., those leading into end nodes 2 and 3):

For end node 2: $\left(\dfrac{1}{6}\right)\left(\dfrac{5}{6}\right) = \dfrac{5}{36}$

For end node 3: $\left(\dfrac{5}{6}\right)\left(\dfrac{1}{6}\right) = \dfrac{5}{36}$

We can use the additive rule to produce

$$P(\text{no more than one 7}) = \frac{5}{36} + \frac{5}{36} + \frac{25}{36} = \frac{35}{36}$$

Note: To shortcut the process, we might have determined the probability of reaching node 1 and subtracted from 1: $1 - 1/36 = 35/36$.

EXERCISE 5

Using the tree below, we need only trace the (two) paths to the silver coin end nodes (shown with checks). Multiplying the associated probabilities for each path produces

Path 1: $(.5)(.75) = .375$

Path 2: $(.5)(.1) = .050$

Applying the additive rule results in

$$P(\text{silver coin}) = .375 + .050 = .425$$

(Clearly this is the equivalent of $1 - P(\text{gold coin}) = 1 - .575$.)

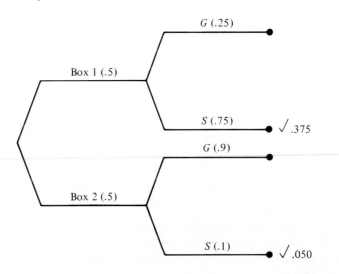

EXERCISE 6

We're provided here with two simple probabilities and one joint probability. Using D to represent drug use and A to represent alcohol use, we have

$$P(D) = .3, \quad \text{a simple probability}$$

$$P(A) = .6, \quad \text{a simple probability}$$

$$P(D \cap A) = .25, \quad \text{a joint probability}$$

The joint probability table below can be easily constructed from the information provided. Using the simple probabilities in the margins and the joint $P(D \cap A)$ in the upper left-hand cell, we have

Alcohol use

	A	A'	
D	.25	.05	.3
D'	.35	.35	.7
	.6	.4	1.0

(Drug use — row label on left)

(As suggested in the hint, we've used $P(D \cap A)$, the one joint probability provided directly, to produce all the joint probabilities involved. We've simply made use of the relationship "sum of the joint probabilities equals the marginal probability.")

(a) $P(D/A) = P(D \cap A)/P(A) = .25/.60 = .417$
(b) $P(A/D') = P(A \cap D')/P(D') = .35/.70 = .50$
(c) $P(D'/A') = P(D' \cap A')/P(A') = .35/.40 = .875$
(d) One form of the test would be: Is $P(D/A) = P(D)$? (In general, is conditional equal to simple?)

Based on our results in (a), the conditional $P(D/A) = .417$. Since the simple probability, $P(D)$, was given as .30, the events are clearly statistically dependent.

EXERCISE 7

Using S to represent the event "respondent supports the change," and 65 to represent the event "the respondent is 65 years of age or older," the given probabilities are

$$P(S) = .60, \quad \text{a simple probability}$$

$$P(65) = .30, \quad \text{a simple probability}$$

$$P(S/65) = .80, \quad \text{a conditional probability}$$

(Be careful. .80 is not a joint probability!) 65' will be used to represent the event "the respondent is under 65 years of age." S' represents the "nonsupporter" event. From the conditional $P(S/65)$ and the simple $P(65)$, we can use the general multiplicative rule to produce the joint probability $P(65 \cap S)$:

$$P(65 \cap S) = P(65)P(S/65) = (.30)(.80) = .24$$

With this one joint probability and the simple probabilities provided, the joint probability table is easy to create:

Age	Opinion: S	Opinion: S'	
65	.24	.06	.30
65′	.36	.34	.70
	.60	.40	1.00

Once constructed, the table can be used to readily produce all the desired probabilities:

(a) The joint probability $P(S \cap 65')$ can be read directly from the lower left-hand cell in the table: .36.

(b) The conditional probability

$$P(65/S') = P(65 \cap S')/P(S') = .06/.40 = .15$$

(c) The conditional probability

$$P(S'/65') = \frac{P(S' \cap 65')}{P(65')} = \frac{.34}{.70} = .486$$

EXERCISE 8

The corresponding Venn diagram is

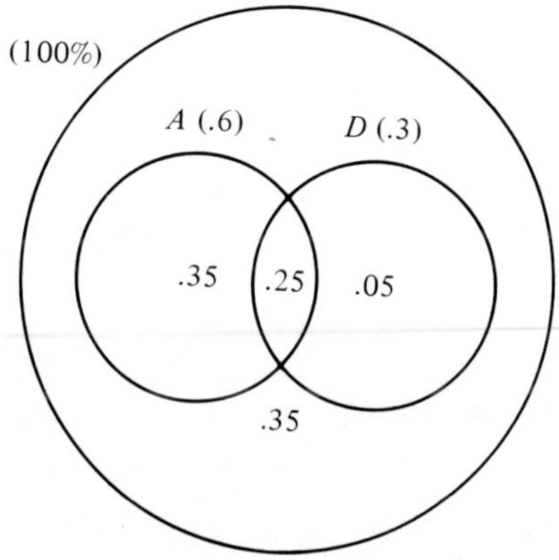

The following are solutions to the chapter-supplement exercises.

EXERCISE S1

We know from our previous work that $P(A \cap B) = .2$ (i.e., the probability of a large blue ball is .2). We also know that $P(B) = .6$ (i.e., the probability of a blue ball is .6). According to our joint/simple rule, then,

$$P(A/B) = \frac{P(A \cap B)}{P(B)} = \frac{.2}{.6} = .33$$

confirming the results of our earlier counting approach (where 2 of the 6 blue balls were large).

EXERCISE S2

We've already established that the overall probability of drawing a large ball ($P(A)$) is .5. We've also established (in Exercise S1) that if we know the ball selected is blue, then the probability that it's large ($P(A/B)$) is only .33. Since the simple and conditional probabilities here are not equal, we have dependent events. The likelihood of occurrence of event A (drawing a large ball) is definitely influenced by the occurrence of event B (drawing a blue ball).

We can also look at the other simple versus conditional comparison:

$$P(B) = .6 \quad \text{and} \quad P(B/A) = \frac{.2}{.5} = .40,$$

implying, as above, statistical dependence.

EXERCISE S3

The diagram below is just one of many possibilities. It shows

$$P(A) = .5 \quad \text{and} \quad P(A/B) = \frac{.2}{.4} = .5 \quad \text{(conditional = simple)}$$

as well as

$$P(B) = .4 \quad \text{and} \quad P(B/A) = \frac{.2}{.5} = .4 \quad \text{(conditional = simple)}$$

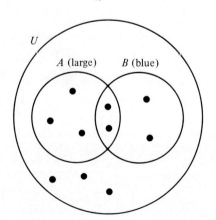

EXERCISE S4

This one's a little easier to create than the statistically independent case in Exercise S3. Just show no overlap (i.e., intersection) of the subset circles. For example,

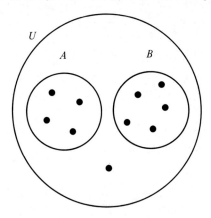

EXERCISE S5

The Venn diagram describes the situation visually. Notice that rather than showing individual points ("dots") within each of the circles, we're showing probabilities directly. We could show a universal set (sample space) total of 100 outcomes (dots), with 80 of the outcomes in the event A circle, 60 in the event B circle, and so on, but it's unnecessary. Everything here is represented on a "percentage" (probability) basis, with the universal set representing 100% of the outcomes.
Using the Venn diagram:

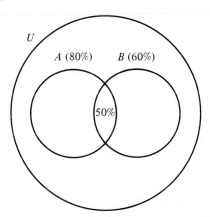

(a) $P(A')$ is simply everything in the universal set (or sample space) outside subset A. Here,
    $100\% - P(A) = 1.0 - .8 = .2$.
(b) Visually, we're interested in what part (percentage) of B is in the intersection of A and B. We can see this as an easy application of the conditional = joint/simple rule.

$$P(A/B) = \frac{P(A \cap B)}{P(B)} = \frac{.5}{.6} = .833 \text{ or } 83.3\%$$

(c) Visually, we need to identify everything that's in circle A along with everything that's in circle B. In rule form, this can be effectively translated to

$$P(A \cup B) = P(A) + P(B) - P(A \cap B)$$
$$= .8 + .6 - .5 = .9 \text{ or } 90\%$$

(d) We can simply repeat the pattern of part (b). We're interested in what part (proportion) of A is in the intersection of A and B.

$$P(B/A) = \frac{P(A \cap B)}{P(A)} = \frac{.5}{.8} = .625 \text{ or } 62.5\%$$

(e) The test for independence goes something like this: Is $P(A/B) = P(A)$? Or, alternatively, is $P(B/A) = P(B)$? (In general, is the conditional equal to the simple?) As we saw in part (b), $P(A/B) = .833$, which is clearly not equal to the .80 value for $P(A)$. Conclusion? Statistical dependence.

As we saw in part (d), $P(B/A) = .625$, which is clearly not equal to the .6 value for $P(B)$. Conclusion? Statistical dependence.

(f) The events are not mutually exclusive, since we are told that the probability that both will occur is .5, not 0, as required for mutually exclusive events. Looking at the Venn diagram, since the A and B circles overlap, we cannot be dealing with mutually exclusive events.

EXERCISE S6

The Venn diagram is

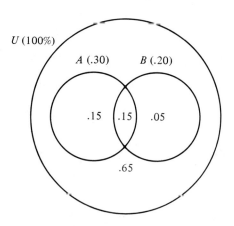

Using the diagram as reference, we have:

(a) $P(A \cup B) = P(A) + P(B) - P(A \cap B) = .30 + .20 - .15 = .35.$

(b) This must be equivalent to the area outside $A \cup B$, so

$$1 - P(A \cup B) = 1 - .35 = .65$$

(c) This must be equivalent to what's in A minus what's in the intersection of the two events:

$$P(A) - P(A \cap B) = .30 - .15 = .15$$

(d) This is the conditional $P(A/B)$. Using the conditional = joint/simple rule, we have

$$P(A/B) = \frac{P(A \cap B)}{P(B)} = \frac{.15}{.20} = .75$$

(e) The events are not independent since the conditional $P(A/B)$ does not equal the simple $P(A)$. That is, $.75 \neq .30$.

EXERCISES

1. For the cases listed below, indicate whether the appropriate probability assessment strategy could best be characterized as classical, relative frequency, or subjective.
   (a) the probability of passing your statistics course.
   (b) the probability that you draw a "full house" poker hand.
   (c) the probability that the Detroit Lions will win the Super Bowl this year.
   (d) the probability that, when your pizza is finally delivered, it will be cold.
   (e) the probability that you choose the "short straw" when it comes time to pay the check.
   (f) the probability that at least one of the eggs in the dozen you've just purchased will be cracked.
   (g) the probability that you win the church raffle (a 1966 Plymouth Duster) if you've purchased 37 of the 200 tickets sold.

2. Which of the following event pairs would seem to be statistically independent?
   (a) exercising regularly, having high blood pressure.
   (b) playing professional sports, making a million dollars.   *No*
   (c) reading at least one book per week, having a full-time job.   *yes*
   (d) reading *Time* magazine, reading *Mad* magazine.   *yes*
   (e) voting Democrat, owning a riding lawn mower.   *yes*
   (f) studying 12 hours a day, sleeping 15 hours a day.   *No*

3. Refer back to the event pairs in Exercise 2. Which would seem to involve events that are mutually exclusive?

4. Two door-to-door salesmen have been sent into a new neighborhood. Their supervisor assesses a .6 probability that salesman A will make a sale,

and a probability of .8 that salesman B will make a sale. She also believes there is a .5 probability that both salesmen will make a sale.
   (a) Construct a Venn diagram to represent the situation.
   (b) What is the probability that at least one of the salesmen makes a sale?
   (c) What is the probability that salesman A will make a sale if salesman B makes a sale?
   (d) What is the probability that both salesmen fail to make a sale?
   (e) Is the sales performance of A independent of the sales performance of B? Explain.

5. As the night editor for the *Daily Planet*, you are waiting for two of your reporters, Abbott and Bocelli, to phone in breaking stories before the paper's 3 A.M deadline. You decide that the probability of Abbott beating the deadline is .6. The probability that Bocelli will beat the deadline is .4. The probability that both will beat the deadline is .10.
   (a) If Abbott beats the deadline, how likely is it that Bocelli will?
   (b) How likely is it that at least one of the reporters beats the deadline?
   (c) How likely is it that neither reporter beats the deadline?
   (d) Are the two events statistically independent? Explain.
   (e) Construct a Venn diagram to represent the situation.

★ 6. In a recent survey of local businessmen and businesswomen, 40% of the respondents said they were regular readers of *Newsweek*, 70% said they were regular readers of *Time*, and 30% said that they read both magazines.
   (a) What percentage of the respondents read at least one of the magazines?
   (b) What percentage of the respondents read neither magazine?

(c) What percentage of the respondents read *Time* but not *Newsweek*? (*Hint:* You might try a Venn diagram to display the problem.)

(d) If a respondent is selected at random, and it turns out that that person is a reader of *Time*, how likely is it that that person is also a reader of *Newsweek*?

(e) Are the events "reader of *Time*" and "reader of *Newsweek*" statistically independent?

(f) Are the events "reader of *Time*" and "reader of *Newsweek*" mutually exclusive?

7. During its next session, Congress is given a 60% chance of taking serious measures to reduce the federal deficit. Congress is also given a 50% chance of voting a substantial pay raise for its members. If you estimate a 40% probability that it will do both,

(a) how likely is it that the Congress will do neither?

(b) if Congress acts seriously on the deficit, how likely is it that it will vote itself a pay raise?

(c) how likely is it that Congress will vote itself a pay raise but will not act seriously on the deficit? (*Hint:* Try a Venn diagram.)

8. GBC, a newly established television network, is introducing two television shows this season. It estimates that "Live and In-Person," an interview show featuring pre-Metallica figures from the world of music, has a 60% chance of being a hit. The other show, "Inside Poultry," a look at the seamier side of the wholesale chicken business, is given a 30% chance of being a hit. The likelihood that both will be a hit is put at 10%.

(a) What is the probability that at least one of the shows will be a hit?

(b) What is the probability that neither show will be a hit?

(c) If "Inside..." turns out to be a hit, how likely is it that "Live . . ." will also be a hit?

(d) Show a Venn diagram to represent the situation here.

★ 9. For an upcoming space probe, you have designed a key electronic component, which, during preflight tests malfunctioned 10% of the time. Consequently you've decided to install an identical unit (i.e., one with a 10% failure rate) as a backup. Based on tests for the two components together, you have found that both the components fail simultaneously about 6% of the time.

(a) What is the probability that at least one of the components will function properly? (*Hint:* The only way this would not happen is if both the components failed.)

(b) What is the probability that the backup component will fail given that the primary component fails?

(c) Is the performance of one component statistically independent of the performance of the other? Explain.

10. Needing to get to Houston tomorrow, you book a flight on two separate airlines, knowing there is a chance that the flights will be overbooked. In fact, you read an article earlier in the day that claimed 6% of all airline flights are overbooked. Construct an appropriate probability tree to answer the questions below. (Assume that flight overbookings are statistically independent.)

(a) What are the chances that both your flights will be overbooked tomorrow?

(b) What are the chances that neither will be overbooked?

(c) What are the chances that exactly one will be overbooked?

★ 11. To promote your movie career, you have hired two theatrical agents. The first of the two has a fairly strong track record. He is successful in finding a movie part for young hopefuls like yourself about 60% of the time. The second agent has been considerably less successful. She is successful in finding parts for young hopefuls only about 20% of the time. Assuming that each agent acts independently, what is the probability that you will end up with at least one movie part? (Construct a probability tree to organize the problem.)

★ 12. Ten people are sitting in IBC's corporate conference room: 6 from the New York office and 4 from the Medford office. If you randomly select a sample group of 3 people,

(a) what is the probability that you would end up with all three people from New York?

(b) what is the probability that your sample would be comprised of one person from New York and two from Medford?

(c) what is the probability that no one from New York is selected?

(*Hint:* Try a probability tree to organize your solution.)

13. You have an important document that has to be

delivered to your Detroit office tomorrow morning. You call a number of overnight delivery services and find that each one claims an 80% probability that it can deliver your document on time. Concerned about your future with the firm, you decide to select two of the services, each of which will be given a copy of the document to deliver. Construct a probability tree to answer the following questions:

(a) How likely is it that your Detroit office will receive the document it needs tomorrow morning?

(b) Still nervous, you decide to go with three services. Now how likely is it the Detroit office receives what it needs tomorrow morning?

(c) How many services should you use if you want to allow no more than 1 chance in 1000 that the document will not be received in time?

14. Three solar batteries are installed in each DPC communications satellite placed in orbit. Testing indicates that each of the batteries is 75% likely to remain in good working order for at least 10 years. The satellite will function as designed if at least two of the three batteries remain in good working order. Assuming the batteries are independent, how likely is it that the satellite will function as designed for at least 10 years?

★ 15. As a resident of Fun City, you naturally feel the need to buy two vicious watchdogs. The dogs' trainer asserts that each dog is 70% likely to detect and "discourage" any intruder.

(a) Assuming that each dog acts independently, how likely is it that an intruder would go undetected?

(b) Still insecure, you are considering two additional measures:

(1) adding one more dog, identical to the others, or

(2) retraining the two original dogs so that each would have a 90% probability of detecting an intruder.

Which alternative would you recommend? Explain.

16. Olympic testing for illegal steroid use by participating athletes involves three independent laboratories. Each specimen submitted by an athlete is separated into three parts. Each part is sent to a different laboratory. If a subject athlete is using steroids, each laboratory is 95% likely to detect the presence of the banned substance. How likely is it that an athlete who is using steroids can escape detection?

★ 17. As project manager, you estimate that there is a 50% probability that completion of your current research program will come within the next month. You concede that it is 30% likely that the program will take two months, and 20% likely it will take a full three months. You also have reason to believe that if the program is completed within one month, there is a 65% likelihood that you can have a significant new product on the market by next Christmas. If the research program takes two months, the chances of meeting the Christmas deadline for new product introduction drops to 40%; and if the research program takes three months, there is only a 10% probability of new product introduction by Christmas. How likely is it that you will not have the new product on the market by Christmas? (*Hint:* Try a probability tree.)

18. Alkaline State University is playing an upcoming four-game series in soccer against archrival Phobic College. Alkaline coaches are convinced that Alkaline should be a 3:2 favorite in each game (that is, Alkaline has a 60% probability of winning each game).

(a) How likely is it that Alkaline will "sweep" the series (win all four games)?

(b) How likely is it that Alkaline will not win a single game?

(c) How likely is it that Alkaline will "split" the series with Phobic (win two and lose two)?

★ 19. You have just received a large shipment of components from your main supplier. Your quality control people have devised a plan for evaluating the shipment by randomly selecting items and subjecting them to rigorous testing: Select one item and test it. If it fails the test (i.e., proves to be defective), send back the entire shipment. If it passes, take a second item to test. If this fails the test, send back the entire shipment. Otherwise, take a third item, and so on. Such testing will continue up to a maximum of five items. If all five items test OK, you will accept the entire shipment and use the components in your productive operations. If the shipment contains just 3% defective components,

(a) how likely is it that you will end up testing exactly three items?

(b) how likely is it that you will eventually decide to accept the shipment?

(c) how likely is it that you will need to test at least three items before you reach a decision either to accept or send back the shipment?

20. You are at the Screaming Slalom Ski Basin and estimate that you have a 25% likelihood of successfully navigating a run down the difficult Blue Slope without a serious fall. Today you plan to make four trips down Blue. However, if you have a serious fall on any one of these downhill trips, that's it—you're done for the day.

(a) how likely is it that you will not make it to the third run?

(b) how likely is it that you will survive the full four-run day without a serious fall?

(c) Suppose you make it through the first three runs unscathed. How likely is it that you will fall on the final run?

(d) Suppose, because of fatigue, the chance of a serious fall increases about 2 percentage points with each run. How likely is it that you will make it through the entire four-run day without a serious fall?

21. The Congressional Budget Office estimates that if the economy continues to grow at the current rate, there is a 95% probability that the federal deficit will be eliminated without new taxes by 1998. If the economy slows, that probability drops to .30, but if it accelerates the probability increases to 99%. Economists generally agree that there is a 60% chance that the economy will continue to grow at the current rate, a 25% chance that it will slow, and a 15% chance that it will accelerate. Use this information to estimate the likelihood that the federal deficit will be eliminated without new taxes by 1998. (Try a tree diagram to organize your solution.)

★ 22. You are trying to assess the likelihood that your company's proposed new "Royal Family" product line will be successful. Your plan is to develop the new line only if the probability of success appears to be more than 60%. Success, you believe, will be linked to two principal factors: general economic conditions over the next two years, and the behavior of your major competitor. If the economy booms, you believe there is an 80% probability that your major competitor will introduce a competing product line. If the economy advances moderately, that probability drops to 40%, and if the economy declines significantly, the probability falls to only 10%. Based on your best analysis, you believe that the economy is most likely to advance moderately over the next 2 years (probability .6), but you also assign a probability of .3 that the economy will boom, and a .1 probability that the economy will decline significantly.

If your competitor introduces its competing product line in a booming economy, you estimate that it is 70% likely that your "Royal Family" line will be successful. On the other hand, if your major competitor introduces its competing line in a declining economy, your company's proposed new line is given only a 1-in-5 chance of succeeding. If the competitor introduces its product line in a moderately advancing economy, you believe that there is a 40% chance that your new line would succeed. Finally, if your competitor decides not to introduce its competing line at all, you anticipate an 80% chance that your firm's "Royal Family" line will be successful, irrespective of the economy. Should your company proceed with plans to develop the new product line? (Try a tree diagram to organize your solution.)

23. You want to estimate your chances of getting tickets to next month's State U.–Acme Tech basketball game. (You don't have the money at the moment to buy the tickets in advance.) You're convinced that the availability of tickets will be affected by such factors as the win-loss record of the two teams and whether the game will be televised. Looking at the respective schedules for the two teams during the month preceding the game, you estimate that currently undefeated State has a .6 probability of not losing a game between now and the Acme matchup. Acme, on the other hand, is given only a .3 chance of being undefeated going into the State game. If only one of the teams is undefeated, you believe there is a 50-50 chance that the game will be on television (hello, Dick Vitale). If both teams are undefeated, the probability of television coverage increases to 90%. If neither team is undefeated, the chance of TV drops to only 20% (so long, Dick Vitale.) If the game is televised, you believe you can reasonably assign a .30 probability that tickets will be available for the game on game day. If the game is not on TV,

then the probability of available tickets decreases to about 10%. How likely is it that tickets will be available on game day? (Try a tree diagram to organize your solution.)

24. The main core at the Crown Mesa nuclear facility has 10 critically important components, each of which must function perfectly for the core to remain within safe boundaries. Each component is 99.7% reliable. If the performance of each component can be considered independent of the others, how likely is it that the core will not stay within safe boundaries?

★ 25. As a news reporter for the *Daily News–Observer–Chronicle*, you are at a political gathering attended by both Republicans and Democrats. In fact, 70% of those in attendance are Republicans and 30% are Democrats. According to best estimates, 90% of all Republicans favor deploying the proposed GHTC defensive missile system, but only 20% of the Democrats favor deployment.

(a) Suppose you plan to choose someone randomly from the crowd to interview. How likely is it that you would choose a supporter of missile deployment? (*Hint:* Try a probability tree.)

(b) Having chosen someone randomly, suppose you now find out that this person is a supporter of deployment. How likely is it that this person is a Democrat?

(c) Suppose the person you have chosen is not a supporter of deployment. How likely is it that that person is a Republican?

26. Seventy percent of your spare parts inventory comes from Mojne Supply, the remaining 30% from Grove-Afton Manufacturing. Over the years, 4% of the Mojne-supplied components have proven defective, and 7% of the Grove-Afton items have been defective.

(a) If you select an item at random, how likely is it that the item will be defective?

(b) If the item you select is defective, how likely is it that the item was supplied by Mojne?

27. In a recent political poll of voters, 40% of the respondents were men and 60% were women. Eighty percent of the men and 60% of the women support the president's proposal for placing a spending limit on national political campaigns.

(a) If a respondent is selected at random, how likely is it that this respondent supports the president's proposal?

(b) If you select a respondent at random and discover that this respondent doesn't support the president's proposal, how likely is it that this person is a male?

★ 28. You are about to implement a drug-testing procedure for company employees. A recent (anonymous) survey revealed that, similar to national figures, about 20% of your employees admitted to using illegal drugs. The random drug-testing procedure you are about to employ is not infallible. In fact, about 5% of the time it will produce a "false positive"—that is, if the person being tested is not a drug user, there is a 5% probability that the test will nevertheless identify that person as a drug user. The test also has a probability of .08 of producing a "false negative"—about 8% of the time a drug user will be identified as a nonuser.

(a) If an employee is selected at random, how likely is it that the drug test will be positive (i.e., how likely is it that the test will identify the person as a drug user)?

(b) If the drug test is positive, how likely is it that the employee is actually a drug user?

29. Forty-five percent of all homes on the local market sell quickly (within 20 days of listing), 35% sell at a moderate rate (within 21 to 50 days), the remaining homes sell slowly (longer than 50 days). Pilot Realty claims that it accounts for 90% of the quick sales (20 days or less), 30% of the moderate sales (21–50 days), and 10% of the slow sales (more than 50 days).

(a) Overall, Pilot accounts for what percentage of local home sales?

(b) Based on the figures above, if Pilot lists a house what is the likelihood that the house will sell quickly?

★ 30. A questionnaire concerning the effect of a recent advertising campaign was sent out to a sample of 500 consumers. Results of the study are reported in the joint probability table

	Saw Ad	Didn't See Ad	
Purchased product	.2	.1	.3
Didn't purchase product	.2	.5	.7
	.4	.6	

(a) If a consumer is selected from the survey at random, how likely is it that he or she either saw the ad or purchased the product (or both)?

(b) How likely is it that a randomly selected consumer in the survey purchased the product?

(c) If a randomly selected consumer in the study purchased the product, how likely is it that he or she saw the ad?

(d) If a randomly selected consumer in the study is known to have seen the ad, how likely is it that he or she purchased the product?

(e) Are the events "seeing the ad" and "purchasing the product" statistically independent? Explain.

31. A survey of 400 graduate and undergraduate students was taken to study student opinions on a number of current social issues. The table below reports responses to the question: "How important is a four-year college education in finding a satisfying and fulfilling career?"

	Very important	Important	Unimportant
Undergrads	80	60	40
Grads	100	50	70

(a) Create from the raw data an equivalent joint probability table.

(b) If you were to randomly select a student from the survey, how likely is it that you would select a graduate student?

(c) If you were to randomly select a student from the survey, how likely is it that you would select an undergraduate student who responded "very important" to the survey question?

(d) If you were to randomly select a student from the survey, how likely is it that the person selected would be a graduate student or someone who responded "important" to the question?

(e) If you were to randomly select a student from the survey, and you learn that this student responded "very important" to the question, how likely is it that this student is an undergraduate?

(f) Does student classification (i.e., undergrad,

grad) appear to be statistically independent of response? Explain.

★ 32. Refer to problem 4. Construct a joint probability table for the situation described there. Use the table to compute
(a) the probability of at least one sale.
(b) the probability of exactly one sale.
(c) the probability that salesman B will make a sale given that salesman A doesn't.
(d) the probability that salesman A does not make a sale given that salesman B does not make a sale.
(e) If you haven't already done so, draw a Venn diagram for problem 4 and compare it to the joint probability table here. Match every entry in the joint table to a corresponding area in the Venn diagram.

33. Refer to problem 5. Construct a joint probability table to compute
(a) the probability that at least one of the reporters will beat the deadline.
(b) the probability that Bocelli will beat the deadline given that Abbott does not.
(c) the probability that Abbott will not beat the deadline given that Bocelli does.
(d) If you haven't already done so, draw a Venn diagram for problem 5 and compare it to the joint probability table here. Match every entry in the joint table to a corresponding area in the Venn diagram.

★ 34. Refer to problem 6. Construct a joint probability table to compute
(a) the percentage of respondents who read neither *Time* nor *Newsweek*.
(b) the percentage of respondents who read *Time* but not *Newsweek*.
(c) the probability that a randomly selected respondent reads *Newsweek* given that the respondent does not read *Time*.
(d) the probability that a randomly selected respondent does not read *Time* given that the respondent does not read *Newsweek*.
(e) If you haven't already done so, draw a Venn diagram for problem 6 and compare it to the joint probability table here. Match every entry in the joint table to a corresponding area in the Venn diagram.

★ 35. Refer to problem 26. If you haven't already done so, construct from your results a full joint probability table.

36. Refer to problem 25. If you haven't already done so, construct from your results a full joint probability table.

37. Refer to problem 27. If you haven't already done so, construct from your results a full joint probability table.

38. Two key guidance control units for the space shuttle have been subjected to repeated testing. Unit Y functioned properly in 96% of the tests. Unit Z functioned properly in 88% of the tests. Curiously, in 90% of the tests in which unit Y failed, unit Z also failed.
    (a) What is the probability that unit Y fails? What is the probability that unit Z fails?
    (b) How likely is it that both units fail? (*Hint:* Be sure you properly interpret the 90% probability given here as a conditional probability.)
    (c) Construct a joint probability table for this situation.
    (d) If unit Z functions properly, how likely is it that unit Y will function properly?
    (e) Are the performance levels of the two units statistically independent? Explain.

★ 39. Local law enforcement authorities are concerned about the growth of teenage gangs in the city. They estimate that 65% of juveniles arrested are gang members. If the crime involved is drug-related, probability of gang membership jumps to 86%. Crime statistics show that 70% of all juveniles arrested are arrested for drug-related crimes.
    (a) If an arrest case involving a juvenile is chosen at random from police files, what is the probability that the case involves a gang member in a drug-related crime? (Be sure you properly identify the 65%, 86%, and 70% probabilities given in the problem.)
    (b) If you learn that the case involves a gang member, what is the probability that the case also is drug-related?
    (c) If you learn that the youth involved is not a gang member, what is the probability that the crime involved is drug-related? (*Hint:* Construct a joint probability table.)

40. Twenty-eight percent of all consumers have made at least one purchase from the Home Shopping Service (HSS), a television shoppers' network. Forty percent of the HSS purchasers are men. If we assume 50% of all consumers are women,

(a) determine the probability that a randomly selected consumer would be a woman who has made at least one purchase from HSS. (*Hint:* Construct a joint probability table. Be sure to properly identify the 28%, 40%, and 50% given probabilities.)
(b) what proportion of male consumers have made at least one purchase from HSS?
(c) determine the probability that a randomly selected consumer would be either a woman or made at least one purchase.

The following problems are based on the material from Chapter Supplement II.

★ 41. Twelve players are waiting on the sidelines to play some pickup basketball at the old Sommers gym. Five players are needed to play the winners of the game currently in progress.
    (a) How many different teams (subgroups) of size 5 could be selected from the group of 12 waiting players?
    (b) Suppose four of the waiting players consider themselves to be guards and the remaining eight consider themselves to be forwards. How many different groups of 5 could you form that would include exactly 2 guards and 3 forwards? (*Hint:* Follow the pattern shown in exercise S9 in the chapter supplement.)

42. Your salesforce consists of 15 salespeople, 10 veterans and 5 rookies. You plan to randomly select six of these people to represent the firm at the January national convention in Duluth.
    (a) Overall, how many different groups of size 6 are possible?
    (b) How many of the groups of 6 would contain exactly 3 veterans and 3 rookies?
    (c) How *likely* is it that the group you select randomly will consist of exactly 3 veterans and 3 rookies?
    (d) How likely is it that the group you select randomly will consist of exactly 2 veterans and 4 rookies?

43. Six different jobs need to be done sometime today. You have a pool of nine available workers. You can assign only one worker to each job and no more than one job to a worker. How many different worker/job arrangements are possible? (*Hint:* Use the permutations expression for your calculation.)

PROBABILITY DISTRIBUTIONS

CHAPTER OBJECTIVES

Chapter 4 should enable you to:

1. Describe the general nature of a probability distribution.

2. Apply the rules of basic probability to create a probability distribution.

3. Differentiate discrete and continuous probability distributions.

4. Describe the role of a random variable in forming a probability distribution.

5. Recognize a binomial situation and produce appropriate binomial probabilities using the binomial function and the associated binomial table.

6. Recognize the bell-shaped normal distribution and use the normal table to produce corresponding probabilities.

7. Identify a Poisson situation and use the Poisson table and the Poisson function to produce appropriate probabilities.

8. Use the normal and Poisson distributions to approximate hard-to-find binomial probabilities. (Chapter Supplement)

9. Use the exponential distribution to produce probabilities for various time, space, or distance intervals between the occurrence of Poisson events. (Chapter Supplement)

Having introduced the elements of basic probability theory in Chapter 3, we plan now to broaden our focus to include a discussion of probability *distributions*. As we proceed, we'll identify the general nature of probability distributions and introduce several special cases in which certain well-defined mathematical functions can be used to produce the probabilities required to form a distribution.

4.1 THE GENERAL NATURE OF A PROBABILITY DISTRIBUTION

Simply stated, a **probability distribution** is nothing more than a full listing of the probabilities assigned to all possible events in a given probabilistic situation or "experiment." (As used here, the term **experiment** refers to any activity that involves uncertain (i.e., "random") outcomes: rolling dice, playing a series of baseball games, taking an exam, trading stocks, etc.) To create a distribution, we'll need both the capacity to identify all possible outcomes and the ability to effectively assess the full array of corresponding probabilities.

In a very real sense, we already have (through our discussions in Chapter 3) the capability to produce precisely the kinds of values necessary to form a full-fledged probability distribution. The example below illustrates the point.

Situation: You supervise a fairly large department within your firm and anticipate that sometime in the next few months there will be two mid-level management positions opening up. You hope to fill the expected vacancies with qualified employees currently working in other jobs within the department. In order to guarantee proper qualifications, you decide that any serious candidate must complete an intensive management training program.

Because the training is particularly rigorous, not everyone who participates in the program is certain to complete it. As a consequence, you plan to send three job candidates—Smith, Jones, and Dunn—in the hope that two, *exactly* two, will succeed. (If fewer than two succeed, then obviously you won't have enough qualified people to fill the vacancies. On the other hand, if all three people successfully complete the program, one will be very unhappy because there are only two positions available.)

Evaluating credentials carefully, you decide that Smith has approximately 3 chances in 4 of successfully completing the course. Jones, too, is given a 3-in-4 chance. Dunn's chances, in your view, are slightly less, about 50-50.

How likely is it that the desired result—exactly two people pass the course—occurs?

Solution: Not surprisingly, we can use precisely the same "layout the ways in which

FIGURE 4.1   A PROBABILITY TREE FOR THE MANAGEMENT TRAINING EXAMPLE

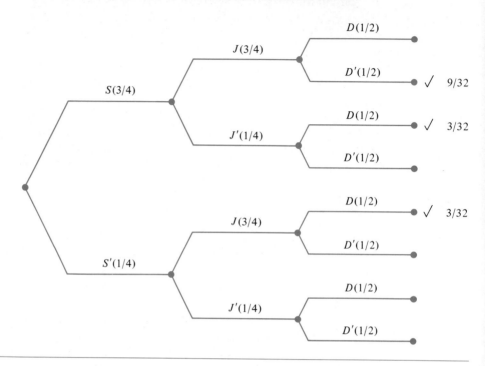

the event can occur" strategy that we saw repeatedly applied in Chapter 3. Exactly how might two people pass the course?

Way 1: Smith passes, Jones passes, Dunn fails

Way 2: Smith passes, Jones fails, Dunn passes

Way 3: Smith fails, Jones passes, Dunn passes

The tree diagram in Figure 4.1 visually traces the possibilities (notice the three pertinent end nodes—those corresponding to the "two pass" event—have been checked off). Assigning probabilities to each of the outcome branches on the tree allows us to compute the appropriate end-node probabilities (9/32, 3/32, and 3/32, respectively) for the two-pass event. Collecting these results in now-familiar fashion produces

$$P(2 \text{ pass}) = \frac{9}{32} + \frac{3}{32} + \frac{3}{32} = \frac{15}{32}$$

Our conclusion? The probability of exactly two of the candidates successfully completing the training course is just under 50%—15 chances in 32.

It's clear that we have the capacity to answer the basic "two-pass" probability question. It should be equally clear that we have the capacity to answer any number of variations on the basic question: the probability that no one passes the course, that all three pass the course, and so on. We have then, the capability to generate the full probability *distribution*.

4.2   CREATING THE DISTRIBUTION

We can begin by formalizing things just a bit.

Defining the Random Variable

To this point in our management training example, we've been describing experimental results (or events) in terms of "number of people passing the course." In the language of probability distributions, this _numerical description of possible events_ could be designated the **random variable** for the experiment. (Had we chosen to describe events here in terms of "number of people failing the course," this numerical description would have replaced "number of people passing" as the designated random variable.)

Identifying Values for the Random Variable

Given our random variable definition, suppose you were asked now to determine the possible *values* for the random variable? Easy enough. In our three-person experiment, the random variable "number of people passing" can take on any of four distinct values: 0 (i.e., no one passes), 1 (exactly one person passes), 2 (exactly two people pass), or 3 (everyone passes).

Assigning Probabilities to Each Value of the Random Variable

Following the same pattern as in the "two-pass" solution, we can readily assign probabilities to each of the four random variable values. (Referring to the tree should enable you to quickly confirm the values shown in the table.)

Number of people passing	Probability
0	1/32
1	7/32
2	15/32
3	9/32

If x is used to represent the "number of people passing" random variable, then the table can be shown as

x	$P(x)$
0	1/32
1	7/32
2	15/32
3	9/32

and labeled the full probability distribution for the management training experiment.

Summarizing a General Distribution-Building Procedure

Based on what we've done so far, we might summarize a general distribution-generating procedure as follows: (1) identify an appropriate random variable to numerically describe results (events) in the experiment, (2) list all possible values for the random variable, and (3) properly assign probabilities to every possible value for the random variable.

> **Note:** As a check of your work, it should always be true that the sum of individual probabilities in a properly defined distribution is 100%, implying that the events listed are mutually exclusive (no "overlapping" events) and collectively exhaustive (the list of events is complete).

4.3 REPORTING AND SUMMARIZING DISTRIBUTION RESULTS

Graphing the Distribution

It may sometimes be useful to show a probability distribution in graphical form. In the Smith, Jones, and Dunn training example, we can create an appropriate graph by drawing a horizontal axis to record values for the random variable x (number of people passing) and a vertical axis to show the $P(x)$ probabilities. Figure 4.2 shows the result. Asked to characterize the shape of the distribution, we might describe it as slightly skewed in the negative direction.

The Distribution Mean

The center of the distribution—the distribution mean—can be easily determined. Noting the similarity between what we're now calling a probability distribution and what we earlier (in Chapter 2) labeled a relative frequency distribution, we'll simply borrow the computational expression

$$\bar{x} = \sum x \cdot P(x)$$

Applied to the current example, the expression produces a distribution mean of

$$0\left(\frac{1}{32}\right) + 1\left(\frac{7}{32}\right) + 2\left(\frac{15}{32}\right) + 3\left(\frac{9}{32}\right) = \frac{64}{32} = 2$$

As a matter of terminology, the mean of a probability distribution is often referred to as the *expected value* of the distribution, or, more precisely, the expected value of the random variable. Having used x to represent the random variable, we'll label the expected value of x as $E(x)$ and show

$$E(x) = \bar{x} = \sum x \cdot P(x)$$

FIGURE 4.2   **GRAPHING THE DISTRIBUTION**

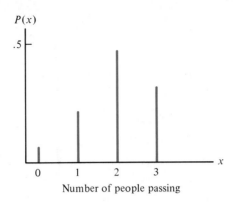

Number of people passing

Interpreting an expected value can be a little tricky. In the training example, we're basically saying that if this three-person experiment were repeated again and again, retaining in each case the same individual chances of passing and failing, then *on average* two people would pass the course.

The Standard Deviation of the Distribution

Consistent with our method for producing the mean of a probability distribution, the standard deviation can be computed by using an approach that we first saw in Chapter 2 to compute the standard deviation of a relative frequency distribution. Specifically,

$$\text{STD DEV}(x) = \sqrt{\sum (x - \bar{x})^2 P(x)}$$

or, substituting $E(x)$ for $\bar{x}$,

$$\text{STD DEV}(x) = \sqrt{\sum (x - E(x))^2 P(x)}$$

For our management training distribution, this means

$$\text{STD DEV}(x) = \sqrt{(0-2)^2 \frac{1}{32} + (1-2)^2 \frac{7}{32} + (2-2)^2 \frac{15}{32} + (3-2)^2 \frac{9}{32}}$$

$$= \sqrt{\frac{4}{32} + \frac{7}{32} + 0 + \frac{9}{32}}$$

$$= \sqrt{\frac{20}{32}} = .79$$

Our work in the management training example, then, has produced a slightly skewed distribution in which the average (or "expected") number of people passing the course is 2. Variability around the expected result is reflected in a distribution standard deviation of .79.

EXERCISE 1 Suppose you plan to toss a coin three times. To make things interesting, assume that the coin has a greater propensity for turning up tails rather than heads. In fact, assume this "loaded" coin has only a 40% likelihood of turning up heads on any one toss (implying, naturally, a 60% chance of turning up tails).

(a) Produce the full probability distribution appropriate to the situation, using as your random variable x the number of heads produced. Start by computing $P(0)$, the probability of no heads turning up.
(b) Graph the distribution.
(c) Compute and interpret the distribution mean.
(d) Calculate the standard deviation of the distribution.

Ans: (a)

x	$P(x)$
0	.216
1	.432
2	.288
3	.064

(b) $E(x) = 1.2$. If this 3-toss experiment were repeated many times, the average number of heads produced would be 1.2.
(c) STD DEV$(x) = .85$

(For a more detailed solution, see the end of the chapter.)

Note: Don't proceed to the next section until you've completed the exercise above. Much of the discussion that follows is built around the results of your work here.

4.4 SPECIAL-PURPOSE PROBABILITY DISTRIBUTIONS

Having seen how the basic rules of probability can be used to create a simple probability distribution, we'll now consider a special set of so-called "theoretical" or special-purpose probability distributions. At the core of each, we'll find a concise mathematical function that can be used to efficiently produce the range of probabilities necessary to describe a full distribution. (Each such function will serve as an effective generator of probabilities only in specific, well-defined circumstances.) While a number of these distributions appear in the statistical literature, we plan to focus on three of the more prominent ones: the **binomial**, the **normal**, and the **Poisson**.

4.5 THE BINOMIAL DISTRIBUTION

Refer to the last exercise, the one in which you were asked to produce the probability distribution for a simple coin-toss experiment. Your approach there was probably fairly conventional, perhaps built around a tree diagram which laid out all the relevant possibilities. We can now introduce an alternative approach that (in the long run, at least) may prove more appealing.

By recognizing in this experiment the presence of four specific conditions (which we'll see shortly), you could have identified the situation as "binomial" in nature, and subsequently made use of the binomial function to produce the full array of "number of heads" probabilities. In the process, you would have streamlined the probability-generating procedure significantly. To see how it works, we'll first have to define the binomial conditions.

The Binomial Conditions

For a situation (experiment) to qualify as binomial, the following conditions need to be met:

1. The situation must involve a number of "trials"—a repetition of the same act again and again. The letter n is frequently used to designate the number of trials involved. (Clearly the coin toss experiment can be cast in these terms. Each toss corresponds to a trial.)

2. Only *two* outcomes are possible on each of the trials. (This is the *bi* part of the "binomial.") One of the outcomes is frequently labeled "success," the other "failure." (In the coin toss experiment, for example, we might designate "heads" as the "success" event and "tails" as the "failure".)

3. The trials must be *statistically independent*. That is, whatever happens on the first trial can't influence what's likely to happen on the second; what happens on the first two trials can't influence what's likely to happen on the third, and so on. (This condition is clearly evident in the coin toss situation.)

4. The probability of success on any one trial must be known, and it must remain constant throughout the entire n-trial experiment. (For example, if the coin in our three-toss experiment has a 40% chance of turning up heads on the first toss, then that 40% probability must hold for every toss. The coin can't change character during the experiment.) We'll normally use p to represent the probability of success on any one trial.

Note: The term *Bernoulli process*, named for the Swiss mathematician Bernoulli, is sometimes used to label the sequence of two-outcome trials that constitute what we're here calling a "binomial situation."

The Binomial Function

Whenever these four conditions are met (i.e., whenever a situation can be identified as binomial), a corresponding mathematical function can be used to generate a full set of related probabilities.

Specifically, if we use n to designate number of trials and p to designate the probability of success on any one trial, the probability of exactly x successes out of n trials in a binomial experiment can be computed as

$$P(x/n, p) = \frac{n!}{(n - x)!x!} p^x(1 - p)^{(n - x)}$$

Coin Tossing One More Time

To illustrate the procedure, suppose we again face the task of computing the probability of tossing exactly two heads in three tosses of our weighted coin—a coin that has a 40% chance of turning up heads on any one toss. Identifying, as we have, the coin toss situation as certifiably binomial, we can use the binomial function to produce a solution.

Setting $x = 2$, $p = .4$, and $n = 3$ gives

$$P(x = 2/n = 3, p = .4) = \frac{3!}{(3 - 2)!2!}(.4^2)(1 - .4)^{(3 - 2)}$$

$$= \frac{6}{(1)(2)}(.16)(.6)$$

$$= 3(.096)$$

$$= .288$$

Interestingly enough, the binomial function is actually replicating the same steps that were required in our earlier solution of the coin toss problem. The first part of the binomial expression,

$$\frac{n!}{(n - x)!x!}$$

which is sometimes called the *binomial coefficient*, is a familiar counting formula (shown in the Chapter 3 supplement) that simply counts the number of ways (3 in this case) in which exactly x successes in n trials can be produced.

The second part of the expression,

$$p^x(1 - p)^{(n - x)}$$

computes the probability of each of the ways (.096 in our current coin toss example). Putting the terms together effectively generates the overall "x successes in n trials" probability ($3 \times .096 = .288$).

EXERCISE 2   Use the binomial function to generate the full set of probabilities (e.g., for 0 heads, 1 head, etc.) in the three-trial coin toss experiment (in which the probability of a head on any toss is .4) and verify that the results match perfectly with the table of results you had produced in Exercise 1.

Ans: See the end of the chapter.

Descriptive Measures for the Binomial Distribution

As we discussed earlier, it may sometimes be useful to summarize a probability distribution by reporting appropriate descriptive measures, usually a mean (expected value) and a standard deviation. In the binomial distribution, the computational effort required to produce such measures can be substantially reduced. In fact, the *expected value* in a binomial distribution can be found simply by multiplying n (number of trials) by p (probability of success on any one trial). For our three-toss coin experiment, then,

$$E(x) = np = (3)(.4) = 1.2$$

The *standard deviation* calculation can be reduced to the square root of n times p times $(1 - p)$. For the coin toss example, this means

$$\text{STD DEV}(x) = \sqrt{np(1 - p)} = \sqrt{3(.4)(.6)} = .85$$

Note: The letter q is frequently used to represent the quantity $1 - p$ in a binomial experiment. The standard deviation expression could be written then as $\sqrt{npq}$.

At this point, you might want to refer to Exercise 1 and confirm that our binomial mean and standard deviation shortcuts have produced the same results as the calculations you made in your original solution to the exercise.

The Binomial Table

Because the binomial function is so widely used in statistical analysis, and because computations may become unwieldy when n is large, tables showing binomial probabilities are available in virtually every basic statistics text. These tables represent the results of someone else performing the necessary binomial computations, eliminating (in most cases) the need for us to do the work of substituting values in the binomial formula.

We can use the coin toss experiment (with a "heads" probability of .40) to establish the usefulness of the tables. Refer to the **binomial table** beginning on page 318. To determine the probability of two heads in three tosses of the 40% coin, we need only (1) reference that section of the table showing an n (first column) value of 3, (2) locate the x (second column) entry of 2, and (3) trace across to the column headed by a p value of .4. At the row and column intersection, you should find the value .288.

As we've confirmed twice before, this .288 represents the probability of exactly two heads in three tosses of our slightly weighted coin. By staying in this same $n = 3$, $p = .4$ section of the table and referencing the complete list of x-value possibilities (0 through 3), we can (and you should) identify precisely the same full set of probability values that we saw earlier for the complete "number of heads" distribution.

EXERCISE 3   Use the table to find the following binomial probabilities:

(a) $P(x = 5/n = 10, p = .3)$
(b) $P(x \geq 7/n = 12, p = .4)$ (i.e., the probability that x is greater than or equal to 7. *Hint:* Use the additive rule for $x = 7$, $x = 8$, $x = 9$, etc.)

(c) $P(2 \leq x \leq 5/n = 15, p = .2)$ (i.e., the probability that x is between 2 and 5 inclusive)

Ans: (a) .1029   (b) .1009 + .0420 + .0125 + .0025 + .0003 = .1582
       (c) .2309 + .2501 + .1876 + .1032 = .7718

The only complication that might arise in using the binomial table occurs when the probability of success (p) is greater than 50%. A glance at the table will reveal that no p values larger than .5 are directly available. (Most binomial tables are laid out in the same way.) How do we get around this limitation? When $p > .5$, we'll need to "rephrase" the problem to create an equivalent problem involving a more accessible p, one with a value less than .5.

To illustrate, suppose we plan to use the table to produce the probability of two heads in three tosses of another weighted coin, in this case a coin that has an *80%* chance of turning up heads on any one toss. In order to use the table, we'll first need to change the focus of the question, since the heads probability of .8 is clearly not listed. Specifically, we can match the two-heads-in-three-tosses result with its corresponding one *tail* equivalent. In other words, we can translate the desired outcome, "two heads in three tosses," into the perfectly equivalent "one tail in three tosses" result. Can you see the advantage?

Symbolically, the change would look like this:

original "heads" question

$$P(x = 2/n = 3, p = .8)$$

equivalent "tails" question

$$P(x = 1/n = 3, p = .2)$$

By changing the emphasis from the heads outcome to the tails equivalent, we're able to switch to a p value of .2 and use the table to solve the equivalent problem. Result? The probability of one tail in three tosses—and so the probability of two heads in three tosses—is (from the table) .3840. (You should check the table to confirm.)

Notice that we didn't subtract the .3840 probability of one tail from 1.0 (i.e., from 100%) in order to produce the probability of two heads, although this is a strong temptation for most beginning students of probability. Once we solve the "tails" problem, we simultaneously solve the original "heads" problem. The value .3840 is the probability of one tail in three tosses and as such is also the probability of two heads in three tosses. The outcomes, although reported from a different perspective (literally, the "other side of the coin"), are equivalent. Try to resist the temptation, in problems like this, to subtract the table value from 100%.

EXERCISE 4   You plan to toss a weighted coin 10 times. The probability of a head on any toss is known to be .7. Use the binomial table to produce

(a) the probability of exactly two heads.
(b) the probability of exactly four heads.

(c) the probability of at least eight heads (*Hint:* At least eight heads is equivalent to no more than two tails.)

Ans: (a) .0014   (b) .0368   (c) .3828

(For a more detailed solution, see the end of the chapter.)

Note: Even in binomial problems where the issue is not explicitly "heads" or "tails," it's often useful to frame the experiment in the simple coin toss setting, especially if you need to negotiate the "successes" to "failures" switch in order to use the binomial table for a solution. The "heads" and "tails" reference provides a clear-cut rudimentary framework for binomial analysis in almost any context.

The Shape(s) of the Binomial Distribution

Shown graphically, the binomial distribution can take on a variety of shapes, depending mostly on the size of the p value involved. For a small p, the distribution will generally appear *skewed* in the positive direction. A p value near .5 establishes a fairly *symmetric* distribution (perfectly symmetric for a p of exactly 50%). A large p will tend to produce a *negatively skewed* result. Figure 4.3 shows some of the possibilities.

Note: As the n value for the binomial gets larger, the distribution tends to look more and more symmetric, even for p values that differ significantly from the "perfect" p of .5. (To convince yourself that this is so, you might, for a p of .2, sketch the distribution for $n = 5$, $n = 10$, $n = 20$, and $n = 30$.) A common rule-of-thumb states that as long as $n \times p \geq 5$, and $n \times (1 - p) \geq 5$, we can consider the corresponding binomial distribution to be essentially symmetric.

4.6   THE NORMAL DISTRIBUTION

We'll take a look now at the **Normal** (or **Gaussian**) distribution, another special-purpose distribution that proves especially useful in sampling theory. (Karl Gauss was a 19th-century German astronomer-mathematician who made extensive use of normal distribution properties in his scientific investigations.) Symmetric in shape (in fact, shaped like a bell), the normal distribution derives its name from the fact that it seems to effectively describe a number of common (or normal) "real-world" phenomena. (For example, the IQ scores of students, the typing speed of office secretaries, the length of fish caught in the river, and the weight of leaves on the oak tree outside could all be expected to follow an approximately normal pattern.)

The Normal Function

As was true for the binomial, the normal distribution can be succinctly described by a mathematical function capable of producing, under appropriate circumstances, the full range of random variable probabilities. Using x to represent the normal random

FIGURE 4.3 **POSSIBLE SHAPES FOR THE BINOMIAL DISTRIBUTION**

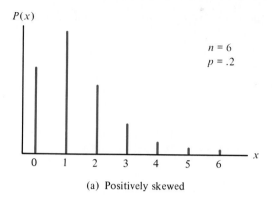

(a) Positively skewed

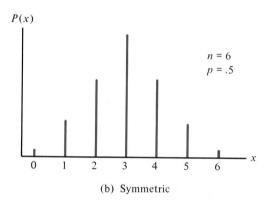

(b) Symmetric

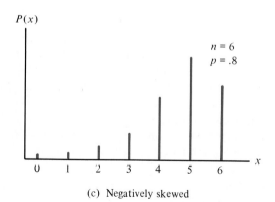

(c) Negatively skewed

FIGURE 4.4   **THE BELL-SHAPED NORMAL CURVE**

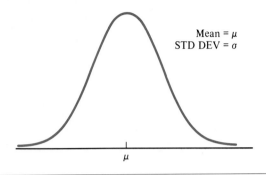

Mean = μ
STD DEV = σ

μ

variable, we can write the normal probability calculator as

$$P(a \leq x \leq b/\mu, \sigma) = \int_{a}^{b} \frac{1}{\sqrt{2\pi\sigma}} \exp\left\{-\frac{1}{2}\left(\frac{x-\mu}{\sigma}\right)^{2}\right\} dx$$

where μ(mu) = mean (or expected value) of the distribution
   σ(sigma) = standard deviation of the distribution
   π = a mathematical constant approximately equal to 3.1417

With proper substitution, it's possible to determine for any set of normally distributed values the likelihood of finding a value within some specified range (i.e., in any *a*-to-*b* interval) defined on the distribution.

You can see that normal probability calculations involve taking the integral of a rather imposing mathematical function. But before you decide to change your major, it should be quickly established that our intention here is simply to acknowledge the functional form for the normal distribution before moving on to a less threatening discussion of a graphical approach to producing normal probabilities. (Translation: We won't be doing any integral calculus here.)

Sketching the Normal Distribution

As we've already noted, the normal distribution follows a symmetrical, bell-shaped pattern. Figure 4.4 shows the characteristic shape. As shown, the distribution is centered on μ (the mean) and has a standard deviation σ. In general, it describes a set of values (*x*'s) that are fairly closely and symmetrically clustered around the mean; as we move farther and farther from the mean in either direction, we would expect to find fewer and fewer values. (Can you see now how this sort of pattern might appropriately describe any number of "normal" or natural, real-world phenomena?)

For our purposes, it's this picture, rather than the underlying normal function, that should come to mind each time the term "normal distribution" is mentioned.

FIGURE 4.5   **INTERVALS ON THE NORMAL CURVE**

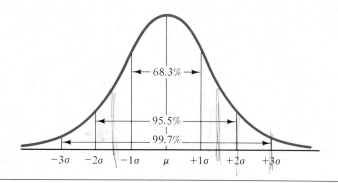

We'll see shortly that, with the help of a table, we can produce from the picture any normal probabilities that might be required.

Normal Distribution Characteristics

Not every symmetrical, bell-shaped distribution is actually a normal distribution. In fact, in order to qualify as normal, a distribution must meet a number of additional—and rather specific—conditions. For example, it will always be true that approximately

> 68.3% of the values in a normal distribution will fall within *one* standard deviation of the distribution mean (or center).

This means that if we were to mark off an interval on the horizontal scale of a normal distribution extending from one standard deviation (i.e., 1σ) below the mean to one standard deviation above the mean, we should expect to find in that interval approximately 68.3% of the values. In more visual terms, this means that 68.3% of the area under any normal curve can be found within the bounds of a -1σ to $+1\sigma$ interval around the mean. (See Figure 4.5.) By implication, then, if we were to randomly select a value from a truly normal distribution, there is a 68.3% probability that the value selected will lie within one standard deviation of the distribution center.

It will also be true that approximately

> 95.5% of the values in a normal distribution will fall within *two* standard deviations of the distribution mean.

This means that 95.5% of the area below the curve can be found within the $\pm 2\sigma$

interval. Consequently, if we select a value randomly from a normal distribution, there's a 95.5% probability that the value selected will lie within two standard deviations of the distribution center.

Finally (at least for the moment),

> 99.7% of the values (nearly all the values) in a normal distribution will be found within three standard deviations of the distribution mean.

This means that 99.7% of the area will be contained within the $\pm 3\sigma$ boundaries, and that there's a 99.7% probability that a randomly selected value will be no more than three standard deviations from the distribution mean.

Note: The total area under the normal curve is of course 100%, with 50% to the right of the mean and 50% to the left. As with any symmetrical distribution, half of the values lie at or below the mean and half lie at or above.

Producing Probabilities for Normally Distributed Values

General normal distribution characteristics like the ones described above can be readily translated to any specific normal distribution (with a known mean and standard deviation) to produce the full range of normal probabilities. To see how this might work, consider a simple experiment:

Suppose I find myself standing next to a large barrel of numbers. The values in the barrel are known to be normally distributed, with a mean of 50 and a standard deviation of 4. I'm going to let you randomly select values from the barrel, but before you do you'll need to answer a few basic questions.

Question 1: How likely is it that your first selection from the barrel produces a number between 46 and 54? *Answer:* For this normal distribution, centered on 50 with standard deviation of 4, the range 46 to 54 converts readily to the kind of $\pm 1\sigma$ (i.e., ± 1 standard deviation) interval that we had referred to above, the sort of interval that should always contain 68.3% of the values in any normal distribution. The probability of selecting a value between 46 and 54 (or $P(46 \leq x \leq 54)$) from this normal distribution is thus .683. See Figure 4.6.

Question 2: How likely is it that the number you select lies somewhere between 42 and 58? *Answer:* Converting these bounds to an equivalent interval measured in standard deviations from the mean, 42 to 58 corresponds to a $\pm 2\sigma$ interval. Since we've established that such a 2σ interval always contains a 95.5% area, we can safely conclude that $P(42 \leq x \leq 58) = .955$. See Figure 4.7.

Question 3: How likely is it that the value selected will lie in the interval from 50 to 54? *Answer:* Given the symmetry of the distribution, the targeted interval is clearly half the $\pm 1\sigma$ (68.3%) interval we dealt with above. Since the normal distribution is always symmetric, splitting the 68.3% area to produce half-areas of about 34.15% gives us the appropriate probability: $P(50 \leq x \leq 54) = .3415$. See Figure 4.8.

FIGURE 4.6 **THE AREA ASSOCIATED WITH P(46 ≤ X ≤ 54)**

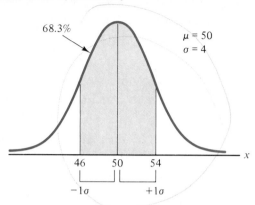

FIGURE 4.7 **THE AREA ASSOCIATED WITH P(42 ≤ X ≤ 58)**

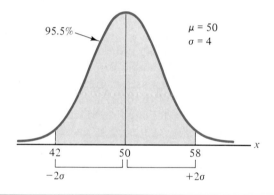

Note: As we move to more complicated questions concerning the calculation of normal probabilities, it's strongly recommended that you develop the habit of sketching the normal distribution involved, centering it on the distribution mean, and identifying the specific area(s) of interest. It's then usually possible to visually piece together a workable solution strategy. Try it out on the next question.

Question 4: How likely is it that the number selected falls between 54 and 58? *Answer:* Working from Figure 4.9, a viable strategy should become quickly apparent: We might (1) find what's between 50 and 58, (2) find what's between 50 and 54, and then (3) subtract the smaller area from the larger area, using the difference as the probability we want. (I'll leave it to you to work through the strategy. From our previous work, you should be able to confirm that $P(54 \leq x \leq 58)$ is $.4775 - .3415 = .136$.)

FIGURE 4.8	**THE AREA ASSOCIATED WITH P(50 ≤ X ≤ 54)**

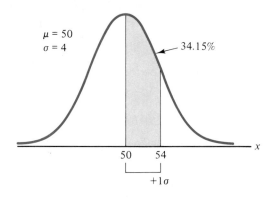

FIGURE 4.9	**THE AREA ASSOCIATED WITH P(54 ≤ X ≤ 58)**

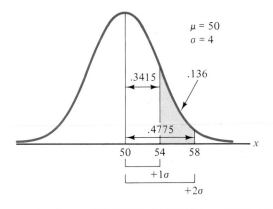

Using the Normal Table

Now for your first real challenge. How likely is it that the value you select from our hypothetical barrel is between 50 and 52? Be careful. There's a strong temptation here, since the interval in question is half of the 50-to-54 interval, to conclude that the corresponding area must be half of the previously computed .3415. Don't give in to the temptation. Figure 4.10 suggests the reason why. While the distribution is undeniably symmetric, it's symmetric about the mean. Although there's every reason to believe that we would find equal areas in the intervals 46 to 50 and 50 to 54, there's no support for the argument that equal areas should be found in the intervals 50 to 52 and 52 to 54. In fact, Figure 4.10 suggests that a substantially larger area lies in the 50–52 segment.

FIGURE 4.10   THE AREA ASSOCIATED WITH P(50 ≤ X ≤ 52)

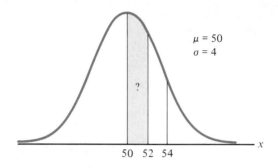

To solve this one, we'll need the help of a table. Turn to the **normal table** on (page 324). The figure at the top of the table shows the nature of the information provided. The focus here is on areas under the normal curve associated with intervals starting at the mean and extending to some number of standard deviations to the right of the mean. Reading specific values from the table requires only that we specify a "number-of-standard-deviations" end point for the interval we want.

For example, suppose we intend to use the table to establish the normal area that lies between the mean and a point 1.2 standard deviations to the right of the mean. Easy enough. Reference the first column of the table and locate the 1.2 entry. (This first column is labeled z, a common normal distribution designator for distances measured in "number of standard deviations from the mean.") By moving your finger one column to the right of the 1.2 z entry, you should find the value .3849. Conclusion? 38.49% of the area in a normal curve will be found in the interval μ to $\mu + 1.2\sigma$.

Now use the same procedure to find the area below the curve between the mean and, say, 2.33 standard deviations to the right of the mean. You should find .4901. (*Hint:* The entries shown across the top row of the table allow you to reference a z score to the hundredths place. A z of 2.33 would consequently be located by first locating a z of 2.3 in the first column, then tracing across to the column headed .03.)

Back to the Barrel

Equipped with the ability to produce these sorts of areas from the table, we can return to the still-pending question: How likely is it that we would select a value from our hypothetical barrel that falls in the interval 50 and 52? (Remember, the values in the barrel have a mean of 50 and a standard deviation of 4.) Converting the distance 50-to-52 to an equivalent z score will give us immediate access to the right normal table entry. We can calculate the necessary z simply by subtracting the mean (50) from the interval endpoint (52) and dividing the result by the size of the standard deviation (4). More succinctly,

$$z = \frac{x - \mu}{\sigma} = \frac{52 - 50}{4} = .5$$

FIGURE 4.11   **USING THE NORMAL TABLE TO FIND P(50 ≤ X ≤ 52)**

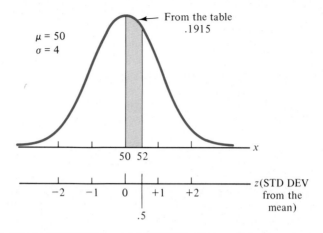

This simply establishes that 52 is .5 standard deviations to the right of the distribution mean of 50. Referencing the table for a z of .5 produces a corresponding area of .1915. Conclusion? The probability of choosing a value that lies in the interval 50–52 (i.e., $P(50 \le x \le 52/\mu = 50, \sigma = 4)$) is about 19.15%. Figure 4.11 shows the basic procedure.

> **Note:** Notice the z-scale—measuring distances in "number of standard deviations" from the mean—drawn parallel to the original x-axis on the graph in Figure 4.1. Extending our earlier recommendation, I would encourage you to produce this sort of sketch for every normal problem you confront. Show the interval(s) of interest on the x-scale and establish the equivalent z-scale marker(s) before proceeding to the table for the appropriate probabilities. Exercise 5 provides an opportunity to try out the approach.

EXERCISE 5   Given a normal distribution of values with a mean of 120 and a standard deviation of 10, find the probability of randomly selecting a value
(a) between 112 and 120 (i.e., $P(112 \le x \le 120)$).
(b) between 120 and 135 (i.e., $P(120 \le x \le 135)$).
(c) between 110 and 138 (i.e., $P(110 \le x \le 138)$).
(d) between 95 and 115 (i.e., $P(95 \le x \le 115)$).
(e) less than or equal to 105. (i.e., $P(x \le 105)$) (*Hint:* Remember, 50% of the values will lie below the mean, 50% above.)
(f) strictly less than 105. (i.e., $P(x < 105)$). (You may want to read the next section before tackling this one.)

Ans: (a) .2881   (b) .4332   (c) .8054   (d) .3023   (e) .0668   (f) .0668

(For a more detailed solution, see the end of the chapter.)

4.7 CONTINUOUS AND DISCRETE DISTRIBUTIONS

Taken together, parts (e) and (f) of exercise 5 raise an issue that we haven't yet discussed. The normal distribution can be broadly classified as a **continuous distribution**, distinguishing it generically from **discrete distributions** such as the binomial. The basis on which this distinction is drawn goes something like this:

In a *discrete* distribution, individual events can be clearly defined as separate and distinct. Graphically, each event can be shown as a discrete point along the horizontal axis of the graph. Probability can be assigned to each individual event and is visually represented by the length of the vertical line segment (or bar) drawn above the corresponding point. Between discrete "point" events, there's only empty space. (You may want to take another look at the binomial distribution in Figure 4.3.)

For *continuous* distributions like the normal, the picture is quite different. Here, every point on the horizontal axis of the graph represents a possible event; there are no "holes" or "breaks"—no empty spaces between points. Individual events tend to lose their identity among the infinite possibilities described by the distribution. As a consequence, probability can't be assigned to any single point event, only to interval events along the horizontal axis. The probability of x being precisely 102.00000 essentially loses its meaning in a continuous distribution. (The probability of any point event here is effectively 0.) Only the probability that x lies within some specified interval has any significance. As a consequence, "area below the curve" replaces "length of the vertical line segment" as the appropriate representation of probability. (In the chapter supplement, we'll take advantage of this contrast in order to effectively use the normal distribution to approximate binomial probabilities.)

What does all this mean? Referring to parts (e) and (f) of exercise 5, we can now conclude that $P(x < 105)$ is precisely the same as $P(x \leq 105)$, since for the continuous normal distribution, $P(x = 105) = 0$.

4.8 THE POISSON DISTRIBUTION

The third and final distribution in our admittedly abbreviated discussion of special-purpose distributions is the **Poisson** (pronounced Pwa-SONE, and named for F. D. Poisson, a 19th-century French mathematician). Like the normal and binomial functions, the Poisson probability function will, under certain conditions, serve as a useful generator of probabilities.

Characteristics of a Poisson Situation

The precise mathematical characteristics of the standard Poisson situation are somewhat complex. As a consequence, we'll give only a general description of the Poisson conditions and leave for another time a more formal treatment of its distinctive mathematical properties. For our purposes, two basic conditions should be met in order for a situation to qualify as Poisson:

1. We need to be assessing probability for a given number of occurrences of some event *per unit time, space, or distance.*
2. The average number of occurrences per unit time, space, or distance must be known and must remain constant; otherwise, the occurrences must be random (i.e., "unpredictable").

Describing some typical Poisson applications should help to clarify the conditions. For example:

(a) Telephone calls coming into a switchboard might meet the Poisson conditions. Suppose the average number of calls per minute is known and constant, but otherwise calls arrive randomly. We might use the Poisson function here to determine the probability that exactly three calls reach the switchboard during any randomly selected 1-minute interval.
(b) Typographical errors on a page of newsprint might well follow a Poisson-type pattern. If the average number of errors per page is known and constant, but otherwise the occurrence of errors is random, we might use the Poisson function to determine, for example, the probability that a randomly selected page contains exactly two typographical errors.
(c) Flaws per square yard of mill-produced fabric may similarly meet the Poisson conditions. If the average number of flaws per square yard is known and constant, but otherwise flaws occur randomly, we could use the Poisson function to determine the probability of finding no flaws (or one flaw or two flaws) in any randomly selected square yard of fabric.
(d) Customers arriving at the checkout counter of the local supermarket may conform to this same Poisson pattern. If the average number of customers arriving per minute is known and constant, but otherwise arrivals are random, we might use the Poisson function to determine the probability that, for example, no more than two people arrive in the next minute observed.

The Poisson Function

In general, we'll use the Poisson probability function to produce the probability of exactly x occurrences of some randomly occurring event, given that the mean number of occurrences (we'll use λ (lambda) to label the Poisson mean) is known and constant. Specifically, the Poisson function is

$$P(x/\lambda) = \frac{\lambda^x e^{-\lambda}}{x!}$$

where λ = Poisson mean
  e = mathematical constant 2.718 . . .

Substitution into the Poisson expression is straightforward. Try it out in Exercise 6.

EXERCISE 6   Arrivals at the checkout counter of your local First Foods Store average two per minute. Aside from this constant average, the arrivals appear to be random. Assuming the appropriateness of the Poisson function, determine

(a) the probability of exactly one arrival in the next minute observed.
(b) the probability of no arrivals in the next minute observed.
(c) the probability of no more than one arrival in the next minute observed.
(d) the probability of exactly three arrivals in the next 2 minutes observed. (*Hint:* If the average arrival rate is two per minute, then the average arrival rate per 2-minute interval must be 4 (i.e., 2×2). Use this revised 2-minute average as the λ value in the Poisson expression and proceed with your calculation.)

Ans: (a) .2707   (b) .1353   (c) .406   (d) .1954

(For a more detailed solution, see the end of the chapter.)

The Poisson Table

As with the binomial and normal distributions, tables are generally available for the Poisson distribution. In fact, the **Poisson table** may be the easiest of the three tables to use. Refer to the Poisson table on page 325 and we can see exactly how it works. Normally all we'll need to do is locate the section of the table associated with the targeted λ value, then trace down the x-column to locate the appropriate row.

EXERCISE 7   Refer to Exercise 6 and use the Poisson table to confirm the results of each of your computations.

Graphing the Poisson Distribution

Displayed graphically, the Poisson distribution is characteristically skewed in the positive (i.e., righthand) direction. Figure 4.12 shows the case for $\lambda = 1$. Notice we're back to sketching a discrete distribution, with the appropriate breaks along the horizontal axis and with probability represented by the length of the vertical line (or bar) above each value of the random variable.

> **Note:** As the value for λ increases, the Poisson looks increasingly symmetrical. To illustrate, try sketching the case in which λ is 10.

Descriptive Measures

The mean (or expected value) of a Poisson distribution is by definition λ. Curiously, the standard deviation of the distribution is the square root of λ. For this reason, the Poisson distribution is sometimes described as a *one-parameter* distribution — know the value of the mean and you immediately know the value of the standard deviation.

FIGURE 4.12   **GRAPHING THE POISSON DISTRIBUTION FOR λ = 1**

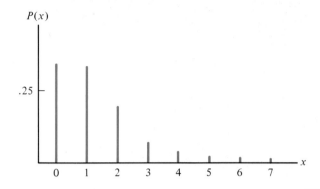

EXERCISE 8 Using Poisson table values, compute the "long way"—with $\bar{x} = E(x) = \sum xP(x)$ and $\text{Var}(x) = \sum (x - \bar{x})^2 P(x)$—the mean and standard deviation for the Poisson distribution when $\lambda = .5$. Confirm that the distribution mean is .5 and that the standard deviation of the distribution is $\sqrt{.5}$.

(For a detailed solution, see the end of the chapter.)

CHAPTER SUPPLEMENT I: APPROXIMATING BINOMIAL PROBABILITIES

Using the Poisson Distribution to Approximate Binomial Probabilities

The Poisson distribution can sometimes be used to effectively approximate binomial probabilities. This stems from the fact that under certain conditions the binomial distribution looks a lot like the Poisson. Specifically, when (1) n for the binomial is large and (2) p for the binomial is small, the binomial distribution takes on a distinctly Poisson-like shape (that is, it's skewed noticeably in the positive direction). Figure 4.13 demonstrates the similarity.

Should we ever find ourselves in a situation in which binomial probabilities are required but are not readily available (for example, if the value for n goes beyond table coverage), we might take advantage of this acknowledged similarity in shape to produce appropriate Poisson approximations. In order to implement the procedure, we can simply compute the mean (expected value) of the binomial involved (np) and select a Poisson with exactly the same center. For example, to identify an approximating Poisson for the binomial case $P(x = 1/n = 100, p = .02)$, we can compute the binomial mean $100(.02) = 2$ and reference the Poisson table

---

FIGURE 4.13   **COMPARING THE POISSON AND THE BINOMIAL**

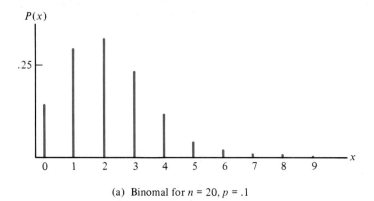

(a) Binomal for $n = 20, p = .1$

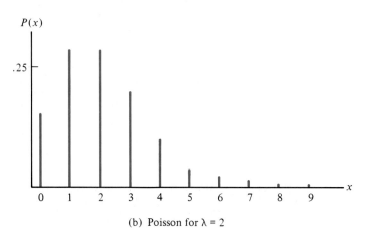

(b) Poisson for $\lambda = 2$

---

for $\lambda = 2$, $x = 1$. The binomial probability $P(x = 1/n = 100, p = .02)$ can then be reasonably approximated with the Poisson "equivalent" $P(x = 1/\lambda = 2) = .2707$.

The Accuracy of Poisson Approximations

We can check the general accuracy of such Poisson approximations by looking at a case or two for which we have both the binomial and the Poisson probabilities available. For example, let $n = 20$ and $p = .05$. From the binomial table, the probability that $x = 2$ is exactly 0.1887. The Poisson equivalent, using $\lambda = 1$, is .1839.

For larger n's and smaller p's, the accuracy of most Poisson approximations should be even better. As a general rule (and given the limitations of the binomial table at the end of the text), we plan to use Poisson approximations for binomial probabilities when the binomial value $n > 30$ and $np < 5$.

Note: It can be shown that the Poisson distribution is actually a limiting case for the binomial function as n gets larger and p gets smaller.

EXERCISE S1   Use the binomial and Poisson tables in the back of the text to compare the binomial probabilities with approximating Poisson values.

(a) $P(x = 5/n = 20, p = .1)$
(b) $P(x \leq 2/n = 18, p = .05)$

Ans: (a) Binomial $= .0319$ Poisson $= .0361$
      (b) Binomial $= .9418$ Poisson $= .9372$

(For a more detailed solution, see the end of the chapter.)

Using the Normal Distribution to Approximate Binomial Probabilities

The normal distribution can also be used to approximate binomial probabilities. In our discussion of the binomial distribution, we established that when n is fairly large and p is not too far from .5, the shape of the binomial is generally symmetric (perfectly symmetric for $p = .5$). Under such conditions, the normal and the binomial look very much alike.

We can sometimes exploit this similarity in shape to produce useful normal approximations for a symmetric binomial. To illustrate, suppose we decide to use the normal distribution to approximate a binomial case in which $n = 100$ and $p = .5$. In order to specify the approximating normal, we need to determine (1) an appropriate mean and (2) a corresponding standard deviation. Easy enough. We'll simply substitute the binomial mean $(np = 100(.5) = 50)$ for the normal mean and the binomial standard deviation $(\sqrt{np(1 - p)} = \sqrt{100(.5)(.5)} = 5)$ for the normal standard deviation. Figure 4.14 shows the match.

It's possible now to approximate binomial probabilities from the normal curve shown. For example, we might approximate $P(45 \leq x \leq 54/n = 100, p = .5)$ by computing $P(45 \leq x \leq 54/\mu = 50, \sigma = 5)$ on the corresponding normal. (The normal approximation here is 0.6294.)

The Accuracy of the Normal Approximation

As we've already suggested, the normal approximation works best when the binomial n is fairly large and p is close to .5. However, for a large enough n, p may differ significantly from .5 and yet still allow for an effective normal approximation. The only requirement is that the binomial involved be sufficiently symmetric.

To be more specific, we can draw on a previous argument. Although it didn't seem important at the time, we had noted in our initial discussion of the binomial that as long as $np \geq 5$ and $n(1 - p) \geq 5$, the shape of the distribution will be nearly symmetric. This premise is at the heart of the test we now plan to adopt to identify

FIGURE 4.14   **MATCHING THE NORMAL AND BINOMIAL DISTRIBUTIONS**
                      ($n = 100$, $p = .5$; $\mu = 50$, $\sigma = 5$)

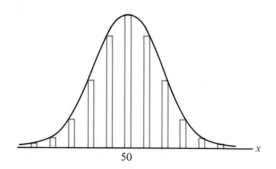

cases in which the normal distribution might be used to approximate the binomial: as long as n is reasonably "large" (at least 30) and both $np \geq 5$ and $n(1 - p) \geq 5$, the normal distribution will tend to produce effective approximations of binomial probabilities.

EXERCISE S2   Use the normal distribution to approximate the following binomial probabilities:

(a) $P(x \geq 110/n = 200, p = .5)$
(b) $P(26 \leq x \leq 35/n = 100, p = .3)$
(c) $P(x \leq 19/n = 30, p = .5)$

Ans: (a) .0793   (b) .6699   (c) .9279

(For a more detailed solution, see the end of the chapter.)

A Final Note on Normal Approximations

Because the normal distribution is continuous and the binomial distribution is discrete, there will, it turns out, always be a slight discrepancy when one distribution is "matched" to the other. In cases where the normal distribution is being used to approximate binomial probabilities, failure to compensate for this discrepancy, particularly where the binomial n is not especially large, may lead to approximations that are not as accurate as they might otherwise be. To deal with this problem, we'll suggest one small adjustment in the basic procedure.

Reconsider part (c) of exercise S2. Here you were asked to produce a normal approximation for the binomial probability $P(x \leq 19/n = 30, p = .5)$. Now compute the actual binomial probability (using the table at the end of the text) and check your answer against the normal approximation. Result? The normal approxima-

FIGURE 4.15   **ADJUSTING THE NORMAL APPROXIMATION TO THE BINOMIAL**

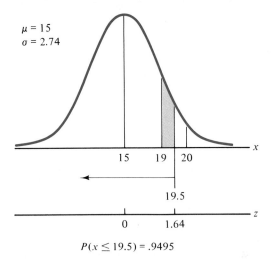

$$P(x \leq 19.5) = .9495$$

tion, while close (.9279 versus .9506), is clearly not perfect. One of the reasons is that the boundary event, $x = 19$, is treated differently in the two distributions. Reconciling that difference should improve our approximation.

As we saw earlier, in the continuous normal distribution (in fact, in *any* continuous probability distribution), every point along the horizontal axis of the distribution graph represents a possible value for the random variable x. For the discrete binomial, on the other hand, this is clearly not true. Between discrete point events in the binomial distribution, nothing is happening, nothing is going on—no area, no probability, no events.

In our example, with $p = .5$ and $n = 30$, the approximating normal shows area (and so, probability) between events 19 and 20; the corresponding binomial shows nothing. In order to ensure that everything in the binomial is matched to a clear counterpart in the normal, we might decide to simply "assign" a part of the normal area between 19 and 20 to the discrete binomial event $x = 19$. How much should we assign? Typically, we'll set a marker halfway between the two values and assign area from the resulting interval. Can you see the implication? The binomial probability $P(x \leq 19/n = 30, p = .5)$ might now (we're suggesting) be more effectively approximated by using the normal table to produce

$$P(x \leq 19.5/\mu = 15, \sigma = 2.74)$$

Rather than simply finding the area below 19 in the approximating normal distribution, we've targeted the area below 19.5. (See Figure 4.15.) In effect, we are (admittedly, a bit arbitrarily) assigning to the discrete point event 19 all the area in the normal curve between 18.5 and 19.5. The probability produced in this slightly modified approach (i.e., $P(x \leq 19.5) = .9495$) does represent a significant improvement over our earlier assessment ($P(x \leq 19) = 0.9279$). (The exact binomial probability was 0.9506.)

As n gets larger, the impact, and so the necessity, of this sort of adjustment diminishes. Exercise S3 will give you a chance to experiment.

EXERCISE S3 Compute the normal approximation to the following binomial probabilities, using the "adjusted" and the "unadjusted" approaches. For parts (a) and (b), compare your results to the actual binomial probabilities.

(a) $P(x \geq 7/n = 12, p = .5)$.
(b) $P(11 \leq x \leq 13/n = 20, p = .5)$.
(c) $P(x \geq 36/n = 100, p = .3)$.

Ans: (a) unadjusted est = .2810   adj est = .3859   binomial = 0.3871
     (b) unadjusted est = .2363   adj est = .3535   binomial = .3542
     (c) unadjusted est = .0951   adj est = .1151

(For a more detailed solution, see the end of the chapter.)

CHAPTER SUPPLEMENT II: THE EXPONENTIAL DISTRIBUTION

Situation: Arrivals at the outside teller window of the First Constitutional Bank appear to follow a Poisson pattern: the average number of arrivals per minute remains fairly constant throughout the day; otherwise individual arrivals appear to be random.

According to our chapter discussion, we could use the Poisson probability function here to determine the probability of exactly two arrivals in the next minute (or four arrivals, or 0 arrivals). It turns out that it's also possible to determine the probability that at least 3 minutes will elapse between the next two arrivals (or 2 minutes, or 10 minutes). In fact, by appealing to the **exponential distribution**—based on a mathematical function derived directly from the Poisson—we can effectively produce the full range of "time-between" probabilities.

The exponential probability distribution can be described by the function

$$P(x \geq t/\lambda) = e^{-\lambda t}$$

where x = time, space, or distance between successive occurrences
    t = any specified value for x
    e = 2.718 ...
    λ = average number of occurrences per unit of time, space, or distance

Any time we have a situation in which the Poisson conditions are met, the expo-

FIGURE 4.16   **THE EXPONENTIAL DISTRIBUTION**

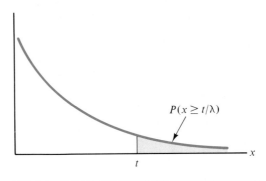

$$P(x \geq t/\lambda)$$

nential distribution is available should we need to determine probabilities associated with intervals of time, space, or distance between successive occurrences of the Poisson event. To illustrate, suppose the average number of arrivals at our teller window is two per minute. The probability that at least 1 minute will elapse between the arrival of one customer and the arrival of the next is

$$P(x \geq 1/\lambda = 2) = e^{-2(1)} = \frac{1}{(2.718)^2} = \frac{1}{7.388} = .135$$

The probability that at least 3 minutes will elapse between the arrival of one customer and the next is

$$P(x \geq 3/\lambda = 2) = e^{-2(3)} = \frac{1}{(2.718)^6} = \frac{1}{403.2} = .0025$$

As you can see, the computations are fairly straightforward. Consequently, no special tables have been provided.

Figure 4.16 shows a representative exponential shape. The "$\geq$" probabilities we've been describing correspond to right-tail areas.

Note: We've established that the Poisson is a discrete distribution. The closely related exponential is, however, a continuous distribution. (Take a minute to consider why this should be true.) If we use λ to represent the mean number of occurrences in the Poisson situation, then $1/\lambda$ is the mean value in the corresponding exponential. The standard deviation is also $1/\lambda$.

EXERCISE S4   On average, the company's switchboard receives six calls per minute. (The calls appear to be random.) A call has just come in.

(a) How likely is it that at least 20 seconds will elapse before the next call comes in?

(b) How likely is it that less than 1 minute will elapse before the next call comes in? (*Hint:* Think 100% minus $P(x \geq 1)$.)

Ans: (a) .135   (b) $1 - .0025 = .9975$

(For a more detailed solution, see the end of the chapter.)

EXERCISE S5   According to a recent study, the average number of serious potholes per mile of Interstate 576 is 4.0. (The occurrences appear to be random.) If your car is currently sitting fender-deep in one,

(a) how likely is it you would have to walk at least a mile along the interstate before you find another pothole?
(b) how likely is it that you would find a pothole no more than a quarter of a mile ahead?

Ans: (a) .0183   (b) $1 - .3679 = .6321$

(For a more detailed solution, see the end of the chapter.)

■ CHAPTER SUMMARY

We developed the ability to create and interpret probability distributions by extending our earlier discussion of elementary probability theory. A probability distribution, in essence, is a full listing of the probabilities assigned to all possible events in a given probabilistic experiment.

We identified the role of a random variable as a numerical description of experimental outcomes. In general, a random variable provides the means to report distribution results concisely and allows for the computation of useful summary measures, such as the expected value and standard deviation.

Probability distributions can sometimes be produced by using special-purpose probability (i.e., mathematical) functions. Each such function can be used to compute probabilities over the full range of possible outcomes (i.e., possible values for the random variable) under a given set of specific conditions. We focused on three of the most widely used of these special-purpose distributions: the binomial, the normal, and the Poisson.

The binomial function is a computer of probabilities in situations that involve a succession of statistically independent, two-outcome trials.

The normal distribution (the familiar bell-shaped curve) shows probabilities associated with common (natural or normal) situations in which most of the outcomes (values) involved are clustered rather closely and symmetrically about the average, with fewer values to be found as we move farther from the center in either direction.

The Poisson function can be an effective generator of probabilities in situations characterized by a constant mean but an otherwise random pattern of occurrences.

We differentiated discrete from continuous probability distributions by focusing on the possible outcomes displayed along the horizontal axis of the distribution graph. If every point along the axis represents a possible outcome, with no breaks to separate one outcome from another, the distribution is continuous and the area under the curve can be used to represent probability. If individual points (discrete, separate, and distinct) are identifiable, the distribution involved is discrete and the length (or height) of a vertical line segment drawn above each point can be used to properly represent probability.

In the chapter supplement, we developed the capacity to use one distribution to effectively approximate hard-to-find probabilities from another: the normal to approximate a symmetric binomial; the Poisson to approximate a binomial that shows positive skewness.

Finally, we introduced the exponential distribution in order to assign probabilities to intervals of time, space, or distance between successive Poisson events.

KEY TERMS

Binomial distribution
Binomial table
Continuous versus discrete distributions
Experiment
Exponential distribution
Normal (Gaussian) distribution

Normal table
Poisson distribution
Poisson table
Probability distribution
Random variable

IN-CHAPTER EXERCISE SOLUTIONS

EXERCISE 1

The tree diagram can be used to structure a solution:

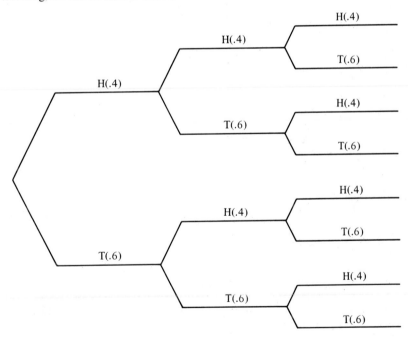

(a) $P(\text{no heads}) = P(x = 0) = (.6)(.6)(.6) = .216$

$P(1 \text{ head}) = P(x = 1) = (.4)(.6)(.6) + (.6)(.4)(.6)$

$+ (.6)(.4)(.6) = .432$

$$P(2\ heads) = P(x = 2) = (.4)(.4)(.6) + (.4)(.6)(.4)$$
$$+ (.6)(.4)(.4) = .288$$

$$P(3\ heads) = (.4)(.4)(.4) = .064$$

(b)

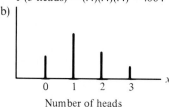

Number of heads

(c) $E(x) = \bar{x} = \sum xP(x) = 0(.216) + 1(.432) + 2(.288) + 3(.064)$

$$= 1.2$$

(d) $\text{STD DEV}(x) = \sqrt{\sum (x - E(x))^2 P(x)}$

$$= \sqrt{(0 - 1.2)^2(.216) + (1 - 1.2)^2(.432) + (2 - 1.2)^2(.288) + (3 - 1.2)^2(.064)}$$

$$= .85$$

EXERCISE 2

$$P(no\ heads) = P(x = 0/n = 3, p = .4) = \frac{3!}{3!0!}(.4)^0(.6)^3$$

$$= 1(1)(.216) = .216$$

$$P(1\ head) = P(x = 1/n = 3, p = .4) = \frac{3!}{2!1!}(.4)^1(.6)^2$$

$$= 3(.4)(.36) = .432$$

$$P(2\ heads) = P(x = 2/n = 3, p = .4) = \frac{3!}{1!2!}(.4)^2(.6)^1$$

$$= 3(.16)(.6) = .288$$

$$P(3\ heads) = P(x = 3/n = 3, p = .4) = \frac{3!}{0!3!}(.4)^3(.6)^0$$

$$= 1(.064)(1) = .064$$

EXERCISE 4

(a) The "heads" question $P(x = 2/n = 10, p = .7)$ can be translated to the equivalent "tails" question $P(x = 8/n = 10, p = .3)$. From the table, with $n = 10$, $p = .3$ and $x = 8$, the resulting probability is .0014.

(b) $P(x = 4/n = 10, p = .7)$ becomes $P(x = 6/n = 10, p = .3)$, which from the table is .0368.

(c) The "heads" question involving 8 or more heads in 10 tosses translates to 2 or fewer "tails"—8 heads equals 2 tails, 9 heads equals 1 tail, 10 heads equals 0 tails. So

$$P(x \geq 8/n = 10, p = .7) = P(x \leq 2/n = 10, p = .3)$$

$$= .0282 + .1211 + .2335$$

$$= .3828$$

EXERCISE 5

(a) $P(112 \leq x \leq 120)$

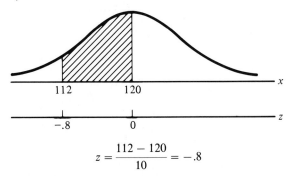

$$z = \frac{112 - 120}{10} = -.8$$

From the normal table, for $z = .8$, the probability is .2881.

(b) $P(120 \leq x \leq 135)$

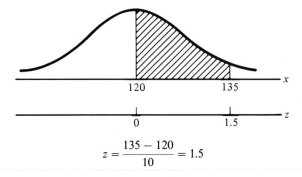

$$z = \frac{135 - 120}{10} = 1.5$$

From the normal table for $z = 1.5$, the probability is .4332.

(c) $P(110 \leq x \leq 138)$

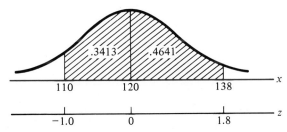

Look up each of the two areas in turn, then sum. Remember, areas in the table all begin at the mean.

$$z = \frac{110 - 120}{10} = -1.0$$

In the normal table for $z = 1.0$, the probability is .3413.

$$z = \frac{138 - 120}{10} = 1.8$$

From the normal table for $z = 1.8$, the probability is .4641. Summing the two probabilities (i.e., areas) provides the solution: $.3413 + .4641 = .8054$.

(d) $P(95 \le x \le 115)$

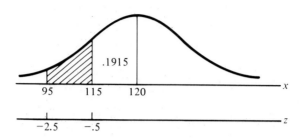

First find the area between 95 and 120; then find the area between 115 and 120. Subtracting the smaller area from the larger area gives the solution.

$$z = \frac{95 - 120}{10} = -2.5, \text{ producing a probability of .4938}$$

$$z = \frac{115 - 120}{10} = -.5, \text{ producing a probability of .1915}$$

$$.4938 - .1915 = .3023$$

(e) $P(x \le 105)$

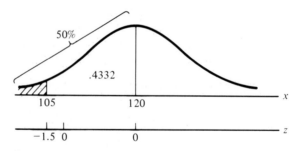

To find the targeted "tail-end" area, find the area between 105 and 120, then subtract it from 50% (.5000). (Remember, half the area in this symmetric distribution must be to the left of the mean and half to the right of the mean.)

$$z = \frac{105 - 120}{10} = -1.5, \text{ producing a probability of .4332}$$

$$.5000 - .4332 = .0668$$

(f) Same as (e).

EXERCISE 6

(a) $P(x = 1 / \lambda = 2) = \dfrac{2^1 (2.718)^{-2}}{1!} = \dfrac{2}{(1)(7.3875)} = .2707$

(b) $P(x = 0/\lambda = 2) = \dfrac{2^0(2.718)^{-2}}{0!} = \dfrac{1}{(1)(7.3875)} = .1353$

(c) $P(x \leq 1/\lambda = 2)$

Use the additive rule to sum $P(x = 0/\lambda = 2) + P(x = 1/\lambda = 1)$. Thus, $.2707 + .1353 = .406$.

(d) $P(x = 3/\lambda = 4) = \dfrac{4^3(2.718)^{-4}}{3!} = \dfrac{64}{(6)(54.5755)} = .1954$

EXERCISE 8

Using probabilities from the table for $\lambda = .5$,

$$E(x) = 0(.6065)$$
$$1(.3033)$$
$$2(.0758)$$
$$3(.0126)$$
$$4(.0016)$$
$$\underline{5(.0002)}$$
$$= 0.5 \text{(the table rounds probabilities slightly)}$$

$$VAR(x) = (0 - .5)^2(.6065)$$
$$(1 - .5)^2(.3033)$$
$$(2 - .5)^2(.0758)$$
$$(3 - .5)^2(.0126)$$
$$(4 - .5)^2(.0016)$$
$$\underline{(5 - .5)^2(.0002)}$$
$$= 0.5$$

$$\text{STD DEV}(x) = \sqrt{VAR(x)} = \sqrt{.5} = .707$$

The following are solutions to the chapter-supplement exercises.

EXERCISE S1

(a) From the binomial table, $P(x = 5/n = 20, p = .1)$ is .0319. Using $np = 20(.1) = 2$ as the mean (λ) for the approximating Poisson produces $P(x = 5/\lambda = 2) = .0361$.

(b) From the binomial table,

$$P(x \leq 2/n = 18, p = .05) = .3972 + .3763 + .1683 = .9418$$

Using $np = 18(.05) = .9$ as the mean (λ) for the approximating Poisson produces

$$P(x \leq 2/\lambda = .9) = .4066 + .3659 + .1647 = .9372$$

EXERCISE S2

(a) To approximate $P(x \geq 110)/n = 200, p = .5)$, use the normal distribution, with

$$\mu = np = 200(.5) = 100 \quad \text{and} \quad \sigma = \sqrt{200(.5)(.5)} = \sqrt{50} = 7.07$$

We're looking then for $P(x \geq 110/\mu = 100, \sigma = 7.07)$.

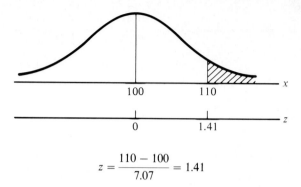

$$z = \frac{110 - 100}{7.07} = 1.41$$

From the normal table for $z = 1.41$, the probability is .4207. Subtracting .4207 from .5000 produces .0793.

(b) To approximate $P(26 \le x \le 35)/n = 100, p = .3)$, use the normal distribution, with

$$\mu = np = 100(.3) = 30 \quad \text{and} \quad \sigma = \sqrt{100(.3)(.7)} = \sqrt{21} = 4.58$$

We're looking then for $P(26 \le x \le 35/\mu = 30, \sigma = 4.58)$.

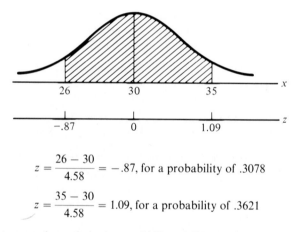

$$z = \frac{26 - 30}{4.58} = -.87, \text{ for a probability of .3078}$$

$$z = \frac{35 - 30}{4.58} = 1.09, \text{ for a probability of .3621}$$

Summing the two produces the estimate: $.3078 + .3621 = .6699$.

(c) To approximate $P(x \le 19/n = 30, p = .5)$, use the normal distribution, with

$$\mu = np = 30(.5) = 15 \quad \text{and} \quad \sigma = \sqrt{30(.5)(.5)} = \sqrt{7.5} = 2.74$$

We're looking then for $P(x \le 19/\mu = 15, \sigma = 2.74)$.

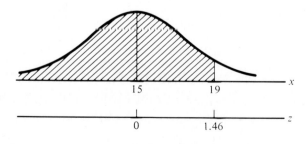

$$z = \frac{19 - 15}{2.74} = 1.46$$

From the normal table for $z = 1.46$, the probability is .4279. Adding .4279 to .5000 produces .9279.

EXERCISE S3

(a) For the unadjusted estimate, we'll produce $P(x \geq 7/\mu = 6, \sigma = 1.73)$ from the normal distribution:

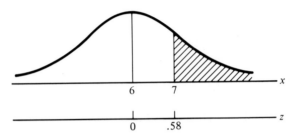

Using $z = \dfrac{7 - 6}{1.73} = .58$, the normal table shows a probability of .4582.

$.5000 - .2190 = .2810$

For the adjusted estimate, we'll produce $P(x \geq 6.5/\mu = 6, \sigma = 1.73)$ from the normal distribution:

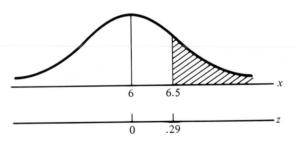

Using $z = \dfrac{6.5 - 6}{1.73} = .29$, the normal table shows a probability of .1141.

$.5000 - .1141 = .3859.$

(b) For the unadjusted estimate, we'll produce $P(11 \leq x \leq 13/\mu = 10, \sigma = 2.24)$ from the normal distribution:

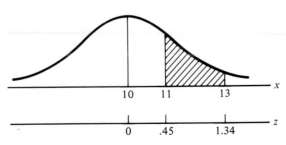

Using $z = \dfrac{11 - 10}{2.24} = .45$, the normal table shows a probability of .1736.

Using $z = \dfrac{13 - 10}{2.24} = 1.34$, the normal table shows a probability of .4099.

Subtracting the smaller from the larger produces $.4099 - .1736 = .2363$

For the adjusted estimate, we'll produce $P(10.5 \le x \le 13.5/\mu = 10, \sigma = 2.24)$ from the normal distribution:

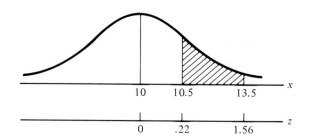

Using $z = \dfrac{10.5 - 10}{2.24} = .22$, the normal table shows a probability of .0871.

Using $z = \dfrac{13.5 - 10}{2.24} = 1.56$, the normal table shows a probability of .4406.

Subtracting the smaller from the larger produces a probability of $.4406 - .0871 = .3535$

(c) For the unadjusted estimate, we'll produce $P(x \ge 36/\mu = 30, \sigma = 4.58)$ from the normal distribution:

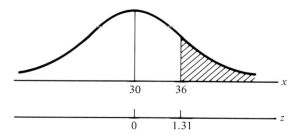

Using $z = \dfrac{36 - 30}{4.58} = 1.31$, the normal table shows a probability of .4049.

$.5000 - .4049 = .0951$

For the adjusted estimate, we'll produce $P(x \ge 35.5/\mu = 30, \sigma = 4.58)$ from the normal distribution:

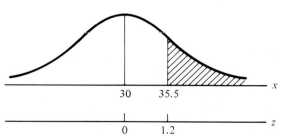

Using $z = \dfrac{35.5 - 30}{4.58} = 1.2$, the normal table shows a probability of .3849.

$.5000 - .3849 = .1151$

EXERCISE S4

(a) $P(x \geq (1/3)/\lambda = 6) = e^{-6(1/3)} = e^{-2}$

$$= \frac{1}{2.718^2} = .135$$

(b) $P(x < 1/\lambda = 6) = 1 - P(x \geq 1/\lambda = 6)$

$$= 1 - e^{-6(1)} = 1 - \frac{1}{2.718^6}$$

$$= 1 - .0025 = .9975$$

EXERCISE S5

(a) $P(x \geq 1/\lambda = 4) = e^{-4(1)} = \dfrac{1}{2.718^4}$

$$= .0183$$

(b) $P(x \leq (1/4)/\lambda = 4) = 1 - e^{-4(1/4)} = 1 - e^{-1}$

$$= 1 - \frac{1}{2.718}$$

$$= 1 - .3679 = .6321$$

EXERCISES

★ 1. You plan to make three sales calls today. You estimate that the chance of making a sale on any one call is about 30%. Construct a probability tree in order to establish the full probability distribution for the random variable (x) "number of sales." Show distribution results in the x, $P(x)$ table format.

★ 2. Refer to exercise 1. Show the distribution graphically and compute both the expected number of sales and the standard deviation of the sales distribution.

★ 3. Use the binomial function to reproduce the set of probabilities you generated in exercise 1. Compute the expected value and the standard deviation of the distribution using the abbreviated computational expressions for the binomial distribution introduced in the chapter (i.e., $E(x) = np$, STD DEV$(x) = \sqrt{np(1 - p)}$).

4. Ajax Construction is currently involved in four major projects. Ajax estimates that each project is 40% likely to be late (i.e., to take longer than originally scheduled). Using a tree diagram, produce the full probability distribution for the random variable "number of late projects."

5. Show your results from exercise 4 in graphical form. Compute the expected value and the standard deviation of the distribution.

6. Use the binomial function to produce the necessary probabilities in exercise 4.

7. Use the binomial table at the end of the text to confirm your results in exercises 1 and 4.

★ 8. Use the binomial table to produce the following binomial probabilities:
(a) $P(x = 4/n = 10, p = .3)$
(b) $P(x = 8/n = 20, p = .6)$
(c) $P(x \leq 12/n = 15, p = .7)$
(d) $P(7 \leq x \leq 13/n = 20, p = .5)$
(e) $P(14 \leq x \leq 18)/n = 30, p = .6)$

9. Use the binomial table to produce the following binomial probabilities:
   (a) $P(x = 5/n = 8, p = .4)$
   (b) $P(x = 8/n = 15, p = .8)$
   (c) $P(x \leq 9/n = 12, p = .9)$
   (d) $P(5 \leq x \leq 8/n = 10, p = .7)$

★ 10. Many educators are concerned that American college and high school students simply do not know basic historical facts. One recent study claimed that 60% of American students do not know which side won the American Civil War. If this 60% figure is accurate, how likely is it that a sample of 30 students will produce
   (a) at least 20 students who know which side won?
   (b) no more than 12 students who know which side won?
   (c) between 15 and 21 students who know which side won?
   (d) fewer than 8 students who know which side won?
   (e) for the random variable "number of students in the sample of 30 who know which side won the American Civil War," compute the mean (expected value) and the standard deviation of possible results.

11. You are to lead a deep-space probe involving extensive scientific experimentation. The guidance system for your space vehicle is obviously critical. The system consists of six independent components, each of which must work as engineered in order for the overall system to function properly. If each component has a 10% chance of failing,
   (a) how likely is it that at least two of the components will fail during the flight?
   (b) how likely is it that no more than three components will fail?
   (c) how likely is it that the system (i.e., all six components) will work properly?
   (d) Suppose you plan to include a backup unit for each of the six components. Tests show that either the backup or the primary component will function properly 95% of the time (i.e., the chance of both a backup and a primary unit failing is only 5%). What is the probability that the system will function properly during the mission?

★ 12. You are about to take a test consisting of 20 true-false questions (a test for which you haven't studied a lick). Your plan is merely to guess randomly on each answer.
   (a) A passing grade on the test is 60%. Given your "pure guess" strategy, how likely is it that you will pass the test?
   (b) To make no worse than a B grade on the test, you would need at least an 80% score. What are your chances?
   (c) Suppose you walk into the exam and discover that the format is not true-false as you had expected, but multiple choice, with four choices for each question. Under your "pure guess" strategy, how likely is it that you pass this exam?

13. You now discover that part (c) of exercise 12 was just a bad dream. The exam is tomorrow, and it will definitely be true-false. Sobered by your nightmare, however, you decide to consider actually doing some studying. You believe that with some studying you can improve your chances of choosing the right answer on any question to .6. If this is the case,
   (a) how likely is it that you will pass the exam?
   (b) how likely is it that you will make a B or better?
   (c) What would your probability of choosing the right answer on any question have to be in order to ensure at least a 95% chance of getting a B (or better) on the exam.

★ 14. Your company has just received a large shipment of GHT rhomboid discriminators. You plan to randomly select and test 20 of the discriminators. Based on test results, you will decide whether the shipment as a whole is satisfactory or whether it should be returned to the supplier. If the overall shipment contains 10% defectives, how likely is it that your sample of 20 will have
   (a) at least three defectives?
   (b) no more than one defective?
   (c) between one and four defectives?

15. Refer to exercise 14. Suppose the firm has a standard policy on shipment quality: If the overall shipment contains no more than 5% defectives, the shipment is considered "good." More than 5% defectives and the shipment should be considered unacceptable and sent back. You will have just one sample of 20 discriminators to test.
   (a) You decide that a sample of 20 with one or more defective discriminators will be con-

sidered sufficient to send back the entire shipment. If this test is applied to a 5% defective batch (i.e., to a "good" batch), how likely is it that you will be led by sample results to mistakenly send back the shipment?

(b) Where should you set the "cutoff" for test results in the sample of 20 in order to ensure that you have no more than a 1% chance of making the mistake of sending back a "good" (5% defective) shipment?

★ 16. The amount of biodegradable waste material produced each day by your productive operations follows a normal distribution, with a mean of 250 tons and a standard deviation of 10 tons. Using x to represent the random variable "tons of waste," find

(a) $P(240 \leq x \leq 260)$ (b) $P(250 \leq x \leq 275)$
(c) $P(235 \leq x \leq 255)$ (d) $P(230 \leq x \leq 240)$
(e) $P(x \geq 234)$ (f) $P(x \leq 226)$

17. Your firm produces YBT Delton sockets that must conform to fairly rigid diameter specifications. Historical evidence shows that actual diameters follow a normal distribution with a mean of 125 mm and a standard deviation of 4 mm. What percentage of diameters will be

(a) between 120 and 130 mm?
(b) between 131 and 137 mm?
(c) less than 136 mm?
(d) more than 132 mm?
(e) less than 116 mm?

★ 18. Annual salaries in a large corporation follow an approximately normal pattern, with a mean wage of $22,300 and a standard deviation of $2400. If an employee is randomly selected, how likely is it that the employee's salary is

(a) between $18,000 and $24,000?
(b) between $16,000 and $20,000?
(c) more than $27,000?
(d) less than $17,000?
(e) It is 68.3% likely that the employee's salary will be between $_____ and $_____. (Make your interval symmetric about the mean.)

★ 19. For a standard normal distribution, use the normal table to determine the z score necessary to construct a (symmetric) interval around the mean in order

(a) to bound 90% of the values in the distribution. (*Hint:* Be sure to draw the normal picture and identify in general the interval

you want to construct, with half the desired 90% area to the right of the mean and half to the left. Locate in the normal table an area as close to .90/2 (.4500) as you can find. Read out the corresponding z score.)

(b) to bound 95% of the values in the distribution.
(c) to bound 99% of the values in the distribution.

20. A newly developed seed has a germination period that follows an approximately normal distribution, with a mean germination time of 16.4 days and a standard deviation of 2.1 days. What percentage of the seeds have a germination period

(a) between 15 and 18 days?
(b) less than 12 days?
(c) 95% of the seeds have a germination period of between _____ and _____ days. (Make your interval symmetric about the mean.)
(d) 80% of the seeds have a germination period of between _____ days and _____ days. (Make your interval symmetric about the mean.)
(e) 99% of the seeds have a germination period of at least _____ days.

★ 21. Monthly demand for your product follows a normal distribution with a mean of 11,000 units and a standard deviation of 1000 units. Suppose you plan to produce 12,000 units for the upcoming month. How likely is it that you will

(a) end up with at least 1500 units of unsold product?
(b) have a shortage of at least 500 units?
(c) end up with no shortage at all?
(d) If you want the chance of a shortage to be no more than 5%, how many units should you plan to produce for the upcoming month?

22. Your Acme Food Filler III is a machine that fills containers with a preset amount of semiliquid product—in this case, a gelatinous substance used in processing day-old dead fish. The average fill weight for the machine can be set and controlled, but there is inevitably some variability in the process. Actual weights, in fact, follow a normal pattern around the mean, with a known standard deviation of .05 ounce.

(a) If the machine is set for a mean fill weight of 8 ounces, what proportion of containers will be filled with less than 8 ounces of product?

(b) If the machine is set for a mean fill weight of 8.05 ounces, what proportion of the containers will be filled with less than 8 ounces of product?

(c) If you want to ensure that no more than 1% of the containers will be filled with less than 8 ounces of product, where should you set the mean fill weight?

★ 23. Your firm produces tires with a useful life that follows an approximately normal pattern. Average life is 36,000 miles, with a standard deviation of 1500 miles. What percentage of the tires will have a useful life

   (a) between 38,000 and 40,000 miles?
   (b) less than 35,000 miles?
   (c) more than 39,000 miles?
   (d) Suppose the company wants to establish a warranty policy. Any tire not lasting as long as the warranty period will be replaced free of charge. What warranty mileage would you suggest to ensure that the firm won't have to replace more than 3% of the tires sold?

★ 24. Use the Poisson function to determine the following probabilities:

   (a) $P(x = 2/\lambda = 3)$
   (b) $P(x = 4/\lambda = 1)$
   (c) $P(x < 3/\lambda = 6)$

25. Use the Poisson table to confirm your answers to problem 24.

★ 26. Use the Poisson table to sketch the Poisson distribution for $\lambda = 2$. Compute the mean and the standard deviation to verify that the mean is λ and that the standard deviation is $\sqrt{\lambda}$.

27. Traffic exiting the Arlington Turnpike at exit 123 conforms roughly to the Poisson conditions. The average number of cars exiting is six per minute. Compute

   (a) the probability of exactly 4 cars exiting during the next minute observed.
   (b) the probability of at least 10 cars exiting during the next minute observed.
   (c) the probability of no more than 1 car exiting during the next minute and a half.
   (d) the probability of exactly 2 cars exiting during the next 30 seconds observed.

28. The appearance of bubbles in the plastic sheets produced by your new extrusion equipment follows a Poisson pattern. The average number of bubbles per square yard of product is 0.2 (otherwise the appearance of bubbles seems perfectly random). Compute

   (a) the probability of finding no bubbles in a randomly selected square yard of plastic.
   (b) the probability of finding exactly two bubbles in a randomly selected square yard of plastic.
   (c) If a full plastic sheet is 6 feet by 9 feet, how likely is it that a randomly selected sheet will show three or more bubbles?

29. Arrivals at the parking lot close to your job follow an approximately Poisson pattern, with a mean arrival rate of 10 cars per hour. You arrive at 9 A.M. and find that there are 5 available spaces (of a total of 20 in the lot). You know that your boss will be arriving at 9:30 and will need a place in the lot to park. Assuming that any new arrival will be in the lot for at least a half-hour and that no one currently in the lot will leave between now and 9:30, compute the probability that your boss will have a place to park at 9:30

   (a) if you decide to park in the lot.
   (b) if you decide not to park in the lot.

30. Equipment malfunctions on the shop floor in your department occur at an average rate of one per hour (otherwise the malfunctions appear to be random). Use the Poisson distribution to determine the probability of

   (a) exactly two malfunctions in the next hour.
   (b) no malfunctions in the next half-hour.
   (c) fewer than three malfunctions in a 4-hour shift.
   (d) between two and five malfunctions in the next hour and a half.

★ 31. Refer to problem 30. Suppose you currently have a maintenance crew capable of effectively handling up to 10 malfunctions per 8-hour shift.

   (a) How likely is it that in any shift, more malfunctions occur than the crew can handle?
   (b) You are considering expanding and upgrading the maintenance crew. If you want to ensure that there is less than a 5% chance that the crew won't be able to handle the malfunction load, you will need a crew that can handle up to _____ malfunctions per 8-hour shift.

32. Typographical errors in the *Daily Times/Mirror* occur at an average rate of .4 per page (otherwise the pattern appears perfectly random). Use the Poisson distribution to determine the probability of finding no errors

   (a) on a randomly selected page.
   (b) in the entire sports section (11 pages).

(c) How likely is it that you will have to read at least five full pages before you come to a page with an error?

The following problems deal with material from the chapter supplements.

★ 33. Use the Poisson distribution to approximate the following binomial probabilities.
   (a) $P(x = 3/n = 200, p = .02)$
   (b) $P(x = 1/n = 50, p = .01)$
   (c) $P(x = 0/n = 1000, p = .002)$
   (d) $P(x \leq 2/n = 100, p = .01)$

★ 34. Use the normal distribution to approximate the following binomial probabilities. (Use the "continuity" correction factor as appropriate.)
   (a) $P(20 \leq x \leq 25/n = 50, p = .5)$
   (b) $P(x \geq 48/n = 100, p = .4)$
   (c) $P(x \leq 85/n = 1000, p = .1)$
   (d) $P(x \geq 230/n = 5000, p = .05)$

35. Use the proper table from the back of the text to determine (or if necessary, approximate) the following binomial probabilities.
   (a) $P(x \leq 5/n = 20, p = .3)$
   (b) $P(22 \leq x \leq 30/n = 60, p = .4)$
   (c) $P(x \leq 4/n = 100, p = .03)$
   (d) $P(x \geq 64/n = 80, p = .7)$

★ 36. Typographical errors in the *Daily Times/Mirror* occur at an average rate of .4 error per page (otherwise the pattern appears perfectly random).
   (a) Use the exponential distribution to determine how likely it is that you would have to read at least five full pages before you come to a page with an error.
   (b) Compare your answer here to your answer in part (c) of exercise 32. Comment on the connection.

37. Accidents on the shop floor serious enough to require medical attention occur at an average rate of 2.0 per 8-hour day (i.e., .25 accidents per hour).
   (a) Use the exponential distribution to determine the probability that at least 2 hours will elapse before the occurrence of the next accident.
   (b) Now use the Poisson distribution to determine the probability that no accidents occur during the next 2 hours. Compare your answer here to your answer in (a) and comment on the connection.

STATISTICAL INFERENCE: ESTIMATING A POPULATION MEAN

CHAPTER OBJECTIVES

Chapter 5 should enable you to:

1. Describe the general nature of statistical inference.

2. Distinguish between random and nonrandom sampling methods.

3. Use a random number generator (or table) to produce a simple random sample.

4. Build a confidence interval estimate for the mean of a population using only sample information.

5. Understand the role of the sampling distribution of means in the construction of a confidence interval.

6. Use the t table to construct confidence intervals.

7. Recognize when and how the finite population correction factor should be used in interval estimation.

8. Determine the appropriate sample size for a simple random sampling plan.

We'll turn our attention now to what we had earlier described as a second major branch of statistical analysis: **inferential** (sometimes called **inductive**) **statistics**. As we'll shortly see, the principles of inferential statistics will add significantly to our ability to translate data into information.

5.1 THE NATURE OF STATISTICAL INFERENCE

Suppose we're confronted with a large data set—call it a population—and we want to know something about a key characteristic (or "parameter") of the data set. (For example, we might want to know the average value, or possibly the proportion of items in the data set that possess a particular attribute (e.g., are blue, or large, or Republican).) Rather than taking a full census (that is, conducting a 100% inspection) of the population, we might choose, for time, effort, or cost reasons, to select only a subset (or sample) of the population values. Our hope is to gain some insight into what would be true of the entire population based on what we see in the sample. Using sample information to draw conclusions (i.e., to make inferences) about a population is the central theme of inferential statistics.

5.2 AN ILLUSTRATION

Assume you find yourself sitting in a classroom with 100 people. Curious about the company you're keeping, you'd like to know more about the prosperity of your classmates. Specifically, you want to assess the average annual income of the group. How might you proceed? You could take a full census of the class, asking each and every class member his or her annual income. Compiling all 100 values, you could then compute the class average and be done. In so doing, you will have effectively generated precisely (assuming everyone has been open and candid) the "population" measure you wanted.

5.3 SAMPLING

The Decision to Sample

Can you think of any reasons, though, why you might not be able or anxious to take a full census of this population? Would time, effort, or cost considerations be relevant? We can certainly conceive of situations in which 100% inspection is simply not an attractive, or even a feasible, option. What recourse is there? How about some form of sampling? You might choose to select only a subset of values from the population and base conclusions about the full population on what you see in the subset.

> **Note:** Other reasons, besides time, cost, and effort, may sometimes suggest or even require taking a sample rather than a census of the full population. For example, in situations where destructive testing is involved, testing an entire population would be practically impossible. (To evaluate the average life of a company's lightbulbs, we couldn't simply wait until they've all burned out and then make our computation.) Or we may find ourselves involved with what amounts to an infinite population (e.g., "all cold sufferers for all time," or "all units that could conceivably be produced by a machine in its current state of adjustment") where we couldn't possibly access all the members of the population.
>
> We might also mention that a sample may actually produce more "accurate" results than would a full census. Careful and controlled examination of a relatively few items may provide more reliable information (that is, it may involve less error) than an exhaustive, and exhausting, look at every member of a very large population.

Assume that sampling is, in fact, the route you plan to take in pursuing the average income problem we've proposed. What issues need to be immediately addressed? Think for a second. What about the issue of sample size? Since you intend to select only a subset of population values, you'll need to decide early in the process just how many values should be included. If you select too few values, you risk an insufficient representation of the population. Choose too many and you defeat the primary purpose of sampling—economizing on the amount of data that needs to be collected. As we'll see later, sampling theory generally offers the means to prescribe an appropriate sample size. For the moment, though, we'll put off discussion of any formal approach and settle the issue simply by choosing a sample size arbitrarily, say 10. Our plan, then, is to select a sample of size 10, take the appropriate measurements and draw conclusions about the population of 100 based on what we see in the sample. (As a matter of notation, we'll use n to represent sample size and N to represent population size. In our example, $n = 10$ and $N = 100$.)

Random Sampling

Having resolved (temporarily) the question of how many students to select for the sample, we can turn to a second key issue: Which ones? Just what sort of *selection procedure* should we use to pick the 10 students we need? It turns out that we have quite a few options, most of which share a common key element—randomness. In order to base our conclusions about the classroom population on firm statistical ground, we'll need to select a **random sample** of the students in the room. And just what's meant by the term *random*?

First of all, it's important to recognize that random is not used here as a synonym for "haphazard." On the contrary, it often takes significant planning and effort to ensure randomness in a sampling procedure. In the simplest form of **random**

sampling (not surprisingly called **simple random sampling** or, sometimes, **unrestricted random sampling**), the term *random* means essentially that every element in the population is given an equal chance of being included in the sample. As we've implied above, ensuring such a condition is not always as easy as it might seem.

Given our simple random sampling definition, could we, for example, form a random sample by picking the first 10 classmates sitting in the first row? Most of us, I think, would be uneasy with this sort of selection procedure. It may be convenient, but is it truly random? I suppose in some sense it is, since conceivably every member of the class had an equal opportunity to choose a front-row seat. But consider the possibility of nonrandom influences: friends tend to sit together, and friendships are often related to common economic backgrounds and social position; or only those who can't afford glasses will be drawn to the front rows of the classroom where they can see the board better. We may be stretching things a bit, but hopefully the point is made. The appearance of randomness can sometimes be misleading.

If selecting the first 10 students seated in the classroom is not such a good idea, exactly how should we produce the random sample we need? How about asking everybody to come to the front of the room? We'll have them mill about smartly while you place a blindfold over your eyes and proceed to tap 10 people on the shoulder. Unfortunately, although this procedure might well produce a satisfactory random sample, it's also likely to produce more confusion than is really necessary. A reasonable substitute? We could begin by assigning a unique identification (ID) number to each member of the class. Since there are 100 students, we might use all the two-digit numbers from 00 to 99. We could then use a **random number table**, such as Table 5.1, to randomly select 10 two-digit numbers. For each number selected, we'll find the corresponding population member and include that member in the sample being formed.

Try it out. Assuming we've already completed the assignment of ID numbers from 00 to 99, we can begin random number selection simply by starting in the upper left-hand corner of the random number table and taking the first two digits encountered. (There are other, more sophisticated ways to initiate random number selection. In fact, people have written dissertations dealing with the subject. We'll try, however, to keep things as simple as possible.) The first two digits, 1 and 6, indicate that population member 16 should be included in the sample. Tracing down the first two columns of the table (we could just as easily have proceeded horizontally or diagonally) produces 76, then 59, and so on. We can continue until all 10 sample members are properly identified.

Sampling with or without Replacement

One issue which may arise at this point deserves comment: Suppose we draw the same two-digit random number a second (or even a third) time as we move through the table? Do we include the corresponding population member more than once, or do we simply discard the second (or third) selection of the same number and continue on? For example, suppose the number 16 (the first one we selected) turned up as the sixth value as well? Does person 16 get included twice in our sample of 10? The answer is, it depends.

TABLE 5.1   **RANDOM NUMBER TABLE (TABLE OF UNIFORMLY DISTRIBUTED RANDOM DIGITS)**

```
1 6 8 7 7 0 4 4 1 9 7 5 9 3 9 9 5 5 6 7
7 6 4 3 1 7 5 4 0 7 5 5 7 6 2 4 1 4 0 3
5 9 2 6 9 3 8 7 9 3 4 4 7 2 2 7 6 7 9 2
5 4 1 4 6 1 3 2 5 8 1 0 0 1 7 5 9 6 5 7
4 1 3 8 9 4 1 8 5 8 1 9 5 6 6 8 3 2 0 3
2 5 1 1 2 5 8 4 1 5 3 4 6 9 3 2 4 8 7 2
4 1 9 8 4 7 5 4 7 8 3 0 4 3 4 6 8 0 6 6
6 7 7 8 2 8 3 8 4 0 3 9 0 7 4 1 1 3 2 0
2 0 0 6 1 5 4 1 8 8 0 3 6 6 9 6 9 8 3 9
1 5 4 5 0 4 4 2 9 9 0 4 6 9 1 4 5 8 5 5
5 8 8 0 2 8 6 6 2 0 0 9 6 8 2 9 9 5 8 3
0 1 3 4 6 4 5 6 9 7 2 6 1 2 0 8 7 9 0 5
9 4 5 0 1 4 3 8 4 9 5 6 4 5 7 7 7 6 0 6
3 7 8 1 6 8 5 0 1 3 6 2 7 9 6 4 7 9 1 1
3 6 5 5 4 4 4 2 7 5 7 6 9 3 8 8 7 8 2 9
7 7 7 4 1 7 2 5 6 7 9 4 1 7 6 3 0 5 8 6
2 2 3 9 8 1 5 4 1 3 9 9 5 4 9 9 8 8 2 5
9 5 1 6 6 1 7 2 5 3 5 6 8 7 7 7 3 5 4 9
5 1 6 1 8 1 4 5 5 2 6 8 8 8 3 5 3 1 7 5
1 3 7 2 3 9 0 4 4 2 9 9 5 4 3 8 6 3 7 9
8 9 3 5 5 1 7 8 9 5 7 5 8 0 0 0 2 1 0 3
4 7 3 2 7 1 7 4 2 5 2 1 5 9 1 0 1 8 7 3
1 3 3 5 8 0 6 9 2 1 6 5 4 4 1 6 8 4 8 1
7 2 4 6 1 4 0 6 7 7 9 8 6 8 7 7 3 9 1 6
0 0 7 5 5 0 7 9 5 2 6 2 2 1 5 4 3 3 8 6
6 1 5 2 0 7 6 5 4 1 7 9 3 9 3 5 9 9 5 9
5 6 9 0 0 5 0 1 7 6 5 1 8 7 0 0 9 3 0 1
3 1 0 5 6 7 1 7 8 7 9 4 3 6 6 7 1 1 4 7
1 5 0 6 6 9 9 3 0 5 7 8 2 6 3 4 7 6 2 5
3 0 2 1 6 2 6 7 8 3 0 8 1 8 2 2 1 2 4 6
0 3 8 9 2 0 1 2 2 4 5 1 5 1 0 8 2 5 9 2
1 8 3 2 3 8 0 3 4 3 1 9 5 8 1 6 4 9 7 7
8 4 8 8 7 8 7 3 6 5 4 5 3 0 6 8 4 3 2 2
7 7 2 9 2 1 4 3 1 8 5 7 0 5 0 9 9 4 5 4
5 3 9 9 4 4 3 2 0 4 9 1 1 8 3 3 7 1 8 2
3 2 0 1 7 6 1 6 4 6 0 9 5 8 3 7 8 7 9 4
8 1 9 3 9 2 8 0 7 8 1 7 2 7 5 5 4 0 0 3
3 9 5 5 4 1 7 1 4 3 7 0 5 6 1 1 2 4 9 7
9 4 2 9 1 8 5 7 7 5 5 3 7 8 4 0 8 2 8 4
5 7 3 8 7 7 5 8 1 1 3 2 8 8 0 2 0 0 4 6
```

We can fine-tune our selection procedure in one of two ways: We can either sample **with replacement** or **without replacement**. Sampling with replacement would mean that should we select the same random number more than once, the corresponding population member would be included more than once in the sample. Conceivably, we could, under this strategy, end up with our sample of size 10 consisting entirely of the same population member represented 10 times—perhaps not a very desirable result (or a very likely one), but under sampling with replacement, it is a possibility. Sampling without replacement, on the other hand, would preclude this kind of extreme outcome. Here, if a random number is repeated, we simply ignore it as we proceed to select 10 unique values for the sample.

In practice, sampling is almost always done *without* replacement. However, we'll begin our sampling discussion assuming all sampling is being done *with* replacement. Why? Most of the computational formulas are considerably simpler for the with-replacement case. Moving to the more widely-used without-replacement approach will involve complicating and extending the basic with-replacement expressions.

A Note on Random Number Tables

A random number table (or, alternatively, a computer function that can be used to generate a sequence of random numbers) is an extremely useful tool in sampling. In order to qualify as a table of random numbers (sometimes labeled a table of "uniformly distributed" random digits), two major conditions need to be met: (1) each of the digits (0 through 9) must have equal representation (that is, 10% of the table entries should be 0's, 10% 1's, etc.), and (2) there must be no apparent pattern to the way in which those entries appear. (Since virtually all such tables are computer-generated, there must be some pattern to the entries, but the pattern is usually so obscure that you or I would not likely be able to detect and exploit it.)

Other Sample Selection Procedures

Back to our example. As we've already implied, simple random sampling is not the only, nor necessarily the best, sampling procedure we might have chosen. For example, a **stratified random sampling** approach might have proven more appealing. To illustrate, suppose you knew beforehand that 30% of your classmates held full-time jobs in addition to going to school, while the remaining 70% were full-time students with, at most, part-time jobs. Would this influence your efforts to produce a sample to represent the population? Very possibly. In this situation, we might pointedly pursue measures that would guarantee representation in the sample for each of the two income-related groups (or strata). We might, for example, make a conscious effort to ensure proportionate representation in our sample of 10 by randomly selecting three members from the "full-time employed" group (stratum) and seven members from the "full-time student" group. Such efforts, characteristic of stratified random sampling, may produce a "better," more efficient sample on which to base conclusions about the population.

Or, for ease of selection, we might have used **systematic random sampling**. With this approach, once a starting point in the population has been randomly selected, population members are chosen at a fixed interval for inclusion in the sample. For our classroom example, only one two-digit random number would be needed to start the sample selection process. That is, we'd need to select a random number in the range 00 to 09 to identify the initial population member to be included in the sample. From then on, we would simply pick every 10th population member to produce the full sample of 10. Once the sample is formed and appropriate measurements are made, sample results can be treated in the same way as those in the simple random sampling case, but sample selection is simplified.

Other sampling procedures, including nonrandom approaches, are available. For example, **judgment sampling**, in which certain "representative" population members are deliberately (rather than randomly) chosen for the sample, might be used. You need to be aware, however, that in using any nonrandom procedure you lose the support of statistical theory when you begin to draw conclusions about the population based on sample results. (Specifically, with any nonrandom sampling procedure, you lose the ability to systematically measure the possible error in your conclusions, a key feature in any of the random sampling approaches.) Unless otherwise noted, we plan to use simple random sampling in our sampling discussions. It serves as the base case from which the other selection procedures can be derived or to which they can be directly compared.

> **Note:** What we've here labeled "random" sampling is also frequently referred to as "probability" sampling or "statistical" sampling.

At Last, The Data

Assume now that you've completed sample selection for the class income example, using simple random sampling with replacement as your selection procedure. Asking each classmate selected to report current annual income produces the following 10 income values (note that population member 41 was included twice, a perfectly legitimate possibility when sampling with replacement):

Population member	Annual income (x)
16	$21,500
76	6,100
59	1,400
54	29,850
41	2,310
25	12,650
41	2,310
67	31,230
20	690
15	2,140

Where do we go from here? Remember, we started out wanting to know the overall (i.e., the population) average income for the class of 100 students. What we have presently, however, is only a sample of 10 observations. The logical next step? Compute the average for the sample. Try it.

Constructing a Confidence Interval

It turns out that the sample mean here is $11,018. (From here on, we plan to use $\bar{x}$ (read x-bar) specifically to denote a sample mean.) Thus we know precisely the mean value for the sample. But does this shed any light at all on our original target—to know the mean value for the population? *Question:* Would you expect the population mean to be exactly the same as the sample mean? That is, would you expect the overall class mean to be precisely $11,018? *Answer:* No, or at least it's not very likely. *Question:* Should we expect the population mean to be somewhere in the neighborhood of the sample mean? *Answer:* Absolutely. And herein lies a key to the inference problem we've posed. Rather than settling for a single best guess of the population mean (a so-called "point estimate") that is very likely to be wrong, we plan to construct an *interval* estimate of the population mean based on sample results.

Next question: If we expect the population mean (we'll begin to use the Greek letter μ (mu) to represent the population mean) to be somewhere in the "neighborhood" of the sample $\bar{x}$, just how large a neighborhood are we talking about? within $2? within $2000? within $20,000? As we'll see shortly, statistical theory plays a crucial role in establishing the right neighborhood size. It will enable us to build an interval (i.e., identify a neighborhood) around the sample mean $\bar{x}$ which we can be *confident* will contain the population mean μ. We'll spend the remainder of the chapter developing the key elements involved in constructing this kind of **confidence interval**.

Factors Influencing Interval Width

Before introducing the theoretical basis for interval construction, we might first find it useful to take a somewhat intuitive approach to the procedure by identifying factors which seem likely to influence the size of the neighborhood (i.e., the width of the interval) around $\bar{x}$ in which we would expect to find μ.

Confidence Requirement We'll begin with a basic question: Would you be comfortable, based on the sample results produced thus far, with the following statement?

"I'm absolutely sure (100% confident) that the population mean will be within 5 cents of the sample mean produced." (That is, μ *must* be in the interval $11,018 \pm .05.)

No? Why not? It seems clear that the interval is simply too small, too narrow, to warrant such a high level of confidence. For an interval in which we might have complete (100%) confidence, we would probably have to specify an extremely large neighborhood ($11,018 \pm $1,000,000?). On the other hand, if we could live with a lower confidence level, a tighter interval (a smaller neighborhood) might be perfectly defensible. If, for example, we wanted to produce an interval statement for which we require only 50% confidence, we could logically propose a much narrower interval than for the 100% (or 99% or 98%) case. The point? The level of confidence required for the interval statement should somehow influence the size of the neighborhood involved: The greater the confidence, the wider the interval.

Sample Size Suppose we had selected a sample of 90 classmates rather than the original 10. Should we expect the resulting interval estimate of μ to involve a significantly smaller neighborhood around the sample $\bar{x}$? It certainly seems like it. After all, with a sample size of 90, we're including in the sample almost the entire population. Consequently, we would expect the sample mean, in all but the rarest cases, to be only a short distance from the population mean. The lesson? Sample size must play a role in influencing interval width (or "precision"): the larger the sample size, all other things equal, the tighter (i.e., more precise) the interval; the smaller the sample size, the wider (i.e., less precise) the interval.

Sample (or Population) Variability   The amount of variability in the sample, insofar as it reflects the amount of variability in the population, is also likely to affect the size of the neighborhood term in our interval estimate. To demonstrate, suppose the income values we produced in the original sample of 10 had shown a distinct pattern: The first income reported was $11,018. Curiously, the second income on the list also turned out to be $11,018, and the third, and the fourth. In fact, each of the 10 sample members reported an income of exactly $11,018. Would this complete lack of variation in sample results influence the width of the interval in which we would expect to find μ? It should. The consistency we see in the sample suggests a corresponding consistency in the population. If the sample values cluster closely together, then it's likely that population values are also closely clustered and that none (or at least not many) of the values lie very far from the mean. If this is true, then the mean of our sample is bound to be close to the mean of the population. It seems inescapable, then, that the degree of variability in sample results (which reflects the degree of variability in the population) has an influencing role to play in interval construction: the less *variability* we see in the sample, the narrower the interval we build.

Summing up, we've suggested three elements which should contribute directly to the formation of a reasonable interval estimate of a population mean based on sample results: (1) the confidence required for the interval (that is, the likelihood that the interval will actually contain the population mean); (2) the size of the sample; and (3) the degree of variability in the population as evidenced in sample results. (This all began when we conceded that sampling won't produce an exact measure of the population mean, but rather will provide an effective and economical way to construct an *interval* estimate of the population mean using only partial information.) In the next section, we'll examine more formally the statistical foundation for confidence interval construction.

5.4   THE KEY TO STATISTICAL ESTIMATION: THE ROLE OF A SAMPLING DISTRIBUTION

In virtually all inferential statistics applications, the notion of a sampling distribution provides the critical link between a sample result and a population characteristic. Understanding the sampling distribution idea is absolutely essential, not only to an understanding of confidence interval estimation, but also to a firm grasp of the entire inference process. Failure to comprehend the sampling distribution idea will almost certainly reduce statistical inference to a garbled mess of formulas to be memorized and tools to be misused. It's important that you take the time to carefully work through the discussion that follows.

Producing the Sampling Distribution of All Possible Sample Means

To understand the notion of sampling distribution, refocus on the class income example. Sample results produced so far show a sample mean of $11,018. Now suppose we were to repeat the sampling experiment. That is, suppose we return the original sample of 10 to the population and use the random number table to produce a second sample of size 10. (Some of the first sample might be included in the second, but by and large we should expect to see different sample members.) Would this second sample show a mean exactly equal to the mean of the first ($11,018)? *Answer:* It's highly unlikely. Let's suppose, for demonstration sake, that the second sample shows a slightly higher mean, say $13,420. Now what? Record the result, return the sample members to the population, and select a third sample of 10, then a fourth, and a fifth, and so on, until we've generated all possible samples of size 10 from the population, recording for each the sample mean, $\bar{x}$. (Needless to say, this could become a pretty tedious procedure.)

> **Note:** There are actually 100^{10} different samples of size 10 that could be produced from this population of size 100. Remember, we had decided that in selecting our sample of 10, we would use sampling with replacement. If sampling were being done without replacement, there would be fewer sample possibilities. How many? Just use the combinatorial formula for 100 things taken 10 at a time: $100!/(100 - 10)!10!$.

Were we to follow the method to conclusion, we would, as we've said, end up producing all possible samples of size 10, together with all the corresponding sample means (i.e., all possible $\bar{x}$'s). This complete list of sample means, arranged in a relative frequency table or perhaps shown in histogram or bar-chart form, is, in fact, the **sampling distribution of means**. It's this distribution that will shortly provide the key to linking any *one* sample mean (for example, our $\bar{x}$ of $11,018) to a statement about the overall population mean, μ.

Predicting Characteristics of the Sampling Distribution of Sample Means

We might stop here for a minute to reflect. Do we really want to (or do we really have to) take the time and expend the energy to generate all these sample means in order to fully appreciate the sampling distribution involved? We could be at it a very long time. The answer, fortunately, is no. As we'll see shortly, statistical theory will enable us to project key distribution characteristics without actually having to produce all the values.

Specifically, statistical theory (mainly in the form of the **Central Limit Theorem**) suggests that the sampling distribution of all possible sample means (1) will follow a bell-shaped pattern which, for larger and larger sample sizes, very closely approximates a normal distribution, (2) is centered on a value exactly equal to the population mean (i.e., the mean of all the sample means in the sampling distribution will be exactly equal to the population mean), and (3) has a standard deviation that is simply and directly related to the population standard deviation. (See Figure 5.1.)

FIGURE 5.1   **THE SAMPLING DISTRIBUTION OF ALL POSSIBLE SAMPLE MEANS ($\bar{x}$'s)**

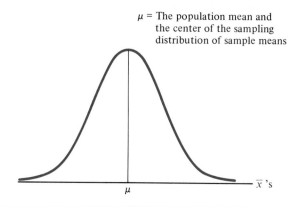

It's these three predictable characteristics (shape, center, and variability) that will allow us to link any one sample mean to the overall mean of the population from which the sample has been randomly selected. We can expand a bit on each:

> **Note:** For the discussion that follows, you may want to refer to Figure 5.2. Here the key role played by the sampling distribution as the link between a population mean and a sample mean is effectively represented.

Shape   Consider first the issue of distribution shape. For large enough sample sizes, the Central Limit Theorem guarantees a near-normal sampling distribution *no matter what the shape of the parent population distribution* (skewed, bimodal, flat, or peaked). In statistical inference, this proves to be an extremely important proposition.

For smaller sample sizes (usually a sample size less than 30 is considered "small") one qualification needs to be added. In the small-sample case, the bell shape of the sampling distribution of means is predictable *only* if the population from which the samples are selected is itself normal. Thus, for our ongoing classroom income illustration, with its sample size of 10, we should be concerned that the overall population of student incomes meets this normal condition. (For purposes of discussion we will routinely make this assumption.) Remember, though, if a slightly larger sample size were involved (say 35), this condition would be un-necessary. In these cases, the emergence of a bell-shaped sampling distribution is ensured irrespective of the shape of the population distribution.

Center   The contention that the center of the sampling distribution of all possible sample means is exactly equal to the population mean appears to be fairly compelling. It seems to make sense that if we were to average out all the sample means, the overall average would be equal to the mean of the population from which the samples were selected, since, collectively, the sample means would actually include every member of the population. This condition holds whether the sample size involved is large or small, or whether the population is normal or not.

FIGURE 5.2   **THE THREE DISTRIBUTIONS IN STATISTICAL INFERENCE**

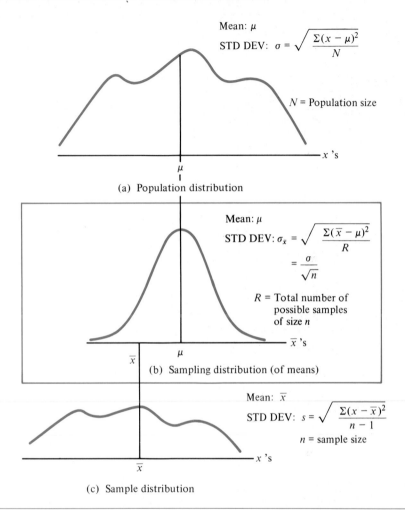

Mean: μ

STD DEV: $\sigma = \sqrt{\dfrac{\Sigma(x - \mu)^2}{N}}$

N = Population size

x's

μ

(a) Population distribution

Mean: μ

STD DEV: $\sigma_{\bar{x}} = \sqrt{\dfrac{\Sigma(\bar{x} - \mu)^2}{R}}$

$= \dfrac{\sigma}{\sqrt{n}}$

R = Total number of possible samples of size n

$\bar{x}$'s

$\bar{x}$ μ

(b) Sampling distribution (of means)

Mean: $\bar{x}$

STD DEV: $s = \sqrt{\dfrac{\Sigma(x - \bar{x})^2}{n - 1}}$

n = sample size

x's

$\bar{x}$

(c) Sample distribution

Variability (Standard Deviation)   Our assertion that variability in the distribution of all possible sample means is directly related to the amount of variability
evidenced in the parent population should also appear reasonable. If there's a great
deal of variability in the population, then sample values, and consequently sample
means, will likely be spread out. If there's very little variability in the population,
then sample values (and sample means) should cluster closely together.

In specific terms, the standard deviation of the sampling distribution (we'll
label it $\sigma_{\bar{x}}$—pronounced "sigma-sub-x-bar") is linked to the standard deviation of
the parent population (which we'll label σ) by the expression

$$\sigma_{\bar{x}} = \dfrac{\sigma}{\sqrt{n}}$$

Note: The above standard deviation expression works if sampling is done with replacement, as it is in our illustrative class income example. The expression becomes a bit more complicated if sampling is done without replacement. This case will be considered later.

5.5 PUTTING THE PIECES TOGETHER

What does all this mean? How can knowing the characteristics of the sampling distribution of all possible sample means have any bearing on estimating the mean of a population using only information from a *single* random sample? (Remember, in statistical estimation we normally have just *one* sample mean, computed from a random sample of *n* observations, to use in estimating the mean of the larger population.)

We can now (and this is important) think of our one sample mean as a value randomly selected from the distribution of *all possible* sample means, a distribution that we now know (1) is approximately normal, (2) is centered on the population mean μ, and (3) has a rather easily computed standard deviation $\sigma_{\bar{x}}$. And what can be said about this sort of value? As with any value randomly selected from a normal distribution, there is a 68.3% chance that this one sample mean will fall within one standard deviation of the distribution center; a 95.5% likelihood that this sample mean will fall within two standard deviations of the distribution center; and a 99.7% likelihood that the value will lie within three standard deviations of the distribution center.

Since we're dealing in this case with a normal distribution whose center is exactly equal to the *population mean*, these statements actually assign a 68.3% probability to the prospect that our randomly selected sample mean will fall within one standard deviation of the population mean; a 95.5% probability that our sample mean will fall within two standard deviations of the population mean; and a 99.7% likelihood that our one sample mean will fall within three standard deviations of the population mean.

We can complete the puzzle by turning things around just a bit. Since we can say that the sample mean is 68.3% likely to fall within one standard deviation of the population mean, we can also conclude that the population mean is 68.3% likely to fall within one standard deviation of the (randomly produced) sample mean. We've simply changed the base point of reference. (If, when you enter your dining room, 68.3% of the time your chair is within 3 feet of the table, it will also be true that 68.3% of the time the table is within 3 feet of your chair.) The 95.5% and 99.7% statements (in fact, any such probability statements) convert in precisely the same manner.

Summing up, we can assign a probability of 68.3% to the prospect that the interval (or "neighborhood")

$$\bar{x} \pm 1 \text{ standard deviation}$$

will actually contain the population mean. Similarly, it is 95.5% likely (or we can be 95.5% confident) that the interval

$$\bar{x} \pm 2 \text{ standard deviations}$$

will contain the population mean. In general, the basic confidence interval expression looks like

$$\bar{x} \pm z \text{ standard deviations}$$

where z represents the proper number of standard deviations for a designated level of confidence (or probability). As we've seen, $z = 1$ for 68.3%, $z = 2$ for 95.5%, etc.

At this point, you might want to check the normal table to verify two other commonly used z values: the one for 95% confidence ($z = 1.96$) and the one for 99% confidence ($z = 2.58$). The z of 1.96 is especially well-used, since 95% confidence serves, in practice, as a kind of "default" level of confidence. In the absence of any other compelling rationale for setting confidence, 95% is most often selected. Most political surveys, for example, typically report results at the 95% confidence level.

Which Standard Deviation?

One additional point needs to be made to complete our general introduction to confidence interval construction. Checking the interval expressions above, you may have noticed that the standard deviation term we've used is not very specific. Depending on the confidence requirement, we've shown intervals involving one standard deviation or two standard deviations or 1.96 standard deviations, etc. But which standard deviation are we dealing with here? the population standard deviation σ? the sample standard deviation (which we'll begin to label s)?

Think about it. You need to remind yourself that the inferences being made and the intervals being constructed are based on the sampling distribution of means. It shouldn't be especially surprising, then, that the standard deviation involved here is $\sigma_{\bar{x}}$, the standard deviation of this sampling distribution (a measure which is often referred to as the **standard error**, or, more completely, the **standard error of the mean**). And how is this value computed? As mentioned earlier,

$$\sigma_{\bar{x}} = \frac{\sigma}{\sqrt{n}}$$

The Magic Formula

We can now show interval statements about μ based on sample results in the general form

$$\bar{x} \pm z\sigma_{\bar{x}}$$

or

$$\bar{x} \pm z\left(\frac{\sigma}{\sqrt{n}}\right)$$

FIGURE 5.3

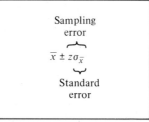

Notice the factors which bear directly on the size of the "neighborhood" term, $z\sigma/\sqrt{n}$: (1) confidence, which determines the value of z, (2) sample size n, and (3) population variability, σ. (This should sound familiar. Think back to our earlier discussion.)

EXERCISE 1 Produce for the class income example a 95% confidence interval estimate of the mean income for the class. (Assume for the moment that the population standard deviation is known to be $5600. Remember, we've already computed the average income for the sample of 10, $11,018.) Interpret the interval you produce.

Ans: $11,018 ± 1.96($5600/$\sqrt{10}$) or $11,018 ± $3471 or $7547 to $14,489

Interpretation: Based on sample results, we can be 95% confident that the sample mean we've produced ($11,018) will be within $3471 of the overall class mean. That is, *there's a 95% probability that the interval $7547 to $14,489 will contain the actual population mean income.* By choosing the 95% confidence level, we're allowing for a 5% chance that the interval may *not* include the class mean. Be sure you understand the interpretation of the interval here. You should be able to apply this sort of interpretation to every interval estimate you produce.

Sampling Error versus Standard Error

Terminology and notation in confidence interval estimation can be confusing. We've used the term standard error for the standard deviation of the sampling distribution of means—symbolically, $\sigma_{\bar{x}}$ or computationally, $\sigma/\sqrt{n}$. A similar term, **sampling error,** is often used when referring to the "plus and minus" or "neighborhood" term in the general confidence interval expression. Figure 5.3 shows the connection.

National surveys often cite sampling error to report the maximum difference we should expect between the sample result reported and the actual population value (i.e., the maximum difference we could expect between the sample mean and the population mean). Following this pattern, we might have reported our sampling results as follows: at the 95% confidence level, our best guess of the class average income is $11,018, with a possible sampling error as large as $3471. In other words, we can be 95% sure that the actual class average income won't differ from the sample average by more than the $3471 sampling error.

5.6 NOTES AND MODIFICATIONS TO THE BASIC METHOD

Having (hopefully) established the basic nature of confidence interval estimation, we can look now at several follow-up issues and minor adjustments to the method. As you read on, it's important not to let details cloud your understanding of fundamental concepts. If you've been able to follow the logic of the procedure to this point, you've established the necessary foundation to understand virtually all of statistical inference. Don't let the twists and turns, and the "ifs" and "buts" of the next few sections threaten that foundation.

Building Intervals When the Population Standard Deviation Is Unknown

Consider again exercise 1 where you produced the 95% confidence interval estimate of average class income. We were told there to assume that the population standard deviation was known to be $5600. In reality, this represents a rather heroic assumption. In most sampling situations, the population standard deviation (σ) is simply not known. (Logically, if the population mean is unknown, it seems unlikely that the population standard deviation would be readily available.) In such situations, we'll need to somehow produce an *estimate* of the population standard deviation, and then adjust our method of confidence interval construction to reflect this added element of uncertainty. (This is especially true in the case of small sample sizes (i.e., sample sizes of less than 30).)

(Intuition would suggest that when σ is unknown and only estimated, the width of the confidence interval should increase (all other things being equal) to reflect the added uncertainty. Put in slightly different terms, if we need to produce an estimate of the population mean when σ is unknown, we should expect to have to hedge a bit more—widening the interval to acknowledge a loss in estimating precision. What we need now is a mechanism to ensure the proper adjustment.)

Using the *t* Distribution

Technically, it's the shape of the sampling distribution of all possible sample means that's directly affected by our not knowing with certainty the population standard deviation. If, as is almost always the case, we choose to use the sample standard deviation to estimate the unknown population standard deviation, theory suggests that the sampling distribution will no longer appear perfectly normal. Rather it will assume the shape of a *t* **distribution**.

FIGURE 5.4 **THE *t* DISTRIBUTION**

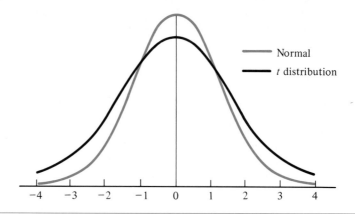

The *t* distribution (sometimes referred to as Student's *t* distribution) is bell-shaped and symmetrical, and, in fact, looks a lot like the normal distribution. (We'll establish shortly that the *t* and the normal distributions are indeed closely related.) The key difference is that, for small sample sizes especially, the *t* distribution appears flatter and broader. Its values are more widely dispersed. (See Figure 5.4.)

Because *t*-distributed values are more widely dispersed than normally-distributed values, intervals constructed on the *t* distribution to collect a given percentage of values will necessarily be wider than comparable intervals on the normal distribution. For example, to establish an interval around the mean that would include 95% of the values in a *t* distribution, we would have to go beyond the normal 1.96 standard deviation limits. (Precisely how far beyond is discussed below.) Not surprisingly, then, confidence interval estimates based on *t*-distributed sample means (appropriate in cases where the population standard deviation is being estimated from sample standard deviation results) will be correspondingly wider (i.e., less precise).

Reading a *t* Distribution Table

Using a *t* distribution table can be a bit tricky. Refer to the *t* table on page 329. Notice especially the shaded area. This shaded area represents the proportion of values in the right tail of the distribution, as distinct from the earlier normal table representation of values between the mean (center) and some number of standard deviations to the right of the mean. These right-tail areas represent the proportion of values that we would expect to find beyond some specified number of standard deviations to the right of the mean.

To demonstrate use of the table, assume we're working with a *t* distribution of sample means and want to determine how far above the center we should establish a boundary to identify, say, the upper 5% of the values in the distribution. The 5% represents a targeted right-tail area. To reference this value in the table, check the entries in the top row. (You should see values of .10, .05, .025, .01, and .005—a set of right-tail probabilities.) Locate the .05 value.

Now look at the first column in the table, the one labeled "degrees of freedom" (df). Briefly described, degrees of freedom are related directly to the size of the sample being used. Without getting deeply involved in a discussion of degrees of freedom (a somewhat complicated statistical concept), we'll simply say that degrees of freedom in a problem such as this can be defined as sample size minus one ($df = n - 1$). If sample size, for example, is 15, then $df = 15 - 1 = 14$.

Assume for the moment that the sample size for our 5% problem is, in fact, 15. To find a boundary that will identify the upper 5% of the values in this distribution, trace down the .05 column to the row showing 14 degrees of freedom. You should find at the intersection the number 1.761. This indicates that 5% of the values in the t distribution with 14 degrees of freedom will be found beyond 1.761 standard deviations to the right of the mean. To check your understanding, find the appropriate t-score boundary for a 1% area, using a sample size of 10. (Answer: 2.821.)

Constructing Intervals on the *t* Distribution

Suppose we now decide to construct around the mean of a t distribution a symmetrical interval that would contain exactly 95% of the distribution values. That is, we want to set an upper and a lower boundary around the center of the distribution in order to bound a 95% area. How could the available t table be used? Remember, the t table shows only right-tail areas.

Given the symmetry of the distribution, our target of 95% would translate to boundaries on either side of the mean that would each account for 47.5% (half of 95%) of the values. That is, we'll need to set a lower bound so that 47.5% of the values in the distribution will lie between the distribution mean and that lower bound, and an upper bound to ensure that 47.5% of the values will lie between the distribution mean and the designated upper bound. (You might, at this point, want to sketch the t distribution here and use it to follow the discussion.)

Consider just the upper boundary for a moment. As we indicated, it should be set in such a way as to include 47.5% of the values from the mean to the bound. We can easily convert this target to a corresponding right-tail area, since 47.5% *within* the bound translates to 2.5% ($.5 - .475 = .025$) *outside* the bound. Having established this area allows direct use of the t table. Enter with a top-row value of .025. If we assume a sample size of 15, we next need to locate 14 in the df row. At the intersection of row and column, you should find a value of 2.145. Conclusion? We should set the upper bound for the 95% interval 2.145 standard deviations above the mean. And the lower bound? Given the symmetry of the distribution, it should be set 2.145 standard deviations below the mean.

This sort of procedure will allow us to adapt our standard approach to confidence interval construction whenever the t distribution is required.

EXERCISE 2 Determine the proper 90% interval boundaries for a t distribution, assuming a sample size of 15. Also do it for the 99% boundaries.

Ans: For the 90% interval, with $df = 14$, $t = \pm 1.761$

For the 99% interval, with $df = 14$, $t = \pm 2.977$

(For a more detailed solution, see the end of the chapter.)

A Return to the Class Income Example

Let's go back to the class income illustration. To this point, we've drawn a sample of 10 and generated the 10 sample incomes shown on page 142. From these values, we've produced a sample mean of $11,018. Suppose we again want to build a 95% confidence interval estimate of the mean of the population from which the sample was selected, but this time we'll assume that the population standard deviation must be estimated from available sample information.

How should we proceed? First of all, we need to recognize that the z score of 1.96 standard deviations associated with the normal distribution 95% interval is no longer appropriate for our new interval estimate. Instead, we'll need to substitute the t-score equivalent, using a t distribution with 9 $(10 - 1)$ degrees of freedom. According to the table, the appropriate t score is 2.262. (Check your t table to verify.)

Next, we need to produce the required sample-based estimate of the population standard deviation. Accordingly, we'll compute the standard deviation of the sample in the usual way, with one minor modification, and use this value as an acceptable proxy for σ.

Refer to the sample incomes and begin the standard deviation computation: take each value minus the mean and square the result, then add the squared differences. To finish the computation, we'd ordinarily have to average out the squared differences by dividing the sum of the squared differences by the number of values involved. Since sample size here is 10, we would expect to divide by 10, take the square root, and be done. In actuality, we'll need to make one small adjustment.

Theory suggests that if we follow the conventional route in computing sample standard deviation (i.e., dividing the sum of squared deviations by n) with the expectation that this measure of sample variability will serve as a useful proxy for population variability, we'll often be disappointed with the results. More specifically, it's been shown that such an approach would lead us to understate rather consistently the population variability (i.e., our estimate of σ would be consistently too low).

To offset the prospect of this consistent understatement, sampling theory has established the proper adjustment: rather than dividing the sum of the squared differences by sample size, we're instructed to divide the sum by sample size minus 1. This will tend to inflate slightly the size of the sample standard deviation, which in turn will, theory confirms, provide a better estimate of the population standard deviation.

Try it out. Use the suggested approach to finish the computation of the sample standard deviation for the 10 income values. Make sure you divide by 9 rather than 10 before taking the square root. Result? The sample standard deviation, which we'll label s, is $12,154 (a rather large value, when you consider that the sample mean was only $11,018).

Note: To probe the issue of precisely why we should divide by $n - 1$ would involve us once again in the notion of degrees of freedom. For an explanation of the degrees of freedom idea, you'll need to see a more advanced text.

Substituting $s = \$12,154$ for the population standard deviation σ in our basic interval expression now produces the appropriate 95% interval.

$$\$11,018 \pm 2.262\left(\frac{\$12,154}{\sqrt{10}}\right) \quad \text{or} \quad \$11,018 \pm \$8,694$$

The Magic Formula: Version II

To reflect the change here, our original expression for confidence interval construction

$$\bar{x} \pm z\left(\frac{\sigma}{\sqrt{n}}\right)$$

becomes

$$\bar{x} \pm t\left(\frac{s}{\sqrt{n}}\right)$$

to accommodate the special case in which the sample standard deviation (s) is used to estimate the population standard deviation σ.

Special note: The only time standard deviation will be computed as the square root of the sum of the squared differences divided by n minus 1 is when the standard deviation for a sample is being computed with the explicit intent of estimating the population standard deviation. In this unique case, we can label the sample standard deviation s and show it as

$$s = \sqrt{\frac{\sum(x - \bar{x})^2}{n - 1}}$$

Don't fall into the common trap of now using $n - 1$ everywhere in your statistical computations (e.g., in computing means, population standard deviations, standard errors, etc.).

The Normal Distribution as an Approximator of the *t* Distribution

The t distribution has been described as more spread out than the normal distribution, and we've already suggested that for any given confidence level the appropriate t score would be larger than the corresponding z score for that same level of confidence. To illustrate, we've shown for 95% confidence and 9 degrees of freedom, a t score of 2.262 versus a corresponding z score (from the normal table) of 1.96. And for 95% confidence and 14 degrees of freedom, we've seen a t score of 2.145.

Watch what happens, though, as we consider larger and larger sample sizes for the t distribution. Locate the appropriate t score for a sample size of 25 (i.e., $df = 24$), maintaining 95% confidence. (Remember, 95% confidence means a .025 right-tail area.) Result? $t = 2.064$. How about for sample size 30? ($t = 2.045$) For sample size 61? ($t = 2.00$) Sample size 121? ($t = 1.98$) Do you see any pattern here?

As sample size, and so degrees of freedom, increases (while confidence is held constant at 95%), the corresponding t scores *decrease*. In fact, they begin to approach the normal-based z score for 95% intervals, 1.96. (Indeed, if you look in the t table using an "infinite" sample size, t and z match perfectly.) The lesson? As sample size increases, the t distribution starts to look more and more like the normal distribution. The normal distribution is actually the "limiting case" for the t distribution as n gets larger and larger.

It's this connection that has moved most authors to suggest that while the t distribution is theoretically the proper sampling distribution to use in cases where the sample standard deviation is substituted for the population standard deviation, if sample size is sufficiently large the more familiar normal distribution will serve as a perfectly satisfactory approximation. And just what's meant by "sufficiently large"? There's general agreement that so long as the sample size is at least 30, the normal approximation holds up pretty well.

A Further Reminder about "Small" Samples

Recall, if you will, another facet of the "small" versus "large" sample issue, one that we noted much earlier in the chapter. When sample size is small ($n < 30$), the shape of the population from which the sample is selected becomes significant. Importantly, for these small sample cases the predictable shape of the sampling distribution (t or normal), on which so much of our inference process depends, is ensured only if the population itself follows a normal distribution. (Statistical tests for population normality exist but will not be pursued here.) If the population distribution is distinctly nonnormal (badly skewed, bimodal, etc.), then the shape of the sampling distribution of means is not easily predicted, and, as a consequence, the inference process we've been describing breaks down. Without a predictable bell shape for the sampling distribution, the sorts of confidence intervals we've been constructing can't be statistically defended.

Fortunately, most of the sampling we'll be doing will involve samples of 30 or more, for which cases the shape of the population distribution is not a significant issue. That is, we'll routinely rely on the general principle that for sufficiently large sample sizes ($n \geq 30$), the shape of the sampling distribution (normal or t) will not be materially affected by the shape of the parent population distribution.

Another Reminder: Don't Lose Sight of the Basics

I realize at this point that we have (as promised) begun to load you up with a number of ifs and thens, and special cases, but I hope these details haven't seriously interfered with your basic understanding of the fundamental interval construction process. Remember, we'll ordinarily have just one sample mean to work with, a sample mean drawn randomly, in effect, from a distribution of all possible sample means. We'll link that one sample mean to the population mean through predictable sampling distribution characteristics (shape, center, and standard deviation).

The shape of this distribution is generally normal—t in special cases where the population standard deviation is unknown and must be estimated from sample values. This characteristic bell shape holds in most cases even if the population itself is not normally distributed. The only exception is for small samples (less than 30), where the population must be approximately normal in order for our conclusions about the shape of the sampling distribution to be theoretically valid.

The Finite Population Correction Factor

For better or worse, we need to add one last detail. It turns out that if sampling is done *without replacement*, the standard error term

$$\sigma_{\bar{x}} = \frac{\sigma}{\sqrt{n}} \quad \text{or} \quad \frac{s}{\sqrt{n}}$$

describing the standard deviation of the sampling distribution of means should be modified slightly. (Remember, to this point in our discussion, we've limited ourselves pretty much to sampling with replacement, although we did allude earlier to a necessary adjustment when sampling is done without replacement.) Specifically, we'll be allowed here to adjust downward (i.e., make smaller) the standard error term by applying something called the **finite population correction (FPC) factor** to the usual standard error expression. The standard error term appropriate in this case becomes

$$\frac{\sigma}{\sqrt{n}} \sqrt{\frac{N-n}{N-1}} \quad \text{or} \quad \frac{s}{\sqrt{n}} \sqrt{\frac{N-n}{N-1}}$$

where $\sqrt{(N-n)/(N-1)}$ is defined as the FPC
and n = sample size
   N = population size

On close inspection, you should be able to see that the FPC term will always take on a value less than 1. Consequently it will always reduce the size of the standard error. A broad justification for this sort of reduction might proceed as follows:

As we noted earlier, when sampling is done with replacement some unsettling possibilities emerge. We could, for example, draw a sample of 10 items from the target population and end up with the same population value being selected again and again. Using the class income illustration, we might conceivably have produced our "representative" sample of 10 by selecting, solely by chance, the highest-income class member 10 consecutive times—not likely, but under sampling with replacement, a real possibility. Such a sample would produce a rather extreme sample mean, one likely to be far from the overall population (i.e., class) mean. In similar fashion, under this with-replacement selection scheme, we might have repeatedly chosen (again, strictly by chance) the lowest-income class member for inclusion in the sample, once again producing a rather extreme sample result.

However, in sampling *without* replacement, a strategy in which the same population member cannot be included more than once in the sample, these sorts of extreme results are simply not possible. Eliminating such extremes translates into less "scatter" or "dispersion" in the sample means distribution. The result is a smaller standard deviation. How much smaller? The FPC calculates the appropriate reduction.

Many textbooks recommend that the FPC adjustment be made only when, in addition to sampling without replacement, we use a sample size that's at least 5% of the population size. The argument here is that in those cases where sample size is *less* than 5% of the population size, the FPC multiplier is so close to 1 that computationally it simply doesn't play a significant role. (Try a few cases to convince yourself.) While the 5% criterion is rather arbitrary, we'll go along in recommending its use when deciding whether the FPC is appropriate. However, it's never "wrong" to use the FPC anytime the population involved is "finite" and sampling is done without replacement.

EXERCISE 3 Reconsider the average class income example. Assume the 10 sample values shown on page 142 were produced using sampling without replacement.

(a) Show the 95% interval estimate of the overall class average income. (Remember, the sample mean was $11,018 and the sample standard deviation was $s = $12,154$.)

(b) Suppose class size had been 1000 rather than 100. Would the FPC play a significant role in reducing the sampling error?

Ans: (a) The interval should look like

$$11,018 \pm 2.262 \left[\frac{12,154}{\sqrt{10}} \sqrt{\frac{100 - 10}{100 - 1}} \right]$$

$$= 11,018 \pm 8694(.953)$$

$$= 11,018 \pm 8285$$

(Notice the value of the FTC term here. It serves to reduce the sampling error, and so the interval width, by about 5%.)

(b) FPC $= \sqrt{\dfrac{1000 - 10}{1000 - 1}} = .995$

a value which would have a negligible effect on sampling error.

<h2>5.7 DETERMINING SAMPLE SIZE</h2>

The With-Replacement Case

Back at the beginning of our average class income example, we raised the issue of sample size and then quickly dismissed it by setting sample size arbitrarily at 10. We can now address the issue more directly.

Three principal factors play a significant role in determining an appropriate sample size for any given situation: (1) confidence level, (2) required interval precision (another term for sampling error), and (3) population variability. Once values for each factor are set (or computed), an appropriate sample size can be easily established. (We could perhaps add a fourth factor—the dollar costs of sampling. In practice, this may frequently be *the* determining factor. Our focus here, however, is on the more purely "statistical" considerations.)

Situation: Suppose we want to estimate, via simple random sampling with replacement, the average age of items currently held in firm ABC's inventory. At the outset, we decide to target a confidence level of 95% for the interval that will eventually be produced. Furthermore, to go along with this 95% confidence requirement, we plan to set a maximum allowable interval width of ± 20 days—that is, we want to be 95% confident that the actual mean age of items in inventory (i.e., the population mean) will be within 20 days of the sample mean computed. How large a sample should we select?

To get us started, consider the basic form of the interval that will eventually be produced

$$\bar{x} \pm z\left(\frac{\sigma}{\sqrt{n}}\right)$$

The precision (or sampling error) term is $z(\sigma/\sqrt{n})$. To ensure a maximum interval width of ± 20 days, we can set this precision term equal to 20:

$$z\left(\frac{\sigma}{\sqrt{n}}\right) = 20$$

The 95% confidence requirement fixes z at 1.96, so that

$$1.96\left(\frac{\sigma}{\sqrt{n}}\right) = 20$$

Can you see where this is taking us? If we just knew σ, we could easily solve the expression for sample size n. But do we know σ? Will we likely know σ at this point in any practical sampling situation? Typically, no. But we're not really stumped.

The standard procedure in a situation like this is to produce an estimate (even if it's only a rough estimate) of σ. For example, we might have had experience with a similar population in the past which could be used to generate an estimate of the standard deviation for the current population. Alternatively, we might decide to select a so-called "pilot sample," a preliminary sample of convenient size, and use sample results to estimate the population σ. To illustrate, suppose we select a pilot sample of 30 items and compute for the sample a standard deviation of 140 days. We can substitute this sample result as an estimate for σ and show

$$1.96\left(\frac{140}{\sqrt{n}}\right) = 20$$

Algebraically solving for n produces

$$n = \left[\frac{1.96(140)}{20}\right]^2 = 188$$

What we've discovered is that a 95% confidence interval estimate of the mean age of the inventory, an interval which can be no wider than ± 20 days, will require a sample size of roughly 188. We use the term *roughly* to take into account the approximating nature of our sample-based estimate (140 days) of the population standard deviation.

Generalizing the computational procedure, we can set

$$z\left(\frac{\sigma}{\sqrt{n}}\right) = \text{PRE}$$

to produce

$$n = \left[\frac{z\sigma}{\text{PRE}}\right]^2$$

where PRE represents the target sampling error (i.e., the desired precision) for the interval that will be constructed.

Changing the value of any one of the three factors in the expression (the target precision requirement PRE, the value of z, or the population σ estimate) will lead to a different sample size recommendation. Tightened precision, for example, would require (all other things being equal) a larger sample size. Less variability in the population would enable us to use a reduced sample size.

EXERCISE 4   Apply the sample size expression to our average class income example. Use the sample results on page 142 as pilot sample results (i.e., use $12,154 as an estimate of the population standard deviation). Stick with 95% confidence, and use a maximum sampling error target of $4000. What n would you recommend?

Ans: $n = \left[\dfrac{1.96(12,154)}{4000}\right]^2 = 35$

Sample Size When Sampling Is Without Replacement

One final comment. Our prescription for determining sample size works anytime sampling is done *with replacement*. If you intend to sample without replacement, the sample size calculation needs to be modified. Here, sample size can be computed according to the following expression:

$$n' = \frac{n}{1 + n/N}$$

where n' = the without-replacement sample size

n = the with-replacement sample size computed according to the standard sample size expression from the preceding section

As you can see, we're simply substituting the with-replacement sample size n into the adjusting expression to produce the without-replacement sample size, n'. Notice that the adjustment will always reduce the sample size requirement since the denominator will always be greater than 1. Thus, for the same confidence and precision targets, sampling without replacement will generally allow us to select a smaller sample than would be required if we were intending to sample with replacement. (This is primarily because we can use the FPC to reduce sampling error whenever a confidence interval is constructed in the without-replacement case. Can you now see why, in practice, sampling is almost always done without replacement?)

We mentioned earlier that many authors dismiss the FPC in cases where the sample size is less than 5% of the population size. In similar fashion, they would have us ignore the $1 + n/N$ sample size adjustment in cases where the recommended n is less than 5% of the population size. The reason is that, computationally, it will have very little effect on results. (Try a few examples to convince yourself.) We'll use this same 5% rule.

EXERCISE 5   Reconsider the ABC inventory example on page 159.

(a) Adjust the sample size requirement to accommodate an assumption that sampling will be done without replacement. Assume ABC has a total of 300 items in inventory. (We had originally prescribed a with-replacement sample size of 188.)

(b) Assume ABC has a total of 10,000 items in inventory. Determine the without replacement sample size.

Ans: (a) $n' = \dfrac{188}{1 + 188/300} = 116$

a reduction in sample size of $188 - 116 = 72$ items

(b) $n' = \dfrac{188}{1 + 188/10,000} = 185$

a reduction of only $188 - 185 = 3$ items

▬▬▬▬ CHAPTER SUMMARY

We saw the basic nature of statistical inference—drawing conclusions about a full set of values (a population) based only on a subset (sample) of representative data. We established the reasons for sampling and outlined alternative sampling pro-cedures. We differentiated between random and nonrandom sampling approaches and argued that only in random sampling could possible errors be systematically measured. Simple random sampling, in which each member of the target population has an equal chance of being included in the sample, was described as the base case for our discussion of statistical inference.

We spent most of our time constructing confidence interval estimates of a population mean (or average), using only the results of a single randomly selected sample. The key theoretical link—the critical connector of sample to population—came in the form of a sampling distribution. Knowledge of this distribution, made up of all possible sample means, was seen as crucial in linking any sample mean to the overall mean of the target population. Curiously, while this distribution is key, it is rarely ever produced. Instead, statistical theory establishes predictable distribution characteristics: the sampling distribution of means will (1) be approximately normally distributed (for sufficiently large sample sizes), (2) be centered on the overall population mean, and (3) have an easily computed standard deviation. By exploiting these three characteristics, we saw how it was possible to relate, in probabilistic terms, a single sample mean to the mean

of the population from which the sample was drawn. The resulting connection followed the standard interval form

$$\bar{x} \pm z\left(\frac{\sigma}{\sqrt{n}}\right)$$

with $z(\sigma/\sqrt{n})$ representing the sampling error, indicating how far from the population mean our one sample mean ($\bar{x}$) is likely to be.

The t distribution was introduced to replace the normal in those cases where the population standard deviation is unknown (and the sample size is less than 30). The finite population correction factor served to reduce standard error when sampling is done without replacement (and the sample size is at least 5% of the population). Finally, a general procedure for establishing a recommended sample size was developed.

Table 5.2 may help you organize some of the chapter details.

TABLE 5.2   **A QUICK-REFERENCE CHART**

When to use the t table versus the normal table for interval construction

		Population standard deviation	
		Known	Estimated
	Large ($n \geq 30$)	Use normal	Normal to approx. t
Sample size			
	Small ($n < 30$)*	Use normal	t distribution

*Note: for small sample sizes, the population distribution must be normal.

When to use the finite population correction factor

	Sample selection is	
	With replacement	Without replacement
$n/N \geq 5\%$	No FPC	Yes FPC
$n/N < 5\%$	No FPC	Optional

When to use the $n - 1$ divisor in computing a standard deviation

For population STD DEV	No
For standard error of the sampling distribution	No
For sample STD DEV when it is to be used to estimate the population STD DEV	Yes

KEY TERMS

Central limit theorem
Confidence interval
Finite population correction (FPC) factor
Inferential statistics
Judgment sampling
Random number table
Random sample
Sampling distribution of means

Sampling error
Sampling with replacement
Sampling without replacement
Standard error
Stratified random sampling
Systematic random sampling
t distribution

IN-CHAPTER EXERCISE SOLUTION

EXERCISE 2

For 90%:

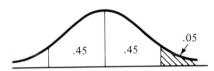

Look in the table for $df = 15 - 1 = 14$ and a tail-end area of .05.
Result: $t = 1.761$

For 99%:

Look in the table for $df = 15 - 1 = 14$ and a tail-end area of .005.
Result: $t = 2.977$

EXERCISES

Assume that sampling is being done with replacement unless otherwise indicated. If the population size is not specified, assume it's very large.

★   1. You want to estimate the average SAT scores for all students who took the Milar SAT preparation course during the past two years. You select a random sample (with replacement) of 100 such students from a comprehensive list of all students

who took the course over the last two years and find that the average SAT score for the sample was 1040, with a standard deviation of 83 points. Produce the 95% confidence interval estimate that would be appropriate here.

★   2. You want to estimate the average years of seniority for employees working for your company. The files of 49 workers are selected at random (sampling with replacement). Average seniority for those in the sample is 13.6 years, with sample standard deviation of 5.2 years.

(a) Construct and interpret a 95% confidence interval estimate of average seniority for the full population of company employees.

(b) Construct the 99% interval. The 80% interval.

(c) For your answers in (a) and (b), identify the sampling error term, and the standard error term in each interval.

3. You want to estimate the average useful life of the new PDR flanges that your company has just put on the market. You take a random sample of 80 flanges to be tested. The test reveals an average life of 1560 hours for the 80 sample flanges (with a standard deviation of 156 hours).

(a) Construct and interpret a 95% confidence interval estimate of average useful life for the full population of PDR flanges produced by your company.

(b) For your answer in (a), identify the sampling error term, and the standard error term.

4. You want to estimate the average time it takes to complete the assembly of a new company product. After allowing ample time for the workers to learn the new assembly procedures, you randomly choose 50 assemblies to observe. For these 50 assemblies, average assembly time is 38.4 minutes, with a standard deviation of 5.6 minutes.

(a) Construct and interpret a 95% confidence interval estimate of average assembly time required for the population of these units now being assembled at your company.

(b) For your answer in (a), identify the sampling error term, and the standard error term.

★ 5. Refer to problem 1. Suppose the sample size had been only 12 instead of 100. Show the 95% confidence interval that would be appropriate here. (What additional assumption about the population of values would have to be made?)

★ 6. Refer to problem 2. Suppose the sample size had been 20 instead of 49. Show the 95%, the 99%, and the 80% intervals that would be appropriate here. (What additional assumption about the population of values would have to be made?)

7. You have just received a large shipment of microbonded applicator units from your principal supplier. To be acceptable, the units should have an average breaking strength of no less than 800 pounds. You select a sample of 49 units

randomly (without replacement) from the shipment and find that the sample average breaking strength is 814 pounds with standard deviation of 35 pounds. Construct a 90% confidence interval estimate of the average breaking strength you could expect to find if all the components in the shipment were tested.

★ 8. In exercise 7, suppose you had chosen a sample of 10 units with the following sample test results:

Unit no.	Break strength
1	840
2	820
3	790
4	850
5	790
6	820
7	880
8	770
9	830
10	850

Produce the appropriate 90% interval. (You can assume that the "population" of breaking strength values for this shipment is approximately normal.)

9. As part of a marketing study, you want to estimate, for customers who have recently purchased a car from your dealership, the average number of dealerships that these customers visited before making their purchase. You randomly select (with replacement) a sample of six customers, with the following results:

Customer	1	2	3	4	5	6
Dealer visits	6	10	4	5	7	4

Use this sample data to construct a 90% confidence interval estimate of the average number of visits for the population being represented here. (You can assume that the population of values is approximately normal.)

★ 10. You want to estimate the average value of current accounts receivable for the firm. In all, there are 2000 such accounts. You take a simple random sample (using sampling with replacement) of 200 accounts and find that the average amount due the firm in the sample is $6500, with standard deviation of $1600.

(a) Show the 95% confidence interval estimate of average accounts receivable.

(b) Suppose the sampling had been done without replacement. Show the 95% interval.

★ 11. Use your result in problem 10(b) to build a 95% confidence interval estimate of the total amount owed the firm by these 2000 accounts. (*Hint:* If the $6500 sample average is your single best estimate of the population average, it would seem reasonable that 2000 × $6500 is your best estimate of the population total. Use the same reasoning to construct the sampling error term here.)

12. You want to estimate the average daily output of all firms (a total of 2300) in your industry. You randomly select, without replacement, 50 such firms and find that the average daily output for the firms in your sample is 1560 units, with standard deviation of 320 units.
   (a) Construct a 99% confidence interval estimate of average daily output for all industry firms.
   (b) Estimate, at the 99% confidence level, the total daily output for the 2300 firms (combined) in the industry.

13. How would your answers in problem 12 change if there were only 150 firms in the industry? Would you need the FPC?

★ 14. You plan to select a simple random sample (without replacement) in order to estimate the average course load for freshman students at Blighton College. From the freshman class of 1600, you randomly choose 16 students, with the following results: sample mean = 16.2 credit hours; sample standard deviation = 5.3 credit hours. Construct a 95% confidence interval estimate of the class average. As you proceed, decide whether you should be using
   (a) the FPC.
   (b) the *t* distribution.
   (c) the assumption that the population distribution is approximately normal.
   (d) division by $n - 1$ rather than n.

15. Refer to problem 2. Which, if any, of the following factors would force you to change your answers? For each factor discuss your reasoning. (Consider each part independent of the others.)
   (a) The size of the company employee population is known to be 5000.
   (b) The shape of the seniority distribution for the population of company employees is decidedly not normal.
   (c) Sampling was done without replacement and the size of the employee population is 10,000.

(d) The standard deviation of the seniority distribution for the population of 5000 company employees is known to be 5.2 years.
(e) Sampling was done with replacement, but you discover that the random selection process ended up including in the sample the same company employee three times.
(f) Sampling was done without replacement and the size of the company's employee population is actually 200.

16. You have taken a simple random sample (without replacement) of 64 students from the campus population in an effort to estimate the average number of miles that students must travel to campus each day. Sample results show an average of 3.4 miles, with standard deviation of 1.6 miles. Answer the following true-false questions. Be prepared to explain your reasoning:
   (a) In order to make a defensible inference about the overall student population average here, the population distribution would have to be approximately normal.
   (b) A 90% interval estimate based on these sample results would be twice as wide as a 45% interval.
   (c) Your best estimate of the population standard deviation here would be 0.2 mile.
   (d) Based on sample results, it would be appropriate to conclude that 95.5% of the students in the campus population would report a driving distance between 3 and 3.8 miles.
   (e) If you repeated this process many times, each time selecting a sample of 64 students and computing a sample mean, you would expect 68.3% of those sample means to be within 1 standard deviation (i.e., .2 mile) of the actual population mean.

★ 17. Refer to problem 1. Suppose you want your 95% interval estimate to have a sampling error no greater than 10 points. How large a sample would be required? (Use the sample standard deviation reported in problem 1 as a reliable estimate of the anticipated population standard deviation.)

18. Refer to problem 2. Suppose you're thinking about taking a larger sample. In fact, you want to use a sample size that will ensure that the sample mean you produce will be no more than one year away from the actual population mean (at the 95% confidence level). How large a sample should you select? (Use the sample standard

deviation reported in problem 2 as a reliable estimate of the population standard deviation.)

19. You plan a study to estimate the average workweek for newly hired accountants at "big five" CPA firms. You intend to take a simple random sample (with replacement) from the population of all accountants recently hired by "big five" firms. You want to ensure a 90% probability that the sample mean you produce will be no more than 1 hour away from the actual population mean. A preliminary estimate, based on previous experience, anticipates a standard deviation of about 12 hours. How large a sample should you select for your study?

★ 20. In exercise 19, suppose your preliminary estimate of the population standard deviation will come from a pilot sample of nine accountants showing the following values:

	1	2	3	4	5	6	7	8	9
hrs/wk	55	68	76	49	45	81	72	60	52

Determine the appropriate size of the full sample to be taken for the study.

21. You plan to select a simple random sample (without replacement) of American workers in order to estimate the average time spent, in minutes per day, travelling to and from their jobs. You initially decide to construct a confidence interval that is no wider than ± 5 minutes. Making a rough estimate of the variability in the population, you determine the required sample size. For each of the cases below, indicate whether the appropriate sample size would increase, decrease, or stay the same. (If the direction of the change cannot be determined without more specific information, so indicate.) Assume any factor not mentioned remains constant.
   (a) You want a higher level of confidence.
   (b) You want a tighter interval.
   (c) The variability in the population values is less than you had initially estimated.
   (d) Your "precision" requirement is loosened, but you want a higher level of confidence.
   (e) You decide that you can live with a lower confidence level. There is more variability in the population of values than you had first estimated.
   (f) You want a tighter interval. You require a higher level of confidence. There is less

variability in the population of values than you had first estimated.

★ 22. You plan to conduct a national survey to determine average monthly salaries for registered nurses with over 10 years of hospital experience. You obtain a registry of 69,100 such nurses and plan to draw a sample, without replacement, from this list. If your intent is to construct a 99% confidence interval estimate in which the sampling error will be no more than $20, how large a sample should you take? (Assume a similar study showed a sample standard deviation of $120.)

★ 23. Suppose you were interested in conducting the study in exercise 22 only for the Portland metropolitan area, where the list of nurses in the targeted category has only 245 names? Recommend a sample size here. Retain the 99% confidence requirement as well as the $20 sampling error goal. Assume the previous sample standard deviation of $120 can be used as a rough gauge of the target group standard deviation.

24. You plan to take a sample (without replacement) of retail outlets nationwide in order to estimate the average selling price of your chief competitor's product. You want to eventually construct an estimate at the 95% confidence level that shows a sampling error no greater than $.50. Previous experience indicates a reasonable estimate of the population standard deviation would be about $2.50. How large a sample would you recommend if the sample will be selected from a population of
   (a) 100 retail outlets?
   (b) 500 retail outlets?
   (c) 5000 retail outlets?
   (d) 20,000 retail outlets?

★ 25. It was stated in the chapter that since we can effectively predict essential characteristics of the sampling distribution of all possible sample means, rarely would we want to actually produce all the values that would make up this distribution. However, we can use the following situation to validate, or at least support, assertions that the sampling distribution of all possible sample means will (1) be bell-shaped, (2) have a center equal to the population mean, and (3) have a standard deviation equal to the population standard deviation divided by the square root of the sample size:

Given the following population of three values
($N = 3$):

Pop member	A	B	C
Value	10	20	30

(a) Compute the mean and the standard deviation of this population.
(b) Now show all possible samples of size 2 ($n = 2$), using sampling with replacement. (You should find nine such samples.)
(c) Compute the mean of each sample and show your nine sample means in a bar-chart (histogram) representation.

(d) Identify the center of the sample means distribution by computing the mean of the nine sample means.
(e) Compute the standard deviation of this set of nine sample means.
(f) Comment on the shape, center, and standard deviation of the sample means distribution as each relates to our sampling distribution predictions.

26. Repeat your work in exercise 25, this time using sampling without replacement. (You should now find only six samples of size 2, three if duplicates are eliminated.) Compare results to those in exercise 25. (*Hint:* Remember the FPC.)

INTERVAL ESTIMATION FOR PROPORTIONS, MEAN DIFFERENCES, AND PROPORTION DIFFERENCES

CHAPTER OBJECTIVES

Chapter 6 should enable you to:

1. Build confidence interval estimates of a population proportion using only sample results.

2. Build confidence interval estimates of the difference between two population means using only sample results.

3. Build confidence interval estimates of the difference between two population proportions using only sample results.

4. Understand the role of the sampling distribution as the key link between sample and population values.

5. Develop an approach for determining appropriate sample sizes in each of these three cases.

6. Recognize special situations in which general interval construction procedures need to be modified and adapted.

To this point, our discussion of confidence interval estimation has been limited to estimates of a population *mean* based on sample results. We now can extend coverage to three additional cases:

- Interval estimates of a population *proportion* (or percentage)
- Interval estimates of the *difference* between two population *means*
- Interval estimates of the *difference* between two population *proportions*

Fortunately, we can apply much of the work done in Chapter 5 and portray each "new" case as little more than a straightforward variation on themes we've already established. As you proceed, concentrate on the similarities among the cases and don't lose sight of the basic inferential principles involved.

6.1 INTERVAL ESTIMATES OF A POPULATION PROPORTION

On occasion, we may find ourselves interested in assessing the proportion (or percentage) of items in a given population which possess a particular attribute or show a particular characteristic: the proportion of voters who support a major political candidate, the proportion of manufactured items in a recently delivered shipment that are defective, the proportion of patients who react positively to a new medication, and so on. In cases where not all population members are easily (or economically) accessible, it's possible (and frequently essential) to turn to sampling theory as a basis for drawing general population conclusions using only partial (i.e., sample) information.

An Example

Suppose we want to establish the proportion (percentage) of local voters (50,000 in all) who would support a proposed change in the state sales tax. Given the size of the voter population, it would be all but impossible to contact and interview every voter in a timely manner. (If it were practically feasible, we could compute precisely the population proportion we want.) How might we proceed? How about taking a sample of local voters, assessing for the sample the sample proportion who support the tax change, and then somehow linking what we see in the sample to what we would expect to see in the population. (Sound familiar?) And just what will enable us to establish the link between sample and population? *Answer:* A clear knowledge of the appropriate **sampling distribution**.

To see how it works here, we'll settle, at least temporarily, two immediate issues: sample size (we'll choose 200 voters for now) and selection procedure (we'll use simple random sampling with replacement). That done, suppose our sample of 200 produces 76 voters (38%) in favor of the tax change. (We can label this sample result $\bar{p}$ (p-bar).) Would you feel comfortable concluding that the overall population proportion (percentage) must be precisely this same 0.38? That is, would you be completely confident that a full census of the local voter population would reveal exactly 38% in favor of the tax change? Not likely.

As was true in the means case of Chapter 5, it's extremely unlikely that the population proportion would be precisely equal to the sample proportion we've produced. But should we expect the population proportion (which we'll begin to label π (pi)) to be somewhere in the *neighborhood* of $\bar{p}$? Absolutely. Our job statistically is to determine just how big a neighborhood is involved. And the key? As mentioned above, it's the idea of a sampling distribution.

The Sampling Distribution of Proportions

Not surprisingly it's not the now-familiar sampling distribution of means that plays the key role here; rather, it's the sampling distribution of all possible sample **proportions**. And the nature of this distribution? If we wanted to produce all the values that make up this particular distribution, we might proceed in the following fashion:

Record the initial sample result ($\bar{p} = .38$) and return all sample members to the population. Randomly select a second sample of the same size and compute a second $\bar{p}$ value. (*Quick Question:* Is this second $\bar{p}$ value likely to be identical to the first? *Answer:* No, even though we might have included some of the same population members in this second sample.) Now record the second $\bar{p}$ (we might label it $\bar{p}_2$ to distinguish it from the first sample $\bar{p}$, $\bar{p}_1$). Return all sample members to the population, select a third sample of 200, record its $\bar{p}$ ($\bar{p}_3$), and so on, until all possible samples of size 200 have been selected from the parent population (a huge number!) and all sample proportions recorded.

This kind of repetitive exercise, if it were actually pursued to completion, would produce the full sampling distribution—a comprehensive list of all possible sample proportions based on samples of size 200 drawn from the population of 50,000. Big deal, you say. Of what possible significance is this listing of sample proportion possibilities? Do we really have to produce this kind of list in order to proceed with the inference process? As I'm sure you suspect, the answer is no.

Just as was true in the means case in Chapter 5, there are predictable characteristics of the sampling distribution here that will allow us to make inferences about the entire population based on a *single* sample result. (See Figure 6.1.) Specifically, statistical theory guarantees that this distribution will (1) be approximately normal in shape (at least for sufficiently large sample sizes), (2) be centered precisely on the population proportion π, and (3) have an easily computed standard deviation (the standard error of the proportions distribution):

$$\sigma_{\bar{p}} = \sqrt{\frac{\pi(1-\pi)}{n}}$$

FIGURE 6.1   **THE SAMPLING DISTRIBUTION OF SAMPLE PROPORTIONS**

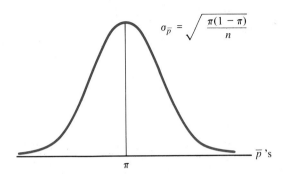

$$\sigma_{\bar{p}} = \sqrt{\frac{\pi(1 - \pi)}{n}}$$

$\bar{p}$'s

π

Given such characteristics, we can establish a confidence interval estimate of the population proportion in precisely the same manner that we had earlier produced interval estimates of a population mean. Viewing any one sample proportion (for example, our sample proportion of .38) as a value randomly pulled from the sampling distribution of all possible sample proportions, and taking advantage of the predictable normal shape, center, and standard deviation of this distribution, enables us to make statements such as

- There is a 68.3% probability that our one randomly produced sample proportion ($\bar{p}$) will fall within one standard deviation of the overall population proportion, π, the center of the sampling distribution.
- There is a 95.5% probability that our one randomly produced sample proportion ($\bar{p}$) will fall within two standard deviations of the population proportion, π, the center of the sampling distribution.
- And so forth.

As before (i.e., in the means case), reversing the reference point enables us to convert these statements to the equivalent

- There is a 68.3% probability (or we can be 68.3% confident) that the overall population proportion π will fall within one standard deviation of any randomly produced sample proportion $\bar{p}$.
- There is a 95.5% probability (or we can be 95.5% confident) that the overall population proportion π will fall within two standard deviations of any randomly produced sample proportion $\bar{p}$.
- And so forth.

Translated to symbols, these interval statements take on the general form

$$\bar{p} \pm z\sigma_{\bar{p}}$$

where $\bar{p}$ = sample proportion

z = number of standard deviations on a normal curve corresponding to a given level of confidence (probability)

$\sigma_{\bar{p}}$ = standard error of the proportions distribution (i.e., the standard deviation of the sampling distribution of all possible sample proportions)

or, equivalently,

$$\bar{p} \pm z \sqrt{\frac{(\pi)(1 - \pi)}{n}}$$

where π = population proportion

You may have noticed one apparent flaw in the argument here. In order to create an interval statement about π using the form above, we need to compute a standard error measure $\sigma_{\bar{p}} = \sqrt{\pi(1 - \pi)/n}$ that seems to assume π is already known. Fortunately, this apparent contradiction can be resolved in a fashion similar to the way we handled a parallel problem in the means case: We'll use sample information in lieu of the required population value in order to compute an acceptable estimate of the necessary standard error term. In other words, we'll substitute the sample $\bar{p}$ for the population π to compute $\sigma_{\bar{p}}$. The general interval expression thus becomes

$$\bar{p} \pm z \sqrt{\frac{\bar{p}(1 - \bar{p})}{n}}$$

As long as sample size is large enough (we'll discuss the sample size issue shortly), this expression will produce a statistically defensible interval estimate of the population proportion π.

EXERCISE 1 Use the standard interval form to create a 95% confidence interval estimate of the population proportion of voters who would support a proposed change in the state sales tax. (Recall that the original sample proportion was .38, based on a sample of size 200.) Interpret the interval you produce.

Ans: $.38 \pm 1.96 \sqrt{\frac{(.38)(.62)}{200}}$ or $.38 \pm .067$

We can state our conclusion as follows: Were we to take a full census of the local voter population (50,000 in all), we can be 95% confident (i.e., there is a 95%

probability) that the interval .313 to .447 (.38 ± .067) would contain the actual proportion of voters in the population who support the measure.

Put in slightly different terms, our best guess of the population proportion of voters who support the tax change is the sample result, .38 (38%). We can be 95% confident that the actual population proportion won't differ from this figure by more than .067 (6.7%).

As was the case in means estimation, the neighborhood term in the interval (.067 in Exercise 1), is commonly referred to as the sampling error and is produced simply by multiplying z by the appropriate standard error value, $\sigma_{\bar{p}}$. (National polls or political surveys would likely report this sampling error amount as the "possible margin of error.")

Determining Sample Size

Having used a sample size of 200 for the voter illustration, we found, through our work in Exercise 1, that this led to a sampling error of 6.7 percentage points at the 95% confidence level. Suppose now we decide that a sampling error this large is simply unacceptable—that the interval estimate which results is just too imprecise to serve a useful purpose. What recourse do we have? One possibility: We can increase the size of the sample.

In fact, we can target a desired sampling error level and prescribe precisely (well, almost) the necessary sample size just as we were able to do in the means case of Chapter 5. To illustrate, assume that a sampling error no larger than 4 percentage points (.04) is desired with no sacrifice in confidence (we still want a 95% confidence level). How large a sample is appropriate?

Drawing on the sampling error term in the general confidence interval expression

$$z \sqrt{\frac{\pi(1 - \pi)}{n}}$$

we can simply set sampling error (or precision) equal to the target level, .04. That is, we'll show

$$z \sqrt{\frac{\pi(1 - \pi)}{n}} = .04$$

For a confidence requirement of 95%, the z term is 1.96, so that

$$1.96 \sqrt{\frac{\pi(1 - \pi)}{n}} = .04$$

Can you see where this is leading? One final substitution, for the as-yet-unknown π, and we can easily solve this sampling error equation for the required n. And the value to be substituted? Why not use the sample result from our initial sample of 200 ($\bar{p} = .38$) as at least a defensible estimate of π, treating the earlier sample as a kind of pilot study? Consequently,

$$1.96 \sqrt{\frac{(.38)(.62)}{n}} = .04$$

Multiplying both sides of the expression by the square root of n, dividing by .04, and then squaring the result gives

$$n = \left[\frac{1.96 \sqrt{(.38)(.62)}}{.04} \right]^2 = 566$$

As long as our pilot sample estimate ($\bar{p} = .38$) of the population proportion π is reasonably accurate, it appears that a sample size of 566 will ensure a sampling error no larger than the targeted 4 percentage points at the 95% confidence level. That is, the sample proportion we produce from a sample of 566 voters is 95% likely to be no more than 4 percentage points away from the actual population proportion.

Generalizing the computation produces the sample size expression

$$n = \left[\frac{z \sqrt{(\pi)(1 - \pi)}}{\text{PRE}} \right]^2$$

where PRE = the desired precision (the tolerable sampling error)
π = the population proportion
z is determined by the confidence requirement

Substituting a sample-based estimate ($\bar{p}$) for the population proportion (π) allows us to write the expression as

$$n = \left[\frac{z \sqrt{(\bar{p})(1 - \bar{p})}}{\text{PRE}} \right]^2$$

EXERCISE 2   For the voter example, suppose a sampling error of no more than .02 is targeted (at the 95% confidence level). Compute the appropriate sample size. Use the earlier sample, with $\bar{p} = .38$, as a pilot for purposes of the calculation.

Ans: $n = \left[\dfrac{1.96 \sqrt{(.38)(.62)}}{.02} \right]^2 = 2263$

A Slight Twist to Determining Sample Size

In the absence of pilot sample information, our approach to the sample size problem needs to be modified slightly. Lacking any preliminary information about π, we can usually fall back on what might be described as a worst-case (or conservative) estimate, producing in the process the maximum required sample size to meet any particular confidence level/sampling error specification. Specifically, we can substitute a value of .5 for π in the sample size expression and then solve for sample size n.

EXERCISE 3   In the voter example, use the 95% confidence requirement and the .04 sampling error target to recompute the appropriate sample size value, using .5 for π. (Remember, we had initially computed a sample size of 566, using pilot sample results ($\bar{p} = .38$) as the estimate for π.)

$$Ans: \quad n = \left[\frac{1.96\sqrt{(.5)(.5)}}{.04} \right]^2 = 600$$

As you can see, the sample size prescribed in Exercise 3 is somewhat larger than the one we produced earlier (600 versus 566). Using .5 as the π value will always produce the *maximum* sample size requirement for a given sampling error/confidence level target. Can you see why? Take a look at the numerator of the sample size expression we've just used:

$$n = \left[\frac{1.96\sqrt{(\pi)(1 - \pi)}}{.04} \right]^2$$

Substituting .5 for π guarantees a maximum numerator value, in turn producing a maximum n. (Try .1, .2, etc., to compare sample size results.) And can you see why using the maximum n might be advantageous? In so doing, no matter what the eventual full sample proportion turns out to be, or no matter what the actual population proportion is (less than .5, greater than .5, or whatever), the corresponding sampling error term will always be within the desired limit (.04 for our exercise).

EXERCISE 4   Reconsider Exercise 2, where, for our voter illustration, the target sampling error was .02 at the 95% confidence level. Use the conservative π estimate of .5 to compute the recommended sample size and compare your result to the 2263 we had computed in Exercise 2.

$$Ans: \quad n = \left[\frac{1.96\sqrt{(.5)(.5)}}{.02} \right]^2 = 2401$$

Footnotes to the Basic Method

Large versus Small Samples in Proportions Estimation   It turns out that the proportions estimation approach we've described here, premised on a normal-shaped sampling distribution, is defensible only in cases where sample size is sufficiently large. In truth, the normal-shaped sampling distribution of proportions that we've been using is, even in the large sample case, only an approximation of an underlying *binomial* distribution. So long as the sample size is large enough, the normal approximation works well. When sample sizes are small, however, this normal approximation can be very unreliable. In such cases, we would need to look directly to the binomial for accurate confidence interval construction. Like most authors of introductory texts, we won't go into the details of just how interval estimates can be produced from the binomial, since this not-too-frequently used procedure can be rather cumbersome. We will need to define, however, what a "large" sample is.

To establish just what we mean by a large versus a small sample in the proportions estimation case, think back to the discussion we had in the supplement to Chapter 4, where we decided that the normal approximation to the binomial is appropriate only when $n \geq 30$ and both $np \geq 5$ and $n(1 - p) \geq 5$. We'll use this same rule of thumb to judge whether a sample qualifies as large or small for purposes of estimating a population proportion. By letting n represent sample size and substituting the sample proportion $\bar{p}$ for p in the np and $n(1 - p)$ expressions, we'll be able to judge whether we have a large enough sample to proceed with our usual normal distribution–based procedure. For example, if $\bar{p}$ is .01, then a sample size of at least 500 would be required (since a sample size of 500 or more would be needed to ensure that $n\bar{p} \geq 5$ and $n(1 - \bar{p}) \geq 5$). If $\bar{p}$ were .9, then a sample size of 50 or more would qualify as sufficiently large.

> **Note:** Many authors recommend that in proportions estimation problems, where the population proportion value is never known with certainty, a sample size of *no less than* 100 should always be used. However, this seems to be an overly conservative condition.

The FPC

In Chapter 5, the **finite population correction** (FPC) factor was introduced as an adjustment to the standard error term for the sampling distribution of means. It finds similar application in the proportions case. Specifically, if (1) sampling is done without replacement, and (2) sample size is at least 5% of the population size, then the standard error for the proportions distribution becomes

$$\sigma_{\bar{p}} = \sqrt{\frac{\pi(1 - \pi)}{n}} \sqrt{\frac{N - n}{N - 1}}$$

where n = sample size
N = population size

Once again, the FPC serves to reduce the standard error.

EXERCISE 5 Suppose you want to estimate the proportion of classmates who plan to spend their junior year abroad. You take a simple random sample (without replacement) of 100 classmates from a total class population of 500. Twelve people in the sample indicate that they plan a junior year abroad. Build a 95% interval estimate of the overall class proportion based on your sample findings.

$$Ans: \quad .12 \pm 1.96 \sqrt{\frac{.12(.88)}{100}} \sqrt{\frac{500 - 100}{500 - 1}}$$

$$= .12 \pm .057$$

6.2   INTERVAL ESTIMATES OF THE DIFFERENCE BETWEEN MEANS

There may be times when the difference between the means of two populations is of significant interest: For example, the difference in average recovery time for two groups of patients being treated with a new medication; or the difference in average test scores for students exposed to two different instructional methods; or the difference in average housing prices across two geographic regions of the country. In such cases, it may prove impossible or impractical to access every member of the respective populations. Consequently, in order to produce a reasonable comparison of the population means, sampling procedures can once again be effectively employed.

An Example

Suppose you want to contrast the average starting salary earned by recent engineering graduates with the salary of recent business school graduates from colleges nationwide. Rather than surveying every recent graduate, you plan to select a simple random sample (with replacement) of 100 recent engineering graduates and 100 recent business school graduates. From the results of your sampling, you hope to estimate the difference in average starting salary.

For the sake of argument, assume that your sampling produces the following results: for the sample of 100 engineers, a sample mean of $27,508, with sample standard deviation of $3400; for the sample of 100 business graduates, a sample mean of $23,392, with sample standard deviation of $5320.

From such results, it's clear that the sample mean difference is $4116 ($27,508 − $23,392). But is it the *sample* mean difference that's of primary interest to us? Not really. What we really want to know is the *population* mean difference. That is, we'd like to identify the difference that would result if we were to include *all* members of the recent engineering graduate population and *all* members of recent business graduate population in the calculation.

Given all that, should we expect the population mean difference to be exactly equal to the sample mean difference we've produced? If you've followed the arguments of the previous section (and the preceding chapter), your answer should be a resounding No. But should we expect the population mean difference to be in the *neighborhood* of our sample mean difference? Absolutely. One more time, our job is to establish just how big a neighborhood is appropriate.

The Sampling Distribution of Sample Mean Differences

Once again the key to establishing a statistically defensible estimate of the targeted population value (here, the difference in population means) comes in the form of a sampling distribution. In this case, it's the sampling distribution of all possible **sample mean differences**. And how would such a distribution be generated?

Record the current sample result ($27,508 − $23,392 = $4116). Return the 100 members of each sample to their respective populations. Take a second sample of

FIGURE 6.2   **THE SAMPLING DISTRIBUTION OF SAMPLE MEAN DIFFERENCES**

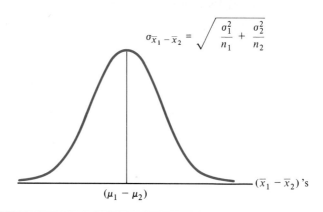

$$\sigma_{\bar{x}_1 - \bar{x}_2} = \sqrt{\frac{\sigma_1^2}{n_1} + \frac{\sigma_2^2}{n_2}}$$

$(\bar{x}_1 - \bar{x}_2)$'s

$(\mu_1 - \mu_2)$

100 engineers and a second sample of 100 business graduates; compute the new sample means and record the sample mean difference. Return these samples, take a third sample from each population, then a fourth, and continue until all possible sample pairs have been produced and all possible sample mean differences have been recorded. The full listing of sample mean differences constitutes precisely the sampling distribution we need.

> **Note:** We'll designate the population of recent engineering graduates as population 1 and the population of recent business grads as population 2, and label the average salary for population 1, μ_1, and the average salary for population 2, μ_2. We'll label the sample means, $\bar{x}_1$ and $\bar{x}_2$, respectively. The sampling distribution, then, comprises all possible $\bar{x}_1 - \bar{x}_2$ sample mean differences, based on samples of size n_1 drawn from population 1 and n_2 drawn from population 2. Sample sizes n_1 and n_2 need not be equal.

Inevitable question: Is it really necessary to generate the full sampling distribution in order to implement our estimating procedure? *Familiar answer:* No. Insofar as statistical theory allows us to anticipate key characteristics of this distribution (we already know what the sampling distribution should look like, or at least we will shortly), we'll need to see only *one* sample mean difference (one $\bar{x}_1 - \bar{x}_2$) in order to produce an effective estimate of the overall population mean difference, $\mu_1 - \mu_2$.

Following a familiar pattern, we would fully expect the sampling distribution of all possible sample mean differences to (1) be approximately normal in shape (for sufficiently large sample sizes), (2) be centered on the population mean difference ($\mu_1 - \mu_2$), and (3) have an easily computed standard deviation, $\sigma_{\bar{x}_1 - \bar{x}_2}$. Statistical theory confirms precisely these three expectations. (See Figure 6.2.)

Importantly, we can view any one randomly produced sample mean difference (including our sample mean difference of $4116) as a value randomly selected from the distribution of all possible sample mean differences. Since this sampling distribution is known to be normal and centered on the population mean difference,

- We can be 68.3% confident that the sample mean difference $(\bar{x}_1 - \bar{x}_2)$ falls within one standard deviation of the population mean difference, $\mu_1 - \mu_2$ (the center of the sampling distribution).
- We can be 95.5% confident that the sample mean difference $(\bar{x}_1 - \bar{x}_2)$ falls within two standard deviations of the population mean difference, $\mu_1 - \mu_2$ (the center of the sampling distribution).
- And so forth.

And as we've seen before, these sorts of statements convert readily to interval statements such as

- We can be 68.3% confident that the actual population mean difference will be within one standard deviation of any one randomly produced sample mean difference.
- We can be 95.5% confident that the actual population mean difference will be within two standard deviations of any one randomly produced sample mean difference.
- And so forth.

Symbolically, such confidence statements translate to

$$(\bar{x}_1 - \bar{x}_2) \pm z\sigma_{\bar{x}_1 - \bar{x}_2}$$

where $\sigma_{\bar{x}_1 - \bar{x}_2}$ = standard deviation of the sampling distribution of sample mean differences (i.e., the standard error of the mean difference)

z = number of standard deviations on a normal curve corresponding to a given level of confidence (probability)

Computing $\sigma_{\bar{x}_1 - \bar{x}_2}$ represents the last piece of the puzzle. Specifically,

$$\sigma_{\bar{x}_1 - \bar{x}_2} = \sqrt{\frac{\sigma_1^2}{n_1} + \frac{\sigma_2^2}{n_2}}$$

where σ_1 = standard deviation of population 1
σ_2 = standard deviation of population 2
n_1 = size of the sample from population 1
n_2 = size of the sample from population 2

We can show the full interval expression, then, as

$$(\bar{x} - \bar{x}) \pm z\sqrt{\frac{\sigma_1^2}{n_1} + \frac{\sigma_2^2}{n_2}}$$

If the population standard deviations are unknown (which will almost always be the case), sample standard deviations s_1 and s_2 can be readily substituted, as long as sample size is sufficiently large (more on sample size shortly). In such cases, the interval expression becomes

$$(\bar{x}_1 - \bar{x}_2) \pm z \sqrt{\frac{s_1^2}{n_1} + \frac{s_2^2}{n_2}}$$

EXERCISE 6   Use sample results from the "average starting salary difference" example to produce the appropriate 90% confidence interval. Recall that the sample means were $\bar{x}_1 = \$27{,}508$ (engineers) and $\bar{x}_2 = \$23{,}392$ (business grads), with sample standard deviations $s_1 = \$3400$ (engineers) and $s_2 = \$5320$ (business grads), and sample sizes $n_1 = 100$ and $n_2 = 100$. Interpret the interval you produce.

Ans: $(27{,}508 - 23{,}392) \pm 1.64 \sqrt{\dfrac{(3400)^2}{100} + \dfrac{(5320)^2}{100}}$

$\qquad = \$4116 \pm 1.64(\$631)$

$\qquad = \$4116 \pm \$1035$

Interpretation: Were we now to survey the entire recent engineering graduate population and the entire recent business graduate population and compute the average starting salary for each, it's 90% likely that the difference in average starting salary would be no more than $1035 away from our sample average difference of $4116.

Alternative interpretation: We can be 90% confident that the interval $3081 to $5151 contains the actual difference in average starting salary for the two populations. (Necessarily, then, there is a 10% chance that the interval will not contain the population difference in average starting salary.)

Footnotes to the Basic Method

Small-Sample Assumptions   When sample sizes are small, two conditions, unnecessary in large sample cases, emerge as statistically important:

1. The populations represented must be normally distributed (paralleling our assumption in the small sample case discussed in Chapter 5).
2. The standard deviations of the two populations represented must be equal.

If either of these two conditions is seriously violated, our standard approach would have to be abandoned. In general, we'll simply make the necessary small-sample assumptions and fine-tune our approach according to the discussion that follows.

Note: How small are small samples here? Conceptually, we should probably return to the familiar small-sample criterion of less than 30 for each of the two samples involved. However, in practical terms, we'll make small-sample adjustments when the *sum* of the two sample sizes is less than 30.

Small-Sample Adjustments

1. *Using the t distribution.*   In those small-sample cases where the population standard deviations (σ_1, σ_2) are unknown and consequently must be estimated from sample results (i.e., from s_1 and s_2), the **t distribution**, not the normal distribution, will effectively describe the corresponding sampling distribution of mean differences. As was true in Chapter 5, this change reflects the added uncertainty of not knowing the population standard deviations precisely.

Using the t table in these cases follows essentially the same pattern as previously discussed, except that degrees of freedom will be computed as $n_1 + n_2 - 2$. This is simply the sum of the individual degrees of freedom, $(n_1 - 1) + (n_2 - 1)$.

When combined sample size is 30 or more, the normal approximation to the t distribution is generally appropriate. That is, for "large"-sample cases, we can go back to the normal table, rather than the t table, even if the population standard deviations are unknown.

2. *Pooling Sample Standard Deviations.*   A second adjustment to the procedure focuses on the "equal standard deviations" assumption required in the small-samples case—an assumption that affects the way in which sample information can be used to estimate the two population standard deviations needed for the standard error term $\sigma_{\bar{x}_1 - \bar{x}_2}$. If past patterns prevail, we would expect a direct and rather matter-of-fact substitution— sample standard deviations s_1 and s_2 for population standard deviations σ_1 and σ_2. In truth, such direct substitution is appropriate only as long as sample sizes are sufficiently large. When small samples are involved, a "pooling" of sample standard deviations is necessary. We can describe the pooling procedure along the following lines:

Question: If, as required in the small-samples case, the two population standard deviations are necessarily equal, but unknown, how could we best estimate their common value? (That is, if we believe, as we must, that the population standard deviations are equal, the question remains: equal to what?) *Answer:* With only sample information available, *combining* (or "pooling") the two sample standard deviations will generally produce the kind of single best estimate we need.

To do this, we'll first average the sample *variances* to estimate the common variance of the two populations. (If population *standard deviations* are equal, population *variances* must also be equal.) We can then take the square root of the average variance to produce the standard deviation. Once computed, this so-called **pooled standard deviation** term will replace the individual s_1 and s_2 values in the standard error expression. (Remember s_1 and s_2 were, in the large-samples case, introduced to estimate σ_1 and σ_2. Now, in effect, an average of these sample values, what we're calling the pooled sample standard deviation, is being used to estimate the common population standard deviation required in the small-samples case.)

To illustrate, suppose a sample of 10 (n_1) from one population shows a sample standard deviation (s_1) of 30, while a sample of the same size $(n_2 = 10)$ from a second population shows a sample standard deviation (s_2) of 50. If we assume that the population standard deviations here are equal, we can produce the pooled *variance* estimate simply by squaring the two sample standard deviations (to produce sample variances) and averaging the results:

$$s^2_{pooled} = \frac{s^2_1 + s^2_2}{2} = \frac{30^2 + 50^2}{2} = 1700$$

Taking the square root of the result produces the pooled standard deviation:

$$s_{pooled} = \sqrt{1700} = 41.2$$

Had we been involved with unequal sample sizes, the averaging procedure used to produce s^2_{pooled} would need to be modified slightly so that the larger sample is given (reasonably, it seems) a greater weight in the computation. If, for example, the size of the second sample in our illustration were 20 rather than 10 as originally proposed, then the prescribed averaging procedure might look like

$$\frac{(10)(30^2) + (20)(50^2)}{10 + 20} = 1966.7$$

or, in general,

$$\frac{n_1 s^2_1 + n_2 s^2_2}{n_1 + n_2}$$

a basic weighted average-type calculation. Taking the square root of this weighted average *variance* would appear to produce the kind of standard deviation estimate we need: $\sqrt{1966.7} = 44.3$.

According to statistical theory, however, one small twist remains. Since we can assume that the individual sample standard deviations s_1 and s_2 were computed using $n_1 - 1$ and $n_2 - 1$ as the appropriate denominator terms (think back to Chapter 5), we plan to use these $n_1 - 1$ and $n_2 - 1$ values to replace n_1 and n_2 as the weights in the weighted average expression. Thus, the pooled variance expression becomes

$$s^2_{pooled} = \frac{(n_1 - 1)s^2_1 + (n_2 - 1)s^2_2}{(n_1 - 1) + (n_2 - 1)}$$

or, equivalently,

$$s^2_{pooled} = \frac{(n_1 - 1)s^2_1 + (n_2 - 1)s^2_2}{n_1 + n_2 - 2}$$

For our example, this translates to

$$s^2_{pooled} = \frac{(10 - 1)30^2 + (20 - 1)50^2}{10 + 20 - 2} = 1985.7$$

Taking the square root of *this* result gives us a statistically sound estimate of the common population standard deviation:

$$s_{pooled} = \sqrt{s^2_{pooled}} = \sqrt{1985.7} = 44.6$$

Using this sort of pooled standard deviation value in the standard error expression gives

$$\sigma_{\bar{x}_1 - \bar{x}_2} = \sqrt{\frac{s_{pooled}^2}{n_1} + \frac{s_{pooled}^2}{n_2}}$$

which, in turn, will produce a confidence interval form that looks like

$$(\bar{x}_1 - \bar{x}_2) \pm t \sqrt{\frac{s_{pooled}^2}{n_1} + \frac{s_{pooled}^2}{n_2}}$$

for the small-samples case.

EXERCISE 7   Return to the average salary difference example, but assume now that the sample sizes involved were 10 from the engineering population and 15 from the business grad population. (The sample means were $\bar{x}_1 = \$27,508$ (engineers) and $\bar{x}_2 = \$23,392$ (business grads), with sample standard deviations $s_1 = \$3400$ (engineers) and $s_2 = \$5320$ (business grads).) Produce a 90% confidence interval estimate of the population mean difference. Be sure to use the appropriate t distribution and the pooled standard deviation term in the standard error computation.

Ans: The pooled sample standard deviation can be produced as follows:

$$s_{pooled} = \sqrt{\frac{(10 - 1)(3400)^2 + (15 - 1)(5320)^2}{10 + 15 - 2}} = 4664$$

Consequently the interval should look like

$$\$4116 \pm 1.714 \sqrt{\frac{(4664)^2}{10} + \frac{(4664)^2}{15}}$$

$$= \$4116 \pm 1.714(\$1904)$$

$$= \$4116 \pm \$3263$$

(Notice t was selected for $10 + 15 - 2 = 23$ degrees of freedom.)

Note: In these small-sample cases, we've suggested proceeding on the *assumption* that the populations are normal and the population standard deviations are equal. In truth, formal tests are available to determine, for any given sample pair, whether these assumptions are statistically viable. We will defer to more advanced texts for details and continue to make the necessary assumptions.

The FPC

The finite population correction factor (FPC) will play its familiar error-reducing role. If (1) sampling is done without replacement, and (2) sample sizes (n_1, n_2) are at least 5% of the population sizes (N_1, N_2), then the standard error term can be properly determined by the expression

$$\sigma_{\bar{x}_1 - \bar{x}_2} = \sqrt{\frac{\sigma_1^2}{n_1}\left(\frac{N_1 - n_1}{N_1 - 1}\right) + \frac{\sigma_2^2}{n_2}\left(\frac{N_2 - n_2}{N_2 - 1}\right)}$$

or, when appropriate,

$$\sqrt{\frac{s_1^2}{n_1}\left(\frac{N_1 - n_1}{N_1 - 1}\right) + \frac{s_2^2}{n_2}\left(\frac{N_2 - n_2}{N_2 - 1}\right)}$$

As is apparent, individual FPCs are applied to each term in the standard error expression.

EXERCISE 8 Suppose you want to estimate the difference in the average number of sick days taken last year by two groups of employees, hourly workers and salaried employees, who work for firm XYZ. From the population of 500 hourly workers you take a sample (without replacement) of 50 and find that the average number of sick days taken was 15.6. For a sample of 50 salaried employees, selected without replacement from a population of 200 salaried employees, the average number of sick days taken is 11.2. The standard deviation turns out to be 1.8 days for both groups. Build a 95% confidence interval estimate of the difference in average sick days taken by the two groups.

Ans: $(15.6 - 11.2) \pm 1.96 \sqrt{\left(\frac{1.8^2}{50}\right)\left(\frac{500 - 50}{500 - 1}\right) + \left(\frac{1.8^2}{50}\right)\left(\frac{200 - 50}{200 - 1}\right)}$

$= 4.4 \pm .64$ days

6.3 INTERVAL ESTIMATES OF THE DIFFERENCE BETWEEN PROPORTIONS

One final case remains before we change themes and move to a new chapter—a case in which we're again involved with two distinct populations. This time it's the difference in population *proportions* that will be of particular interest: the difference, for example, between the proportion of male voters who consider themselves to be politically conservative, and the proportion of female voters who would so describe themselves; or the difference between the proportion of all recent buyers of American-built cars who report general satisfaction with their purchase, and the proportion of all recent buyers of Japanese-built cars who would make a similar report.

As was true previously, when access to every member of the population(s) is either impossible, impractical, or too costly, we'll rely on sampling procedures to draw population conclusions from sample information.

An Example

As chief of medical research at Lathrop Research Labs, you have developed a promising new "cure" for the common cold, one that you believe will normally relieve all cold symptoms in 48 hours or less. You randomly select a sample of 150 cold sufferers to whom you will administer the new medication (call this the "test" group) and 150 cold sufferers who will be given a harmless placebo (call this the "control" group). (For purposes of our illustration, we can consider the test group to be a random sample drawn from the entire population of cold sufferers who would take the new medication. Similarly, the control group can be considered a random sample drawn from the population of all cold sufferers who take no medication at all.)

Your job is to estimate the difference in 48-hour recovery rates between the two populations represented. That is, you'll want to contrast the proportion of 48-hour recoveries in the test group sample (call it $\bar{p}_1$) with the proportion of 48-hour recoveries in the control group sample (call it $\bar{p}_2$) in an effort to estimate the population proportion difference ($\pi_1 - \pi_2$) that would prevail were you to run the test for *all* cold sufferers.

Now, for the sake of argument, suppose you find that 90 members (60%) of the test group sample and 75 members (50%) of the control group sample report complete recovery within 48 hours. What can we conclude? Could we expect, for example, that a full testing of the two populations would reveal a *population* proportion difference ($\pi_1 - \pi_2$) precisely equal to the 10% *sample* proportion difference ($\bar{p}_1 - \bar{p}_2$) we've uncovered? Not likely. But would we expect the population proportion difference to be somewhere in the *neighborhood* of the 10% sample result? Absolutely. The only real question is, how big a neighborhood? As before, sampling theory, here in the form of a sampling distribution of all possible **sample proportion differences**, will help determine the appropriate size of the neighborhood involved.

The Sampling Distribution of Sample Proportion Differences

At this point, you should be able to fill in all the relevant blanks in a very familiar argument. Our one sample proportion difference ($\bar{p}_1 - \bar{p}_2$) can be viewed as one value randomly selected from the sampling distribution of *all possible* sample proportion differences, using sample sizes of n_1 (150) and n_2 (150).

And how might this distribution be generated? Record the sample porportion difference ($\bar{p}_1 - \bar{p}_2$) of .10 that we've just produced. Replace the two samples of 150 and 150. Draw two new samples of similar size. Record the corresponding ($\bar{p}_1 - \bar{p}_2$) difference. Return these two samples. Draw two new samples, and so on.

As in all the cases before, the predictability of sampling distribution characteristics eliminates the need to actually produce all the values. As Figure 6.3 indicates, we can predict the distribution's

- Shape—normal (at least for sufficiently large sample sizes)
- Center—the population proportion difference ($\pi_1 - \pi_2$), and

FIGURE 6.3 **THE SAMPLING DISTRIBUTION OF SAMPLE PROPORTION DIFFERENCES**

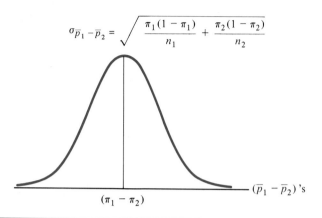

$$\sigma_{\bar{p}_1 - \bar{p}_2} = \sqrt{\frac{\pi_1(1 - \pi_1)}{n_1} + \frac{\pi_2(1 - \pi_2)}{n_2}}$$

$(\pi_1 - \pi_2)$

$(\bar{p}_1 - \bar{p}_2)$'s

- Standard deviation

$$\sigma_{\bar{p}_1 - \bar{p}_2} = \sqrt{\frac{(\pi_1)(1 - \pi_1)}{n_1} + \frac{(\pi_2)(1 - \pi_2)}{n_2}}$$

If we select (as we have in our $\bar{p}_1 - \bar{p}_2$ of .10) one value randomly from this distribution, how likely is it that this value will fall within one standard deviation of the population proportion difference? Within two standard deviations? If our past arguments have made any sense at all, the general confidence interval expression appropriate here should be readily apparent:

$$(\bar{p}_1 - \bar{p}_2) \pm z \sqrt{\frac{\pi_1(1 - \pi_1)}{n} + \frac{\pi_2(1 - \pi_2)}{n}}$$

And if π_1 and π_2 are unknown, the interval form becomes

$$(\bar{p}_1 - \bar{p}_2) \pm z \sqrt{\frac{\bar{p}_1(1 - \bar{p}_1)}{n_1} + \frac{\bar{p}_2(1 - \bar{p}_2)}{n_2}}$$

to accommodate $\bar{p}_1$ as an estimate of π_1 and $\bar{p}_2$ as an estimate of π_2 in the standard error term.

EXERCISE 9 For the cold remedy problem we've proposed, construct the 90% confidence interval estimate of the population proportion difference in recovery rates. Interpret the interval you produce.

Ans: $(.60 - .50) \pm 1.64 \sqrt{\dfrac{(.60)(.40)}{150} + \dfrac{(.50)(.50)}{150}}$

$$= .10 \pm 1.64(.057)$$

$$= .10 \pm .093 \quad \text{or} \quad .007 \text{ to } .193$$

Interpretation: If we were to extend the experiment to all cold sufferers, we can be 90% confident (there's a 90% probability) that the interval .007 to .193 will contain the actual difference in recovery rates for those given the new medication and those given only a placebo. (Admittedly a rather wide interval, suggesting perhaps the need for larger sample sizes.)

Putting things in slightly different terms, we can be 90% confident that the actual population proportion difference would be no more than 9.3 percentage points away from the sample difference of .10.

The FPC

The finite population correction factor can be applied under the usual conditions. When appropriate, the standard error becomes

$$\sigma_{\bar{p}_1 - \bar{p}_2} = \sqrt{\frac{(\pi_1)(1 - \pi_1)}{n_1}\left(\frac{N_1 - n_1}{N_1 - 1}\right) + \frac{(\pi_2)(1 - \pi_2)}{n_2}\left(\frac{N_2 - n_2}{N_2 - 1}\right)}$$

or, when population proportions are unknown,

$$\sqrt{\frac{(\bar{p}_1)(1 - \bar{p}_1)}{n_1}\left(\frac{N_1 - n_1}{N_1 - 1}\right) + \frac{(\bar{p}_2)(1 - \bar{p}_2)}{n_2}\left(\frac{N_2 - n_2}{N_2 - 1}\right)}$$

EXERCISE 10 Suppose you want to estimate the difference between the proportion of male employees in your firm who smoke and the proportion of female employees who smoke. You take a simple random sample of 100 male employees (without replacement) from the population of 500 male employees who work for the firm: 27% of the males in the sample are smokers. You take a similar sample of 100 female employees (out of a total of 800 women who work for the firm): 21% of the females in the sample are smokers. Construct a 90% confidence interval estimate of the actual difference in the male-female proportions across the entire firm.

Ans:

$$(.27 - .21) \pm 1.64 \sqrt{\frac{(.27)(.73)}{100}\left(\frac{500 - 100}{500 - 1}\right) + \frac{(.21)(.79)}{100}\left(\frac{800 - 100}{800 - 1}\right)}$$

$$= .06 \pm .09$$

TABLE 6.1 **A SUMMARY OF KEY EXPRESSIONS IN STATISTICAL ESTIMATION**

For a population mean (μ)	For a population proportion (π)
The basic interval form:	The basic interval form:
$$\bar{x} \pm z\sigma_{\bar{x}}$$	$$\bar{p} \pm z\sigma_{\bar{p}}$$
The standard error term:	The standard error term:
$$\sigma_{\bar{x}} = \frac{\sigma}{\sqrt{n}}$$	$$\sigma_{\bar{p}} = \sqrt{\frac{\pi(1-\pi)}{n}}$$
$$\left(\text{or approx. } s/\sqrt{n}\right)$$	$$\left(\text{or approx. } \sqrt{\frac{\bar{p}(1-\bar{p})}{n}}\right)$$
If sampling w/o replacement:	If sampling w/o replacement:
$$\sigma_{\bar{x}} = \frac{\sigma}{\sqrt{n}}\sqrt{\frac{N-n}{N-1}}$$	$$\sigma_{\bar{p}} = \sqrt{\frac{\pi(1-\pi)}{n}}\sqrt{\frac{N-n}{N-1}}$$
$$\left(\text{or approx. } \frac{s}{\sqrt{n}}\sqrt{\frac{N-n}{N-1}}\right)$$	$$\left(\text{or approx. } \sqrt{\frac{\bar{p}(1-\bar{p})}{n}}\sqrt{\frac{N-n}{N-1}}\right)$$
In the small samples case ($n < 30$):	In the small samples case ($n < 30$ or $n\bar{p} < 5$ or $n(1-\bar{p}) < 5$)
$$\bar{x} \pm t\frac{s}{\sqrt{n}}$$	
$$\text{where } s = \sqrt{\frac{\sum(x-\bar{x})^2}{n-1}}$$	not covered
$$df = n - 2$$	

A Final Note

As we did in the earlier proportions estimation case, we'll sidestep the complications of the small samples issue in the current proportions difference case. So long as $n \geq 100$ and the $np \geq 5$, $n(1 - p) \geq 5$ criteria are met for each of the two samples involved, the procedure we've described in this section (based on the normal distribution) will generally be appropriate. (You should notice that we've used 100 as the minimum sample size here rather than the usual 30. While this may be a bit conservative (i.e., smaller sample sizes may be admissible), we'll recommend taking no chances that the normal shape of the sampling distribution is jeopardized by using samples that are too small. As a general rule, the larger the n, the firmer our statistical footing for making the sorts of estimates we've been making.)

■■■■■■■ CHAPTER SUMMARY

We extended the confidence interval estimation procedure to address three additional problem types: (1) estimating a population proportion (estimating the proportion or percentage of items in a population that possess a particular attribute or characteristic), (2) estimating the difference between two population means, and (3) estimating the difference between two population proportions. In each case, an appropriate sampling distribution played the key role in establishing a link between the results of a

For the diff. between means ($\mu_1 - \mu_2$)	**For the diff. between proportions ($\pi_1 - \pi_2$)**
The basic interval form:	The basic interval form:
$$(\bar{x}_1 - \bar{x}_2) \pm z\sigma_{\bar{x}_1 - \bar{x}_2}$$	$$(\bar{p}_1 - \bar{p}_2) \pm z\sigma_{\bar{p}_1 - \bar{p}_2}$$
The standard error term:	The standard error term:
$$\sigma_{\bar{x}_1 - \bar{x}_2} = \sqrt{\frac{\sigma_1^2}{n_1} + \frac{\sigma_2^2}{n_2}}$$ $$\left(\text{or approx. } \sqrt{\frac{s_1^2}{n_1} + \frac{s_2^2}{n_2}}\right)$$	$$\sigma_{\bar{p}_1 - \bar{p}_2} = \sqrt{\frac{\pi_1(1 - \pi_1)}{n_1} + \frac{\pi_2(1 - \pi_2)}{n_2}}$$ $$\left(\text{or approx. } \sqrt{\frac{\bar{p}_1(1 - \bar{p}_1)}{n_1} + \frac{\bar{p}_2(1 - \bar{p}_2)}{n_2}}\right)$$
If sampling w/o replacement:	If sampling w/o replacement:
$$\sigma_{\bar{x}_1 - \bar{x}_2} = \sqrt{\frac{\sigma_1^2}{n_1}\left(\frac{N_1 - n_1}{N_1 - 1}\right) + \frac{\sigma_2^2}{n_2}\left(\frac{N_2 - n_2}{N_2 - 1}\right)}$$ $$\left(\text{or approx. } \sqrt{\frac{s_1^2}{n_1}\left(\frac{N_1 - n_1}{N_1 - 1}\right) + \frac{s_2^2}{n_2}\left(\frac{N_2 - n_2}{N_2 - 1}\right)}\right)$$	$$\sigma_{\bar{p}_1 - \bar{p}_2} = \sqrt{\frac{\pi_1(1 - \pi_1)}{n_1}\left(\frac{N_1 - n_1}{N_1 - 1}\right) + \frac{\pi_2(1 - \pi_2)}{n_2}\left(\frac{N_2 - n_2}{N_2 - 1}\right)}$$ $$\left(\text{or approx. } \sqrt{\frac{\bar{p}_1(1 - \bar{p}_1)}{n_1}\left(\frac{N_1 - n_1}{N_1 - 1}\right) + \frac{\bar{p}_2(1 - \bar{p}_2)}{n_2}\left(\frac{N_2 - n_2}{N_2 - 1}\right)}\right)$$
In the small samples case ($n_1 + n_2 < 30$):	In the small samples case ($n_1 < 100$ or $n_2 < 100$, etc.):
$$(\bar{x}_1 - \bar{x}_2) \pm t \sqrt{\frac{s_{\text{pooled}}^2}{n_1} + \frac{s_{\text{pooled}}^2}{n_2}}$$ where $s_{\text{pooled}} = \sqrt{\frac{(n_1 - 1)s_1^2 + (n_2 - 1)s_2^2}{n_1 + n_2 - 2}}$ $df = n_1 + n_2 - 2$	not covered

single sample and the population from which the sample was selected.

The sampling distribution of all possible sample proportions, normal-shaped (in large sample cases), centered on the overall population proportion and showing an easy-to-compute standard deviation, enabled us to connect a single sample proportion to a corresponding proportion in the larger population.

The sampling distribution of all possible sample mean differences, normal-shaped (in large sample cases), centered on the difference between the two population means and showing an easy-to-compute standard deviation, provided the key to relating the difference between two sample means to the difference between the means of the two populations from which the samples were selected.

The sampling distribution of all possible sample proportion differences, normal-shaped (in large sample cases), centered on the difference be-

tween the two population proportions and with an easy-to-compute standard deviation, connected the difference between two sample proportions to the difference between the proportions which characterize the two populations from which the samples were drawn.

Paralleling our discussion in Chapter 5, we saw some of the adjustments to the basic method required in the presence of certain special conditions: the t distribution in small sample cases when the population standard deviation is not directly known; the finite population correction factor when sampling is done without replacement.

Finally, we identified a strategy for computing appropriate sample sizes to meet target interval specifications. In general, if a desired confidence level is set, and a maximum tolerable error range (sampling error) is established, sample size computations are relatively straightforward.

Table 6.1 provides a useful summary reference.

KEY TERMS

Finite population correction
Pooled standard deviation
Sampling distribution
   of proportions

   of the difference between means
   of the difference between proportions
t distribution

EXERCISES

Assume that sampling is being done with replacement unless otherwise indicated. If the population size is not indicated, assume it's very large.

(Exercises 1 through 14 deal with confidence interval estimates of a population proportion.)

★ 1. *World Review*, the monthly newsmagazine, has recently changed its format (more holiday recipes and Gary Larson cartoons, less political commentary) and is interested in subscriber response. In a randomly selected sample of 200 subscribers, 64% (128 subscribers) said they preferred the old format. Construct a 95% confidence interval estimate of the proportion of all *World Review* subscribers who favor the old format. Interpret the interval you produce.

2. In a recent random sample of 1350 registered Democrats across the country, 525 of those surveyed expressed opposition to a government proposal to relax trade restrictions with Cuba. Use the sample information to build the 95% confidence interval estimate of the proportion of all Democrats nationwide who would oppose relaxation of trade restrictions.

3. The FAA is concerned about the increasingly high proportion of Acme Airlines (formerly Air Naugatuck) flights that arrive more than 30 minutes after their scheduled arrival times. You select a simple random sample of 100 flight records for the past year, and discover that for 40 of these sampled flights, the planes were late (beyond the 30-minute limit).
   (a) Construct and interpret the 98% confidence interval estimate of the overall late arrival rate.
   (b) Construct and interpret the 85% confidence interval estimate of the overall late arrival rate.

★ 4. There is concern that too many newly created jobs in the economy are low-paying service industry positions. In a random sample of 1000 newly created positions (out of a total of approx-

imately 130,000 new jobs), 610 could be accurately classified as low-paying service industry positions.
   (a) Use this sample result to construct the 95% confidence interval estimate of the overall percentage of new jobs that could be so classified.
   (b) Identify the standard error term and the sampling error term in the interval.

★ 5. You are planning to sample area television viewers in order to estimate the proportion of all TV sets that were tuned to channel 4 to watch the latest CBS miniseries.
   (a) If you intend to produce a 95% confidence interval that will be no wider than ±4 percentage points, how large a simple random sample would you recommend? (The results of a small pilot study showed a sample proportion of .12.)
   (b) How large a sample would be required for a 90% interval?
   (c) Suppose no pilot study results were available. Recompute the recommended sample sizes for parts (a) and (b).

6. Tri-State Bank wants to determine the proportion of its customers who would take advantage of its proposed new interest-bearing checking account, and plans to select a simple random sample to make an appropriate estimate. It wants a sample size that will ensure a 95% probability that the actual population proportion will not differ from the sample proportion by more than 3 percentage points (±.03).
   (a) Assume you have no prior information allowing you to estimate what the sample result might be, and you are not interested in doing any sort of pilot study. How large a sample would you recommend?
   (b) Assume a preliminary study suggests that the population proportion is approximately 20%. How large a sample would you recommend for the survey?

★ 7. Your firm wants to monitor the response of customers to its new flavor of toothpaste ("like a soft breeze on a cool Palm Springs evening"). Suppose you target a sampling error no larger than .02 for the proposed 95% interval estimate. How large a sample would you recommend? (Assume you have no prior information about the likely population proportion.)

8. Your company has just received a large shipment of components commonly used in your productive operation. You randomly select a sample of 120 of these components and test each one. Of the 120 components tested, you discover 15 of them are defective. Produce the 95% confidence interval estimate of the percentage (proportion) of defective items in the entire shipment, then explain how your interval estimate would change under the following conditions (consider each condition independent of the others):

(a) The shipment contained only 500 components. Sample size was 120. Sampling was done without replacement. Fifteen defectives were found.

(b) The shipment contained 5000 components. Sample size was only 10 rather than 120. Sampling was done with replacement. Three defectives were found.

(c) The shipment contained 5000 components. All 5000 of the components were inspected. Fifteen defectives were found.

★ 9. Refer to problem 8. Assume that the shipment consisted of 500 components (and that sampling was done without replacement). Based on the reported sample results,

(a) Produce and interpret the 95% confidence interval estimate of the total number of defective components in the shipment.

(b) Produce and interpret the 90% confidence interval estimate of the total number of defective components in the shipment.

10. You want to estimate the proportion of students at Pelton College (enrollment 1200) who favor a change in the pass/fail grading system. A random sample of 200 students is selected without replacement. Eighty-five students in the sample express support for the change.

(a) Construct a 90% confidence interval estimate of the population proportion here.

(b) Construct a 90% confidence interval estimate of the total number of students at Pelton who favor the change.

11. W. D. "Skip" Broccoli works on the production line at Magnus Industries. As his supervisor, you randomly monitor his activity throughout the 40-hour workweek. Checking at 50 randomly selected times during the week, you find that Skip is on break away from his job 11 times.

(a) Using a 95% confidence level, estimate the proportion of time Skip was on break during the week.

(b) Convert your 95% estimate of the proportion (percentage) of time Skip was on break to an equivalent estimate of the total time, out of the 40-hour week, he spent on break.

★ 12. You are interested in estimating the proportion of taxpayers who took advantage of the newly modified (and controversial) deduction for business travel expenses on their most recent tax returns.

(a) If you want your estimate to be within 4 percentage points of the actual population proportion (at the 95% confidence level), how large a sample should you select? (Assume no prior information.)

(b) Suppose you want to reduce the sampling error term to 2 percentage points, but you don't want to increase the sample size beyond your recommendation in (a). What reduction in confidence level would you have to settle for?

★ 13. We have established that three predictable characteristics of the sampling distribution of all possible sample proportions enable us to make statements about a population proportion using the results of a single sample. While it is generally unnecessary to generate the full array of all possible sample results (the entire sampling distribution of sample proportions) in a standard proportions estimation problem, we can try a small example to demonstrate the three key characteristics that sampling theory would predict for this distribution (i.e., shape, center, and standard deviation). Consider a population consisting of four family members. Each member is asked whether he or she plans a career in advertising. Below are their responses:

Amy	Bob	Carla	Dan
yes	no	no	yes

(a) Determine the proportion of population

members who plan an advertising career. (Label this value π.)

(b) List all possible samples of size 2 that can be selected (sampling with replacement) from this population of 4. (You should produce 16 such samples.)

(c) For each of the 16 possible samples, show the sample $\bar{p}$, the proportion who plan an advertising career.

(d) Construct a relative frequency histogram (bar chart) to display your list of sample proportions ($\bar{p}$'s). Comment on the shape of this sampling distribution.

(e) Compute the average of the 16 sample proportion values and show this value as the center of your sampling distribution of proportions histogram. Compare this value to the population proportion (π) computed in part (a).

(f) Compute the standard deviation for your list of 16 sample proportions using the general square root of the average squared deviations approach. (Label this value $\sigma_{\bar{p}}$.)

(g) Compute the standard deviation of the sampling distribution of proportions using the expression

$$\sigma_{\bar{p}} = \sqrt{\frac{\pi(1-\pi)}{n}}$$

Compare the result produced here with the result you produced in (f) and comment on the significance of the comparison.

14. Refer to problem 13. Assume now that sampling will be done without replacement. Work your way through all parts of the problem for this modified situation and report the similarities and differences you find in your results.

(Exercises 15 through 26 deal with confidence interval estimates of the difference between two population means.)

★ 15. Your firm is testing two order-processing systems to assess late delivery times for orders processed under each. System 1 processes orders on essentially a "first come, first served" basis. System 2 uses a more sophisticated "Nowakowski prioritized mapping" technique. Using a sample of 50 orders processed under system 1 and 50 orders processed under system 2,

you find the average "late" time for the system 1 sample is 29.2 days, with standard deviation of 4.5 days. For the system 2 sample, average late time is 13.1 days with standard deviation of 3.5 days. Build and interpret the 90% confidence interval here to estimate the difference in late times for the two systems.

16. It has sometimes been suggested that people who reach top management positions (chief executive officers, etc.) are, on average, taller than those persons who stay at mid-level positions. You take a simple random sample of 400 top-level managers and 400 long-time mid-level managers, and find the average height for the top-level sample to be 73.2 inches, with standard deviation of 1.8 inches. For the mid-level managers, average height is 68.7 inches, with standard deviation of 2.7 inches. Construct and interpret the appropriate 99% confidence interval estimate of the difference in average heights for the two populations represented here.

★ 17. A survey of wage scales for workers of comparable skill levels was conducted in different regions of the country. In the sample of 500 office workers in the Northeast, the average hourly wage was $9.84, with standard deviation of $1.23. In the Southeast, a sample of 400 office workers showed an average of $7.48, with standard deviation of $2.04. Construct the appropriate 90% confidence interval estimate of the overall difference in average hourly wage for the population of office workers in the two regions.

18. Two different arthroscopic surgery techniques have been developed for correcting serious knee injuries. (Call the surgical techniques A and B.) You are interested in assessing the difference in average healing times for the two procedures. You select 50 patients to undergo surgical technique A and 50 patients to undergo surgical technique B. Average recovery time for technique A patients turns out to be 95 days, with standard deviation of 12 days. For technique B patients, average recovery time is 88 days, with standard deviation of 8 days. Treating the two patient groups as simple random samples, construct the 95% confidence interval estimate of the difference in average recovery times for the population of all technique A patients and the population of all technique B patients.

★ 19. It has been proposed that there is a difference in

the number of hours of TV watched per week by good students (B average or better) versus poor students (D average or worse). You take a simple random sample of 1400 good students and find that the average weekly TV hours for the sample is 22.4, with sample standard deviation of 4.7 hours. For a sample of 1000 poor students, the weekly average is 31.5, with a standard deviation of 6.1 hours.

(a) Build and interpret the 99% confidence interval estimate of the population average difference in TV-watching hours for good students versus poor students.

(b) Suppose sample sizes were in fact 14 good and 10 poor students. Revise your interval and discuss the changes.

(c) For part (b), what additional population assumptions are required in order to produce the interval in the manner prescribed.

20. Margaret St. Cloud, marketing director for New World Concepts, a fashion retailer, wants to establish the difference in average in-store purchase amounts for customers under the age of 25 and for customers 25 or older. She selects a simple random sample of 50 customers under 25 and 50 customers 25 and older. For the under-25 sample, purchase amounts average $163.50, with standard deviation of $45.80. For the 25-and-over sample, purchase amounts average $146.30, with a standard deviation of $27.00.

(a) Build and interpret the 95% confidence interval estimate appropriate here.

(b) Suppose only eight customers were included in each sample. Revise your interval and discuss the differences.

(c) For part (b), what additional population assumptions are necessary in order to produce the interval in the manner prescribed.

21. You have recently purchased 1000 LKP fusion units from Rochester Industries and 1100 similar units from Beaverton Technology. You select (randomly, without replacement) 12 of the Rochester units and 12 of the Beaverton units. The Rochester sample averaged 1092 hours of useful life, with standard deviation of 29.2 hours. The sample of Beaverton units averaged 984 hours, with standard deviation of 31.4 hours. Using a 90% confidence level, construct and interpret an appropriate interval estimate of the difference in average life for the two populations

represented. (Assume that the two populations represented here are normally distributed and that they have equal standard deviations.)

★ 22. Your boss has expressed interest in assessing the difference in average meltdown temperature for two competing brands of DB conical resonators. You take a simple random sample of five brand W and five brand F devices with the following results:

Brand W		Brand F	
Sample member	Meltdown temps (°F)	Sample member	Meltdown temps (°F)
1	156	1	145
2	144	2	141
3	148	3	146
4	152	4	139
5	150	5	149

(a) Compute the sample means ($\bar{x}_1$ and $\bar{x}_2$) and sample standard deviations (s_1 and s_2) here.

(b) Compute the pooled estimate of population standard deviation.

(c) Using the pooled estimate in (b), construct and interpret the 95% confidence interval estimate of the population mean meltdown temperature difference. Be sure to use the appropriate t score for your interval. (You can assume that the populations represented here are normally distributed with equal standard deviations.)

23. Refer to problem 21. Indicate how things would change under the following conditions. Unless otherwise indicated, consider each condition independent of the others:

(a) Sampling was done with replacement.

(b) Sampling was done without replacement. Total shipment size was 100 from each supplier.

(c) Shipment sizes were 15,000.

(d) All 15000 components in each shipment were inspected.

★ 24. Political editorial writer D. Bunker Fish is interested in campaign spending by current members of the state legislature. Specifically, he is interested in comparing the spending levels of Republican versus Democratic members. Mr. Fish has taken a simple random sample (without replacement) of 40 Republican members of the state legislature (from a total of 102 Republican members) and a similar sample of 40

Democratic members (from a total of 110 Democratic members). Carefully checking expenditures, Fish found that in the last election the Republicans in the sample spent an average of $7800, with standard deviation of $1400, while Democrats spent an average of $6250, with standard deviation of $1100. Build and interpret the 95% confidence interval to estimate the difference in average campaign expenditures for the two populations represented here.

★ 25. Ivy Bound, Inc. offers a home study program to prepare students for the college entrance exams (the SATs), suggesting that test scores will be better for those students who follow the 10-week, $250 home study program. To evaluate the effectiveness of the program, you take a sample of 200 students who took the home study course and 200 students who did not. The average SAT score for the sample who took the course was 1130, with standard deviation of 56 points. The average score for the sample who did not take the course was 1080, with standard deviation of 84 points.

(a) Construct and interpret the 95% confidence interval estimate of the "population" mean difference in scores.

(b) What are the two "populations" involved here?

(c) Suppose you wanted to produce an interval estimate that had a sampling error term no larger than 10 points at the 95% confidence level. Assuming that you plan to take samples of equal size from each population (i.e., $n_1 = n_2$), how large a sample from each population should you take? (Use the results above as a pilot sample.)

26. You are planning a study to estimate the difference in the average amount spent weekly on compact audio discs (CDs) by two different groups of consumers: teens under the age of 15 (population 1) and adults over the age of 30 (population 2). You want your interval estimate of the mean difference in purchases to show a sampling error of no more than $.10 at the 90% confidence level. Assuming you will take samples of equal size from the two populations ($n_1 = n_2$), what sample size is required? A preliminary sample showed sample standard deviations of $1.23 for the population 1 sample and $2.04 for the population 2 sample.

(Exercises 27 through 33 deal with interval estimates of the difference between two population proportions.)

★ 27. In a recent national survey of 800 men and 800 women (randomly selected), 64% of the men and 43% of the women reported that they have absolutely no interest in the outcome of the next Wrestle-Mania extravaganza. Construct and interpret the 90% confidence interval estimate of the population proportion difference here.

28. The National Collegiate Athletic Association (NCAA) is concerned about the graduation rate for student athletes involved in revenue-producing sports, especially football and basketball, versus that for student athletes involved in nonrevenue sports like swimming, lacrosse, and the like. From a list of freshmen athletes who entered college five years ago, a random sample of 500 football and basketball players and a sample of 500 participants in other collegiate sports is selected. It turns out that the sample of football and basketball players showed a graduation rate of 38%, while the sample of other student athletes showed a graduation rate of 43%. Based on these sample results, produce a 90% confidence interval estimate of the difference in graduation rates for the two populations represented here. What conclusions (if any) can be drawn?

★ 29. You have received shipments of 15,000 components from each of your two main suppliers, Alcott Industries and Brookside, Inc. You take a simple random sample (without replacement) of 150 of the components from the Alcott shipment and find that 15 of them fall outside acceptable tolerance levels (they are defective). For a similar sample of 150 components from the Brookside shipment, 9 of the components are defective. Construct and interpret the appropriate 95% confidence interval estimate of the overall difference in percent defectives for the two shipments.

30. Refer to Problem 29. Suppose each of the two shipments contained 1000 components. Revise your 95% interval.

31. You set up a market test in two separate regions of the country to examine the effects of a new wholesale pricing policy. In the first region, a sample of 100 industrial customers is selected randomly (without replacement) from a list of

1100 regular accounts and asked to react to the new policy. Twenty-two of the customers in the sample react negatively to the new policy. In a similar sample of 120 industrial customers, drawn (without replacement) from a list of 1400 accounts in the second region, 18 react negatively. Build and interpret an appropriate 99% confidence interval estimate of the difference in negative reaction rates for the two customer populations represented here.

★ 32. To establish a comparison of customer satisfaction levels, you have taken a sample of 200 recent buyers of an American-made car, and 200 recent buyers of a Japanese-made auto. One hundred and twenty of the American-car buyers and 142 of the Japanese-car buyers expressed a very high level of satisfaction with their purchases.

(a) Construct the 90% confidence interval estimate of the difference in the population proportion of American- versus Japanese-car buyers who would express this level of satisfaction.

(b) Suppose you wanted your 90% interval to have a sampling error no larger than .04. Assuming sample sizes will be equal, how large a sample from each of these popula-

tions would you recommend? (Use the results from the original samples of size 200 as pilot study results.)

(c) Suppose you had no prior sample information. Reconsider your sample size recommendation from part (b) and recalculate the appropriate sample sizes.

33. There is general concern (and disagreement) over the apparent link between smoking and lung cancer. In a simple random sample of the medical records of 1000 deceased smokers and 1000 deceased nonsmokers, it was found that 26 of the smokers and 20 of the nonsmokers died of lung-cancer-related causes.

(a) Build and interpret the 95% confidence interval estimate of the difference in lung cancer-caused mortality rates for the two populations represented here.

(b) Can you draw any broad scientific conclusions from these sample results? Explain.

(c) Suppose we wanted to ensure a sampling error for a 95% confidence interval estimate of the population difference that is no larger than 0.5%. How large a sample would you recommend? (Assume sample sizes will be equal. Use the results of the original sample here as pilot study results.)

STATISTICAL HYPOTHESIS TESTING: HYPOTHESIS TESTS FOR A POPULATION MEAN

CHAPTER OBJECTIVES

Chapter 7 should enable you to:

1. Describe the general nature of hypothesis testing.

2. Understand the key role of the sampling distribution in statistical hypothesis testing.

3. Devise a hypothesis test for a population mean.

4. Define the significance level for a hypothesis test.

5. Measure the chance for error in hypothesis testing.

6. Determine the appropriate sample size for any hypothesis test involving a population mean.

7. Produce the operating characteristic curve for a hypothesis test of the population mean. (Chapter Supplement)

Having spent a great deal of time describing confidence interval estimation, we now turn to a second side of statistical inference: statistical hypothesis testing. Usefully, most of the principles involved in our discussion of interval estimation can be readily translated to this new inferential setting. (We'll see, for example, that the crucial task of linking sample evidence to population characteristic is once again accomplished by using an appropriate sampling distribution.) We'll simply build on the foundation that we've already established.

7.1 CONTRASTING INTERVAL ESTIMATION AND HYPOTHESIS TESTING

As we saw in Chapters 5 and 6, interval estimation procedures essentially involve selecting a sample from a larger population and using sample results to estimate the value of some population characteristic. In statistical hypothesis testing, the procedure is slightly different. Here, we'll start out by making a statement about the population, then select a sample to see whether that statement will hold up in light of the sample information produced.

To put things a little more formally,

> In hypothesis testing, a statement—we'll call it a *hypothesis*—is made about some characteristic of a subject population. A sample is then drawn from the population in an effort to establish the truth or falsity of the statement.

And just how is the truth or falsity of the statement established?

> If sample results are sufficiently unlikely under an assumption that the statement is true, then the statement is judged to be false.

A simple example should help to clarify things:

Suppose you arrive home from the store with a bushel basket of strawberries. The grocer has assured you that the berries are of finest quality and that no more than 5% of them will show even the smallest blemish. You take a random sample of 20 berries from the basket and find that all 20 are blemished. What do you think now of your grocer's contention? Have you produced sample results that severely challenge the grocer's statement? I suspect that even to the casual observer, the answer appears obvious. It seems highly unlikely that this sort of result could have been produced had the grocer's statement been true. In fact, our result could fairly be judged so unlikely that we would now almost certainly have to reject the grocer's claim and conclude that more than 5% of all the berries in the basket are blemished.

Plainly stated, we put the grocer's statement (hypothesis) to the test and rejected it as indefensible in light of compelling sample evidence. Case closed. Or is it? Are there issues to be explored here? Absolutely. For example, we probably need to discuss whether a sample of 20 is sufficient for a judgment about the entire population (bushel) of berries. And we certainly need to establish what's meant by "sufficiently unlikely" (or "highly unlikely"). Would 19 blemished berries in the sample have been "unlikely" enough to force us into rejecting the grocer's contention? How about 16? 10? 5? 2? 1? Just where do we draw the line between the kinds of sample results that would lead us to reject his assertion, and those that would make his assertion appear believable? It's these sorts of issues that will provide the focus for the remainder of the chapter.

7.2 MONTCLAIR MOTORS: An Example

At this point, let me establish a slightly more ambitious example for our discussion:

Montclair Motors, a major automobile manufacturer, has been troubled by persistent charges from a respected consumer group that its 1989 models may have had a serious design flaw which, after a few years of ordinary road wear, could cause the front axle to buckle under normal pressure. The original design called for an average breaking strength of 5000 pounds per square inch (psi), with a standard deviation of 250 psi, and Montclair has issued repeated reassurances that road wear has not eroded this standard.

Montclair is convinced that the axles remain solid (with a current average breaking strength of 5000 psi, as designed), but nevertheless plans to recall a random sample of 50 cars (out of a total of 10,000 1989 models sold) in an effort to answer the questions that have been raised.

If sample results provide evidence that the average breaking strength is now unsatisfactory (i.e., that it's now below 5000 psi), a full recall of all 1989 models will be conducted, at a cost to the company of nearly $10 million. Our job is to construct an appropriate hypothesis test to effectively evaluate Montclair's claim.

7.3 FORMALIZING A TESTING PROCEDURE

Establishing the Hypotheses

To formally initiate a hypothesis testing procedure, we first need to identify the two competing positions inherent in the situation. We'll then choose one of the two as the position to be tested directly, and label it the **null hypothesis**. We'll label the other position the **alternative hypothesis**. Here, our two competing positions might briefly be described as:

Position A: Montclair's claim. The axles are perfectly OK; were we to examine the full population of axles, we would find the average breaking strength to be 5000 psi (precisely as designed).

Position B: The consumer group's concern. The axles are not OK; an examination of all 10,000 axles in the population would reveal an average breaking strength of less than 5000 psi.

Importantly, whichever position we designate as the *null* hypothesis will assume a real prominence in the test we set up. As a general rule, we'll hold fast to the null position unless or until convincing sample evidence to the contrary is produced. (Think back to the strawberries example. There the grocer's claim served as the null hypothesis. We were willing to believe the grocer until strong sample evidence to the contrary was produced.)

The *alternative* hypothesis generally represents a kind of "fall-back" position. Should evidence force us to abandon the null position, we would then embrace the alternative. (Could you now identify the implied *alternative hypothesis* in the strawberries illustration?)

For our initial discussion, we'll adopt Montclair's position as the null hypothesis (and put off until later a discussion of how we might generally decide which position should be designated as the null). We can now succinctly represent the competing positions as

$$H_0: \mu = 5000 \quad \text{(The null hypothesis)}$$
$$H_1: \mu < 5000 \quad \text{(The alternative hypothesis)}$$

with μ representing the mean breaking strength for the population of all 10,000 Montclair axles. (We'll later revise the null hypothesis just slightly.)

Notice we've used H_0 (*H*-sub-zero or *H*-nought) to label the null hypothesis and H_1 (*H*-sub-1) to label the alternative. Notice, too, that both hypotheses focus on μ, the population mean in question, not on a sample result. The null and alternative hypotheses are *always* statements about a subject population, never about a sample. It's important that you remember this as we start to develop the details of the test.

The Role of Sampling

Our plan now is to sample 50 axles and use sample results either to reject or accept (some authors prefer the term "fail to reject" rather than "accept") Montclair's claim (which we've now designated the null hypothesis). And what kind of sample evidence would be considered sufficient to reject H_0? According to an earlier argument, compelling evidence should appear in the form of a sample result "so unlikely under an assumption that the null hypothesis is true," that the null hypothesis must consequently be judged false. The trick now is to clearly identify those "unlikely" results.

A Preliminary Step: Anticipating Possible Sample Results

Following the pattern set in the strawberries illustration, consider your instinctive response to some of the sample results possible in the Montclair Motors case.

Suppose, for example, we tested the 50 cars in the sample and found that the average breaking strength for the front axles examined was only 3 or 4 or 5 psi. Could we reasonably continue to accept Montclair's position? It wouldn't take a rocket scientist to conclude that in light of such strong sample evidence, we should find it exceedingly difficult to hold fast to the belief that average breaking strength for the overall population of front axles is 5000 psi. Why? Even though all we have is partial (i.e., sample) information, this sort of sample result seems far too unlikely to have come from a population that has a mean of 5000. Getting this sort of result would almost certainly cause us to *reject* H_0 as no longer believable. (It's the strawberries case all over again.)

But how about sample results of 4997 or 4999 or 5001 psi? Would we consider any of these to be compelling evidence that should cause us to reject Montclair's position? Are these such unlikely sample results (under an assumption that the population mean is an acceptable 5000 psi) that they should lead us to dismiss the claim that the overall population average is as designed? My instincts say probably not. With a population mean of 5000, I shouldn't be at all surprised to find a sample mean slightly lower (or higher) than 5000. (I certainly wouldn't expect that every sample mean would be *exactly* 5000.) Consequently, I would tend to regard these sorts of sample results as perfectly compatible with a continued belief in H_0.

What have we discovered so far? A sample average of 3 or 4 or 5 psi would likely lead us to abandon Montclair's position (i.e., to reject H_0), while a sample average of 4997 psi probably would not. It would seem, then, that somewhere between 3 (or 4 or 5 psi) and 4997 psi there's a line to be drawn, a line which will serve as a boundary between (1) sample results that would be judged compatible with a continued belief in Montclair's position, and (2) sample results that would be judged sufficiently strong to challenge (and reject) that belief. We'll try to establish a sound and standardized approach to producing just such a line.

The Key: The Sampling Distribution of Means

Think back to our Chapter 5 discussion of sampling distributions and consider what the sampling distribution of means for the Montclair Motors case would look like *under the assumption that the null hypothesis here is true*. Predictably, this distribution (call it the *null distribution*) should be approximately normal and, to be consistent with the null hypothesis, centered on 5000 psi. (See Figure 7.1.)

Question: Is this null distribution likely to yield values in the vicinity of 3 (or 4 or 5) psi? *Answer:* Hardly. If the scale in Figure 7.1 is anywhere near accurate, these are the kinds of values (i.e., sample means) we should expect to find far down in the left tail of the distribution—a long, long way from the hypothesized center— making them highly *unlikely* values. How unlikely? Knowing what I know about the characteristics of a normal distribution, I'd say that such results have less than one chance in a million of ever being selected from this particular sampling distribution.

Question: Suppose I now selected a single random sample of size 50 from the axles population, and the sample showed a mean breaking strength of 3 (or 4 or 5) psi. Could I reasonably conclude that such a mean came from the null distribution we've shown in Figure 7.1 (i.e., the sampling distribution associated with a

FIGURE 7.1   **THE NULL SAMPLING DISTRIBUTION OF SAMPLE MEANS**

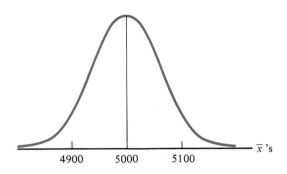

true null hypothesis)? No way. It's simply too unlikely a sample mean (we've assigned a chance of less than 1 in a million) to have come from a distribution centered on 5000 psi. Given this kind of sample mean, I would almost certainly have to conclude that it came from a different sampling distribution, one centered on a value below the hypothesized μ of 5000 (and therefore consistent with a true *alternative* hypothesis). In short, this sort of sample mean would almost certainly cause me to *reject* the null hypothesis.

Choosing a Significance Level

In discussing the act of rejecting the null hypothesis, we've used the term *too unlikely* a sample result to come from the null distribution. But generally speaking, just how "unlikely" does a sample result have to be before we can use it as sufficient (or "significant") evidence to reject a null hypothesis? In our example, 1 chance in a million (.000001) seemed to qualify; so, I'd suspect, would 1 chance in 500,000 (.000002). But how about 1 chance in 1000, or 1 in 100, or 1 in 10?

The fact is that no absolute standard for "unlikely" sample results exists. Consequently, we might ourselves set the standard and demand, for example, that only a sample mean that has "no more than a 1% chance of being selected from the null distribution" will be judged as sufficient sample evidence to reject the null hypothesis. (Or, as we'll see later, we could rely on certain conventional measures used in various scientific fields to set the standard.)

Technically speaking, the standard for unlikely sample results, no matter how it's established, is commonly labeled α (**alpha**) and referred to as the **significance level** for the test. (Statistical hypothesis tests are often called *tests of statistical significance* or simply *significance tests*.) In a very real sense, then, a

> Significance level defines just what we mean by a sample result "so unlikely" under an assumption that the null hypothesis is true that such a result would cause us to reject the null hypothesis and accept the alternative.

Note: The most commonly used significance levels in scientific hypothesis testing are 5% and 1%, but the specific nature of the situation may well suggest other α values. Later we'll examine the consequences and the implications of choosing a significance level.

Drawing the Line

Suppose we were to choose a significance level (an α value) of .05 for the Montclair Motors test, meaning we plan to hold fast to a belief in Montclair's claim that the axles are OK ($\mu = 5000$) unless or until we randomly produce a sample mean that has no more than 1 chance in 20 (i.e., no more than a 5% chance) of emerging from the null distribution of Figure 7.1. Could we at this point begin to specify just where these null-challenging "5%-or-less-likely sample results" are to be found? Absolutely.

Since it's disturbingly low sample means that will discourage us from believing Montclair's claim, we'll isolate the lower 5% of possible sample means by drawing a boundary line in the left-hand tail of the null distribution. Checking the normal table shows that a mark set approximately 1.64 standard deviations below center on a normal curve will effectively establish a boundary separating the lower 5% from the other 95%. (Five percent in the left tail means 45% between the center and the boundary. Entering the normal table with a "target" probability of 45% results in a z-score of 1.64 standard deviations.) We'll label this 1.64 standard deviation cutoff z_c. (See Figure 7.2.)

Our "test of hypotheses," then, should look something like this:

Select a sample of fifty 1989 Montclair cars and test each front axle. If the average breaking strength for these axles lies more than 1.64 standard deviations below 5000, reject Montclair's claim that its population of axles (10,000 in all) still has a mean breaking strength of 5000 psi. If the sample result falls above the z_c cutoff of 1.64 standard deviations, continue to believe (accept) Montclair's claim. (See Figure 7.2.)

Processing Sample Results through the Test

Let's use the test we've devised in order to process a couple of illustrative sample results. Suppose, for example, we found in our random sample of 50 axles a sample mean (i.e., an $\bar{x}$) of 3 psi, a sample result that we've already instinctively established as one that should signal us to reject the null hypothesis. Our job now is to determine how far (in standard deviations) 3 psi falls below the null hypothesis mean of 5000 in the null sampling distribution. If it's more than 1.64 standard deviations, we reject H_0. A simple computation will produce what we need.

$$z_{sample} = \frac{3 - 5000}{\sigma_{\bar{x}}}$$

or, more generally,

$$z_{sample} = \frac{\bar{x} - \mu}{\sigma_{\bar{x}}}$$

FIGURE 7.2   **SETTING THE Z-SCORE BOUNDARY ON THE NULL DISTRIBUTION**

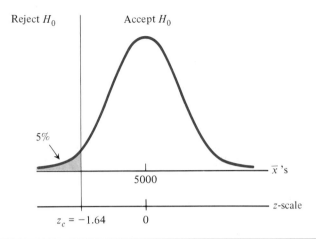

Notice that the denominator here is $\sigma_{\bar{x}}$, the standard deviation (standard error) of the sampling distribution of means shown in Figure 7.2. In *any* test of hypotheses, we will invariably find ourselves measuring distances on the sampling distribution. Consequently, it's the dispersion (i.e., the standard deviation) *here*, not in the population or in any individual sample, that's relevant. The computation of $\sigma_{\bar{x}}$ involves the usual formula

$$\sigma_{\bar{x}} = \frac{\sigma}{\sqrt{n}}$$

where σ = standard deviation of the parent population
     n = sample size

If we assume (as we will for the moment) that the standard deviation of the population is still the original 250 psi, then

$$\sigma_{\bar{x}} = \frac{250}{\sqrt{50}} = 35.35 \text{ psi}$$

Substituting this value in the z_{sample} expression gives a z score for 3 psi of $(3 - 5000)/35.35 = -141.35$, indicating that this sample result would fall approximately 141 standard deviations below the null hypothesis mean of 5000, obviously far beyond the test's $z_c = -1.64$ standard deviation cutoff. Conclusion? We reject H_0 and conclude that Montclair's claim simply is not true.

EXERCISE 1   Given the test we've set up for Montclair Motors, suppose the sample mean turned out to be 4945 psi. Compute the associated z score, using the pattern we've established, and judge whether such a result would lead us to reject

FIGURE 7.3   **SETTING THE BOUNDARY ON THE ORIGINAL SCALE**

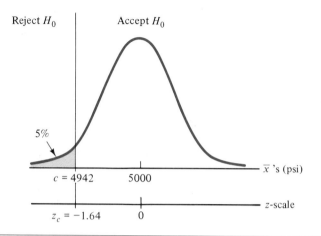

or accept H_0.

$$Ans: \quad z_{sample} = \frac{4945 - 5000}{35.35} = -1.55$$

Since z_{sample} (-1.55) is inside z_c (-1.64), we can't use this sample evidence to reject Montclair's claim, so we accept the null hypothesis.

A Slight Variation

The test we've described clearly establishes a z-score cutoff, z_c, below which we reject and above which we accept (i.e., don't reject) the null hypothesis. Figure 7.2 shows this sort of a z-score boundary. It may sometimes be preferable to establish an equivalent boundary on the "original scale" of the problem. For the Montclair Motors case, for example, we may want to translate the -1.64 standard deviation cutoff to a corresponding mark on the psi scale of the null sampling distribution. To accomplish this, we can simply multiply 1.64 by $\sigma_{\bar{x}}$ ($1.64 \times 35.35 = 58$ psi) and set the boundary marker (call it c) this distance to the left of center (i.e., 58 psi below 5000). Our cutoff, then, would be set at $c = 5000 - 58 = 4942$ psi, and the test could be restated as follows:

> Take a random sample of 50 axles. If the mean breaking strength for the sample is less than 4942 psi, reject Montclair's claim. If the sample mean exceeds this bound, accept Montclair's claim.

Figure 7.3 shows the test.

> **Note:** We'll tend to favor this "original scale" procedure for boundary setting, principally because it allows for an easier, less technical communication of test specifics. Either approach, however, will always produce the same conclusion.

EXERCISE 2   Construct the appropriate test if we were to choose to test Montclair's claim ($\mu = 5000$) at the 10% significance level ($\alpha = .10$). Draw the null sampling distribution and show the test boundary as both a z-score cutoff (z_c) and as a marker (c) on the original psi scale. Clearly label the accept/reject regions for the test. How would this test process a sample mean of 4940 psi? Would it lead you to accept or reject Montclair's claim?

Ans: The z_c cutoff should be set at -1.28 on the z scale, which translates to a c of 4954.75 psi on the original scale. With a sample mean of 4940 psi, we should reject the null hypothesis (i.e., reject Montclair's claim).

(For a more detailed solution, see the end of the chapter.)

A Note on Statistical Significance

We might take a moment here to clarify the term *statistical significance*. As a matter of terminology, any sample result that leads us to reject the null hypothesis is commonly referred to as a statistically significant sample result. We can say that whenever a sample result (in this case, a sample mean) falls in the reject H_0 region of a hypothesis test, the null hypothesis is rejected because a *statistically significant* difference exists between the sample result and the hypothesized population value. Seen in these terms, the value of α sets the boundary for what is and is not statistically significant. Thus we can label sample results significant at the 5% significance level, or significant at the 1% significance level, and so on. (For example, the sample result of 4940 psi in Exercise 2 can be labeled statistically significant at the 10% significance level. You should check to see if it's statistically significant at the 5% level or at the 1% level as well.)

7.4   THE POSSIBILITY OF ERROR IN HYPOTHESIS TESTING

Whenever we make a judgment about an entire population of values based only on partial (i.e., sample) information, we may end up being wrong. In hypothesis testing, in fact, it's possible to identify two types of potential errors, labeled (not too creatively) **Type I error** and **Type II error**. We can define each in fairly simple terms.

> Type I error: rejecting a true null hypothesis
> Type II error: accepting a false null hypothesis

An effective hypothesis test will always attempt to keep the likelihood of such errors within acceptable limits.

The Probability of Type I Error

Consider again the Montclair Motors case and assume for the moment that the null hypothesis is true (although of course, we wouldn't know it unless we saw all

10,000 axles). If this is the case (i.e., if $\mu = 5000$ psi) and we proceed to draw a sample of 50 axles, then we're very likely to see sample results in the vicinity of 5000, the center of the null sampling distribution. We may, however, occasionally produce a sample result (i.e., a sample mean) far from this center. Some of the sample means, in fact, will be so far from the center of 5000 that, according to the test we devised above, they would cause us to reject Montclair's claim. But—and this is important—if, as we've now assumed, Montclair's claim is correct, rejecting that claim is exactly the wrong thing for us to do! We would be making a mistake. More specifically, we'd be committing a Type I error.

And how likely is it that our test would lead us to make such an error? Interestingly enough, it turns out that the significance level (α) gives us precisely the measure we want. Earlier, we described the significance level as a measure of how unlikely a sample result would have to be, under an assumption that the null hypothesis is true, before we would use that sample result as evidence that the null hypothesis is *not* true. With a significance level of 5%, for example, we would use as evidence to reject the null hypothesis any sample mean that had less than a 5% chance of being randomly selected from the sampling distribution of means associated with a true null hypothesis (i.e., the null distribution). Critically, however, this doesn't mean that the null hypothesis *couldn't* be true given such a sample result, only that this sort of sample result is extremely unlikely under an assumption that the null hypothesis is true.

In effect, we're admitting in our 5% test that if the null hypothesis is true, we could nevertheless get a sample result that, under the cutoff rule we've established, would lead us to *reject* the null hypothesis (and therefore, make a Type I error). How likely? For our test, 5%. In general? It's the significance level α.

This is tough stuff. Let me try a slightly different explanation. If the null hypothesis is true, then the null sampling distribution will be the source of our one randomly selected sample mean. But we've decided to draw a boundary on this distribution, below which a sample result will be regarded as evidence that should lead us to *reject* the null hypothesis. With a significance level of 5%, we're saying that any sample result out there in the 5% tail of the null distribution will signal us to not believe that this sample result came from the null distribution at all. Consequently, in 5% of the cases in which the null hypothesis is true, we will be led by sample results to reach the wrong conclusion—to believe that the null is false. Result? We make a Type I error.

> **Note:** Keep in mind that α, which we're now describing as the probability of a Type I error, is actually a *conditional* probability; *given* that the null hypothesis is true, α measures the likelihood that we will produce a sample result that would have us incorrectly reject the null hypothesis. (In some cases, α can better be described as the *maximum* probability of a Type I error. More on this idea later.)

The Probability of a Type II Error

Now suppose that the average breaking strength for the full population of Montclair's axles is actually 4900 psi, well below the null hypothesis mean of 5000. (In a sampling situation, of course, we wouldn't know this without a full inspection of all 10,000 axles.) If this is the case, then clearly the null hypothesis we've been working with since the very beginning of our illustration is false and the alternative hypothesis is true.

But just how well would the test we've devised (with $\alpha = 5\%$) perform in this circumstance? Remember, our test identified a cutoff at 4942 psi; any sample result below 4942 psi would lead us to reject Montclair's claim, any sample mean above 4942 psi would lead us to accept Montclair's claim. Figure 7.4(a) shows the test in visual terms.

Focus on the region of "accept H_0" sample results. If the population mean is actually 4900 psi, could we possibly select a sample showing a sample mean above the cutoff value of 4942, putting it in the accept H_0 region of the test? That is, if μ is only 4900, could we nevertheless produce a sample whose mean would, according to our test, lead us to mistakenly accept Montclair's claim that $\mu = 5000$? My immediate instinct is to say yes. But to effectively respond to the question, we really need to consider related sampling distribution characteristics.

Refer to Figure 7.4(b). Here we've simply superimposed on the test axis of Figure 7.4(a) the sampling distribution of means (based on a sample size of 50) appropriate to a population of axles whose overall mean breaking strength is 4900 psi. (Clearly, we're showing here a sampling distribution consistent with a true alternative hypothesis. Consequently, we'll label this the *alternative* (as opposed to the null) sampling distribution.) Under our assumption that the population mean is actually 4900 psi, it's from this alternative sampling distribution of means (the one centered on 4900) that our one sample mean will actually be selected.

Again, the question: Is it possible that this distribution could yield a sample mean that would lead us to accept Montclair's claim (i.e., to accept the original null hypothesis)? It sure looks like it. Check the (shaded) right-hand tail of this alternative distribution. It's clear that this portion of the curve falls in the accept H_0 region of our test. Here we'll find exactly the sorts of sample means that would lead us into a Type II mistake. These are sample mean possibilities that come from the alternative distribution but, under our test (with its cutoff of 4942), are treated as indicators that the *null* hypothesis is true and so should not be rejected! Bingo! Type II error.

And how likely is it that if the population mean is 4900, we would nevertheless produce a random sample mean which would lead us to mistakenly accept Montclair's claim? All we need to do is use the normal table to find the proportion of values that are associated with the shaded area in the alternative distribution (i.e., the values that lie above 4942 psi).

The mechanics of the procedure are fairly straightforward. If we assume that the population standard deviation is still 250 psi, then the alternative distribution is normal and centered on 4900, with a standard deviation $\sigma_{\bar{x}} = 250/\sqrt{50} = 35.35$. We can compute the appropriate z score for 4942 on the alternative distribution as $z = (4942 - 4900)/35.35 = 1.19$. Checking the table for $z = 1.19$ produces an area of .3830. Subtracting .3830 from .5000 gives us what we need: $.5000 - .3830 = .1170$. (This is the area we've labeled β (beta) in Figure 7.4(c).) Conclusion? If the population mean breaking strength is only 4900 psi, our test has an 11.7% probability of leading us into a Type II error. In other words, if our test is applied to

FIGURE 7.4   **COMPUTING β**

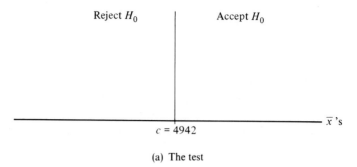

(a) The test

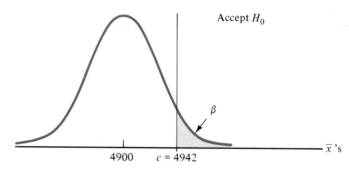

(b) Superimposing the alternative distribution
on the test when $\mu = 4900$

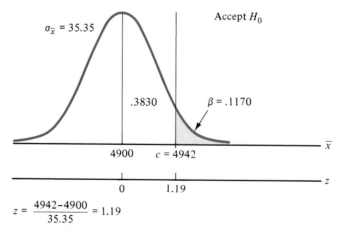

$$z = \frac{4942-4900}{35.35} = 1.19$$

(c) Computing β

a population of axles whose true mean breaking strength is only 4900 psi, there is an 11.7% probability that we will mistakenly conclude that Montclair's claim is correct.

As a general rule the probability of a Type II error (which we're now calling β) will be found by identifying the proportion of values in the alternative distribution that fall in the accept H_0 region of the established test. (Like α, β is a conditional probability; it's the probability of Type II error *given* a specific alternative μ.)

EXERCISE 3   Compute the value of β for the Montclair Motors test ($n = 50$ and $c = 4942$) if the actual population mean breaking strength were 4920 psi. As you proceed:

(a) Show the basic test we've devised on the original ($\bar{x}$) axis.
(b) Superimpose onto the test axis the alternative distribution involved (i.e., the one centered on 4920).
(c) Determine the area in the alternative distribution that lies in the accept H_0 region of the test.

Ans: If $\mu = 4920$, $\beta = .2676$.

(For a more detailed solution, see the end of the chapter.)

The Relationship between α and β

Let me pose a question that may well have occurred to you as you considered our discussion of Type I error. If α represents the probability of making a Type I error, then why use a test that sets this risk as high as 5%? or even 1%? Why not construct the test using an α level closer to 0? (In fact, if we let $\alpha = 0$, we'd never make a Type I error because we would never reject the null hypothesis!) Can you see the potential problem? Choosing a test with a near-zero chance of Type I error will typically result in a test with a high risk of Type II error.

To illustrate, suppose we decided in the Montclair Motors case to use a significance level of 1% rather than the original 5%. As an immediate consequence, the accept/reject boundary c will be pushed lower, to approximately 4918 psi (see Figure 7.5(a)). This lower value for c will cause β to rise. In fact, we can easily track the precise change in β. If we assume that the population mean is actually 4900 psi (rather than Montclair's claim of 5000), we need only find the proportion of values in the corresponding alternative sampling distribution (the one centered on 4900) that lie above the new $c = 4918$ psi boundary. Result? For $\mu = 4900$, β now is 0.305. (See Figure 7.5(b).)

Reducing α from 5% to 1% has increased our chances of making a Type II error from 11.7% to 30.5% (if the population mean is actually 4900 psi). In the absence of any adjustment to sample size, we can always expect this sort of effect: lowering α will raise β.

FIGURE 7.5

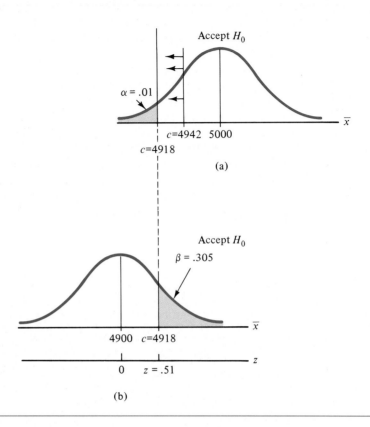

(a)

(b)

Choosing an α Level

Knowing what we now know about possible errors in tests of hypotheses, we can further expand our discussion of the issues involved in selecting a suitable significance level for such tests. Arguably, if we choose not to defer to convention in choosing α (we've already mentioned 5% and 1% as the most commonly used α values), a careful consideration of error consequences may provide a useful guide to selecting an appropriate significance level. In choosing, as we did initially, an α of 5% for the Montclair Motors test, we now realize that we were implicitly assigning to our test a 5% risk of Type I error—a 5% chance that if Montclair's claim is true, the test would nevertheless lead us to incorrectly reject that claim. But what are the consequences of such a mistake?

According to the problem statement, should sample evidence lead us to reject Montclair's null hypothesis, a full recall of all (10,000) 1989 models will be conducted, at a cost to the company of $10,000,000. To incur such a cost unnecessarily—which is precisely what would happen if a Type I error were made—obviously represents a severe penalty to Montclair. In light of this, is 5% too great a risk of error? Understandably, Montclair might well prefer a test with a considerably lower α level.

Before moving too quickly to a test with a lower α, however, we might stop to consider the other side of the error "coin." What consequences would there be in the Montclair Motors case should a Type II error be made? Accepting Montclair's claim that the axles are OK when in fact they're not would result in no recall, meaning that a potentially serious safety problem would go uncorrected, at least for the moment. I imagine any owner of a 1989 Montclair model (or for that matter, any member of Montclair's legal staff) would be inclined to prefer a test for which β is quite small, even if this means a substantially higher α.

What should we conclude? In attempting to establish a proper significance level for any test of hypotheses, we might do well to seek out a test that properly balances α and β levels to reflect the relative costs and consequences of Type I and Type II errors. Unfortunately, finding just the right balance may be more easily said than done. As a consequence, many scientific tests are, by default, designed around a commonly accepted significance level of 5% (or occasionally 1%).

Determining Sample Size

It's possible in hypothesis testing to set target values for α and β and to subsequently establish a sample size n (and a cutoff value c) to meet these targets. That is, rather than deciding on a sample size arbitrarily (we assumed a sample size of 50 for Montclair Motors), we might instead choose desired α and β levels and subsequently determine the sample size necessary to produce such error characteristics in the test.

To illustrate, suppose, having considered the relative consequences of Type I and Type II errors in the Montclair Motors case, we decide that (1) if Montclair's claim (the null hypothesis) is true, we want only a 1% chance of incorrectly rejecting that claim (i.e., we want $\alpha = .01$); and (2) if the average breaking strength of the population of axles is only 4900 psi, we want to allow no more than a 5% chance of incorrectly accepting Montclair's claim (i.e., we want $\beta = .05$). How large a sample should we select?

Figure 7.6(a) shows the appropriate *null* sampling distribution, with its center set at 5000 psi. In order to accommodate the targeted α value of .01, we need to establish a boundary, c, 2.33 standard deviations below 5000. (Recall the rationale. Check the normal table if you're not sure of the proper z score.) Figure 7.6(b) shows the *alternative* distribution centered at 4900 psi. In order that β not exceed 5%, we need to set boundary, c, no less than 1.64 standard deviations above 4900. (Make sure you're convinced.) Since c actually represents the same point in both instances, it must be true that

$$5000 - 2.33\left(\frac{\sigma}{\sqrt{n}}\right) = c = 4900 + 1.64\left(\frac{\sigma}{\sqrt{n}}\right)$$

where $\dfrac{\sigma}{\sqrt{n}}$ is the standard deviation of the two sampling distributions involved.

If we only knew or at least had an estimate for σ, the population standard deviation,

FIGURE 7.6   **DETERMINING SAMPLE SIZE FOR SPECIFIC α/β TARGETS**

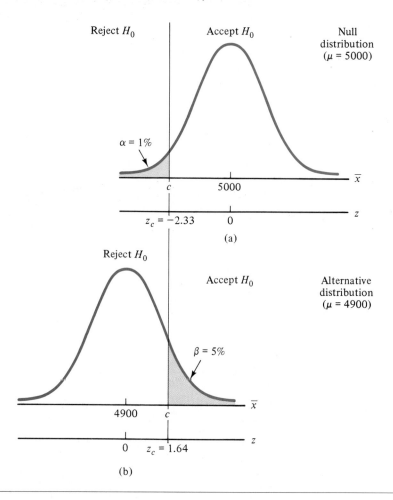

(a)

(b)

we could simply equate the two sides of the c expression and solve for n. In practice, we might look to a pilot study or to previous experience with a somewhat similar population for such an estimate, but here let's continue to assume that σ is 250 psi. Thus,

$$5000 - 2.33\left(\frac{250}{\sqrt{n}}\right) = 4900 + 1.64\left(\frac{250}{\sqrt{n}}\right)$$

Solving the equation for n produces $n = 99$. Substituting 99 for n in either side of the above expression for c establishes $c = 4941$ as the recommended cutoff for the test (by coincidence a value quite close to the c we found earlier). (Be sure you can confirm that the prescribed test, with $n = 99$ and $c = 4941$, actually produces the desired α and β levels.)

EXERCISE 4   Suppose we had set an α target of 2% and a β target of 1% (if $\mu = $ 4900 psi) for the Montclair Motors test. Determine the appropriate sample size and boundary value c. (Assume $\sigma = 250$ psi.)

Ans: $n = 120$ $c = 4953.2$

(For a more detailed solution, see the end of the chapter.)

7.5   ADDITIONAL ISSUES IN HYPOTHESIS TESTING

Having hopefully made a reasonable pass at a number of fundamental hypothesis testing ideas, we can now go back and fill in some of the details we skirted the first time through.

A Closer Look at the Hypotheses

Recall our selection of the null and alternative hypotheses for the Montclair Motors ease:

$$H_0: \mu = 5000 \qquad \text{(Montclair's claim)}$$
$$H_1: \mu < 5000 \qquad \text{(the consumer group's concern)}$$

H_0 was intended to reflect Montclair's claim that its axles are still perfectly OK (that is, were we to test all 10,000 of its 1989 models, we would find that average breaking strength continues to be 5000 psi). On reflection, however, it appears that the null hypothesis we've proposed is probably a little too specific. Rather than arguing that its axles have a mean breaking strength of exactly 5000 psi, Montclair is more likely claiming that its axles show an average breaking strength of *at least* 5000 psi. (Clearly neither the firm nor the consumer group would be alarmed if the axles exceeded design specifications.) If this is so, we need a slightly different set of competing hypotheses:

$$H_0: \mu \geq 5000 \qquad \text{(the population mean is at least 5000)}$$
$$H_1: \mu < 5000 \qquad \text{(the population mean is less than 5000)}$$

Fortunately, this minor modification changes little else in our hypothesis testing procedure; the null distribution used to establish a proper cutoff will still be centered on 5000 psi (now the lower bound of the null position), and we'll still look to the lower tail (below 4942 for the 5% test) of the distribution for the sort of sample results that would lead us to reject Montclair's claim.

> **Note:** α, the significance level for the test and a measure that we had previously described as the probability of Type I error, could now be better described as the *maximum* probability of Type I error. Can you see the reason for this slight change in emphasis? (See Exercise 27 at the end of the chapter.)

FIGURE 7.7 **A TWO-TAILED HYPOTHESIS TEST**

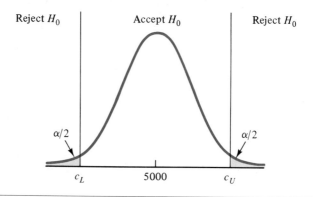

Two-Tailed Hypothesis Tests

Had Montclair truly intended to claim that its axles show a mean breaking strength of *exactly* 5000 psi, implying that variation from this standard *in either direction* would be alarming, some test specifics would have to change. The hypotheses, for example, would be more properly stated as

$H_0: \mu = 5000$ (the population mean is exactly 5000)

$H_1: \mu \neq 5000$ (the population mean is not exactly 5000—it's either less than *or* greater than 5000)

and our boundary-setting task would involve setting *two* cutoffs: a lower bound (below which a sample mean would cause us to reject Montclair's claim) and an upper bound (above which a sample mean would also cause us to reject Montclair's claim).

Figure 7.7 illustrates this **two-tailed** procedure (versus our earlier **one-tailed** approach). For any chosen significance level, we simply split α in half and distribute the resulting half-probabilities to each of the two tails of the null distribution. We can then use the normal table to establish the cutoff values, c_L and c_U, the lower and upper bounds, respectively, for our test. A sample result outside either bound will lead us to reject the null hypothesis; a sample result within the bounds allows us to continue to accept the null.

EXERCISE 5   Using an α of 5%, construct a test of Montclair's claim ($\mu = 5000$) for the two-tailed case. Use a sample size of 50 and continue to assume that the population standard deviation is 250 psi. Report the upper and lower boundaries for your test. Compare your results to our previous one-tailed test.

Ans: The boundaries should be set 1.96 standard deviations to either side of the null hypothesis mean (5000), at 4930.7 psi and 5069.3 psi.

(For a more detailed solution, see the end of the chapter.)

In most problems, a close look at the situation will suggest which of the two test types is most appropriate. If the issue is whether the population mean differs from a particular (hypothesized) value, and it seems immaterial whether the difference is to the low side or the high side of that value, a two-tailed test should be used. If the specific *direction* of the difference is important, then a one-tailed test should be set up. In cases where the issue is unclear, the two-tailed test is usually the choice by default.

The Bias of the Null Hypothesis

Throughout our discussion of the Montclair Motors case, in both the one- and two-tailed versions, we've left unchallenged our designation of Montclair's claim as the null hypothesis. It should now be recognized, however, that in selecting Montclair's claim as the null, we've introduced a substantial **bias** into the testing process.

We've already acknowledged that, once chosen, the null hypothesis assumes a distinct prominence in any subsequent test. As we've seen, we will characteristically hold fast to a belief in the null proposition unless or until significant sample evidence to the contrary is produced. Only then will we abandon the null and embrace the alternative. By choosing Montclair's position as the null hypothesis, we are implicitly establishing a predisposition to accept its claim—putting the alternative position at a distinct disadvantage.

Can you see the problem? In a very real sense, we've "loaded the dice" in favor of Montclair's claim—placing the burden of proof, in effect, on those espousing the alternative position. Had we chosen to reflect the concern of the consumer group in the null hypothesis, a substantial change in emphasis would emerge. (Consider how different our justice system would be if the principal premise (null hypothesis?) were "guilty until proven innocent.") Exercise 6 gives you a chance to try out this idea.

EXERCISE 6   Construct a one-tailed test for the Montclair Motors case, using the consumer group's concern as the null hypothesis. That is,

$$H_0: \mu \leq 5000 \qquad \text{(the consumer group's concern)}$$
$$H_1: \mu > 5000 \qquad \text{(Montclair's position)}$$

Use a significance level of 5% and a sample size of 50, and assume the population standard deviation is still 250 psi. Report your cutoff c as a mark on the psi scale. (*Hint:* Set c in the upper tail of the null distribution centered on 5000, since it is now especially high sample means that will move us away from a belief in the new null hypothesis.)

(a) Compare the conclusion you would reach were you subsequently to produce a sample mean of 4950 psi (1) under this test and (2) under an equivalent test in which the null hypothesis is chosen to reflect Montclair's position (the original test in our earlier discussion).

(b) Make the same sorts of comparisons you did in part (a) for a sample mean of 4990 psi and for a sample mean 5030 psi.

(c) If you represented the consumer group, which form of the test would you prefer (i.e., which null hypothesis would you want to use)?

Note: We've shown the null here as $\mu \leq 5000$. I suppose we could argue about whether the "equal" part of this "less than or equal to" condition should be included in the null position or as part of the alternative hypothesis. The convention is to include it in the null. In practical terms, since we're typically dealing with a continuous distribution of values, it actually doesn't matter.

Ans: The accept/reject boundary should be set at 5057.9 psi.

(a), (b) For each of the sample results cited (4950, 4990, and 5030 psi) we would, according to the test proposed here, accept the *consumer group's* claim. Under the original test, in which we would accept Montclair's claim for any sample result above the accept/reject cutoff of 4942 psi, these same sample results would lead us to accept *Montclair's* claim.

(c) If the consumer group is looking to put Montclair on the defensive, it would undoubtedly prefer to use $\mu \leq 5000$ as the null. This test is biased in favor of the consumer group's position. Montclair would have to show a sample mean well above its claimed population mean of 5000 before its claim would be accepted.

(For a more detailed solution, see the end of the chapter.)

Choosing a Null Hypothesis

Given the kind of inherent bias we've described, it's clear that care must be taken in deciding which of the competing positions should be designated as the null hypothesis. A number of selection strategies intended to prevent exploitation of the natural bias toward the null hypothesis have been proposed. Here are some of the possibilities.

1. The "good sport" approach: Use as the null hypothesis the position *opposite* the one you would like ultimately to be proved true. (In essence, the "other side" is given the advantage of being the null hypothesis.) Here you assign to yourself, or to those supporting your case, the burden of proof to produce sample evidence which will overwhelm a contrary or disagreeable null position. Examples: our product has an average useful life of no more than three years; our competitors offer a lower average price for their products than we do for ours; our sales performance has diminished since last year.

2. The skeptic's "show me" approach: In order to test claims of "new and improved" or "better than" (or "different from") what is currently or normally the case, we'll use as the null hypothesis the skeptic's view which

essentially contends that "new is no better than (or no different from) old." This sort of healthy skepticism is characteristic of the general scientific community when it evaluates new treatments, medications, or claims of scientific breakthroughs. Examples: the proposed new headache remedy acts no faster than commonly used treatments; the new metal alloy is no stronger than those currently in use; the proposed new accounting system is no more error-free than the system that has been used for years.

3. The "status quo" or "if-it's-not-broken-don't-fix-it" approach: Here we'll use the status quo (i.e., no change) position as the null hypothesis. Compelling sample evidence to the contrary must be produced before any change in the status quo is recognized. Examples: the machine continues to function properly; the productive process is running as it was designed to run; the recent shipment contains no more than the usual number of defective items. (You can probably see similarities to (2) above.)

Whether you choose to follow one of these guidelines or not, careful thought should be given to forming your null and alternative hypotheses. At the very least, the favorable bias enjoyed by the null position should be clearly understood before you proceed with the test (or, for that matter, before you interpret the results of a test conducted by someone else).

7.6 USING THE *t* DISTRIBUTION IN HYPOTHESIS TESTING

For the same reasons discussed in Chapter 5, if the population standard deviation σ is unknown and must be estimated by the sample standard deviation s, then the *t* distribution (*not* the usual normal distribution) appropriately describes the sampling distribution of means required to set up a hypothesis test. (Hypothesis tests are, in fact, often referred to simply as "*t* tests.") Here the *t* table, with $n - 1$ degrees of freedom, should be used to set proper test boundaries. (This means we'll replace our normal z_c boundaries with t_c equivalents.) For large enough sample sizes (i.e., $n \geq 30$), the normal approximation to the *t* distribution is, as before, perfectly acceptable.

Note: As in Chapter 5, in the small sample case ($n < 30$), an additional assumption is required to legitimize our hypothesis testing procedure: the population of values must be normally distributed.

EXERCISE 7   For the Montclair Motors case, assume that sample size is 15, rather than 50, 1989 models, and that we no longer can assume $\sigma = 250$ psi. Instead, we plan to use the sample standard deviation, s, to estimate the current population standard deviation, σ. If the sample of 15 shows a sample mean of 4922 psi and a standard deviation of 220 psi, construct a test of the hypotheses

$$H_0: \mu \geq 5000 \quad \text{(Montclair's claim)}$$
$$H_1: \mu < 5000 \quad \text{(the consumer group's concern)}$$

using a significance level of 5%, and report your conclusion. (Assume that the population of values here is approximately normally distributed.)

Ans: The accept/reject cutoff t_c should be set at 1.761 standard deviations below the mean on the t scale or at 4900 psi on the original scale. A sample mean of 4922 psi would lead us to accept Montclair's claim.

(For a more detailed solution, see the end of the chapter.)

7.7   THE FPC IN HYPOTHESIS TESTING

The finite population correction (FPC) factor also plays a role in hypothesis testing, just as it did in interval estimation. If sampling is done without replacement (as is normally the case) and if the sample size is at least 5% of the population, then the standard deviation of the sampling distribution of means (i.e., the standard error term) should be adjusted with the FPC factor

$$\sqrt{\frac{N-n}{N-1}}$$

so that

$$\sigma_{\bar{x}} = \frac{\sigma}{\sqrt{n}} \sqrt{\frac{N-n}{N-1}}$$

EXERCISE 8   Suppose the Montclair population of 1989 models consisted of only 500 automobiles. You plan to take a sample (without replacement) of 50 of these models to test the hypotheses

$$H_0: \mu \geq 5000$$
$$H_1: \mu < 5000$$

Assuming that we know the population standard deviation to be 250 psi, show the proper cutoff value c for your test at the 5% significance level. What conclusion would be reached if your one sample of 50 showed a mean of 4940?

Ans: $c = 5000 - 1.64(35.35)\sqrt{\dfrac{500-50}{500-1}} = 4945$ psi

A sample mean of 4940 should cause you to reject Montclair's claim.

(For a more detailed solution, see the end of the chapter.)

A Final Note

Beginning students sometimes find it difficult to decide, for one-tailed tests, in which tail of the distribution the test should be conducted. The key is in the direction of the inequality shown in the alternative hypothesis. If the inequality "arrow" points in the right-hand direction ($>$) then the boundary for the test will be set in the right-hand (or upper) tail of the distribution. If the inequality arrow points in the left-hand direction ($<$), then the boundary will be established in the left-hand (or lower) tail.

CHAPTER SUPPLEMENT: THE OPERATING CHARACTERISTIC CURVE

The Range of β Values

In describing Type II error, we outlined the basic procedure for determining β, the likelihood that any given hypothesis test could lead the user into mistakenly accepting a false null hypothesis. However, the calculation of β depended very much on the presumed center of the alternative distribution.

For example, in the Montclair Motors illustration (with Montclair's "$\mu \geq 5000$ psi" claim as the null hypothesis), we had computed β under the assumption that the population mean breaking strength was 4900. But what if μ were 4950? or 4895? In each of these cases, Montclair's claim would clearly be false, yet, I suspect, the test we established could lead us to conclude incorrectly that Montclair's claim was true. In fact, each of the alternative μ possibilities will produce its own distinct β value.

EXERCISE S1   To illustrate, reconsider the Montclair Motors hypothesis test for

$$H_0: \mu \geq 5000 \quad \text{(Montclair's claim: the axles are OK)}$$
$$H_1: \mu < 5000 \quad \text{(the axles are not OK)}$$

where we had used a significance level of 5% and a sample size of 50. This is the test, you may recall, in which we had established a cutoff at 4942 psi—the sample mean boundary below which we will reject Montclair's claim, above which we will accept it. For each of the following possible population means (except for two), we've calculated β, the probability that the test would lead the user to mistakenly accept the null hypothesis. Fill in the two remaining β values—the one for $\mu = 4920$ (actually, you were asked to compute this value in exercise 3) and the one for $\mu = 4980$. (For each case, assume $\sigma = 250$ psi.)

The "real" population mean	The probability of our test accepting H_0
4850	.0047
4890	.0708
4920	
4950	.5910
4980	
4999	.9463

Ans: For $\mu = 4920$, $\beta = .2676$

For $\mu = 4980$, $\beta = .8577$

(For a more detailed solution, see the end of the chapter.)

FIGURE 7.8

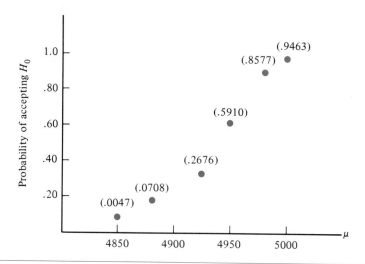

Figure 7.8 shows a plot of β values from Exercise S1.

Sketching the OC Curve

"Connecting the dots" in Figure 7.8 produces a substantial portion of the so-called **operating characteristic (OC) curve** for our test (Figure 7.9(a)). In general, an OC curve for this sort of test describes the full range of "accept H_0" probabilities associated with *all* population mean values (not just for population means that would make the null hypothesis false). The exercise below provides the kind of values we'll need to complete the OC curve for our Montclair Motors test.

EXERCISE S2   We've added to the list of population means in Exercise S1 the following μ possibilities, and computed for two of the three cases the corresponding accept H_0 probabilities. Compute the remaining accept H_0 probability. (Note: Just as before, you need to find how much of the alternative sampling distribution lies in the accept H_0 region of the test.)

The "real" population mean	The probability of our test accepting H_0
5005	.9625
5010	
5030	.9936

FIGURE 7.9

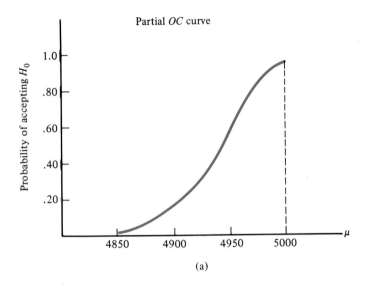

(a)

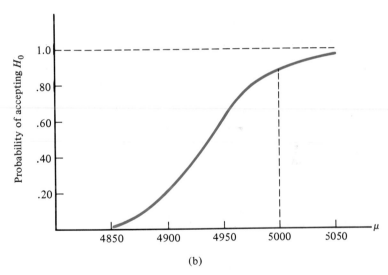

(b)

Ans: For $\mu = 5010$, the probability that the test would lead us to accept H_0 is .9726. (For a more detailed solution, see the end of the chapter.)

Clearly, the three acceptance probabilities produced in Exercise S2 could not properly be labeled β values, since accepting H_0 (Montclair's claim) is, in each case, the right thing to do. In fact, for all population means at or above 5000 psi, such probabilities simply represent the likelihood of correctly accepting a true null hypothesis. In any event, plotting these additional points (Figure 7.8(b)) completes the sketch of the full OC curve.

As a comprehensive picture of performance, the OC curve is often used to compare the potential effectiveness of two or more candidate tests for a given hypothesis testing situation.

▌▌▌▌▌ **CHAPTER SUMMARY**

Our discussion focused on hypothesis testing, the second side of statistical inference. Initially describing the general hypothesis testing rationale, we spent most of our time dealing with a procedure designed to test statements concerning a population mean. That standard procedure went something like this:

A statement (hypothesis) is made about the mean of a target population. A sample is then selected and a sample mean computed in order to judge the truth or falsity of the original statement. Should the sample mean turn out to be substantially different from what one would expect if the statement (hypothesis) were true, then the statement (hypothesis) is judged to be false.

We saw in this case the central role played by the same sampling distribution of means that had been introduced in Chapter 5. Knowing key sampling distribution characteristics—shape, center (if the null hypothesis were true), and standard deviation—gave us the capacity to assign probabilities to various possible sample results. By identifying the most unlikely of these results, we were able to establish a boundary for the sorts of "unusual" sample mean values that would lead us to reject the statement (the null hypothesis) being tested.

One of the crucial elements in the testing procedure was the significance level assigned to the test. Significance level was described as a probability which defines just what we mean by "unlikely" or unexpected sample results—the kind of results that would lead us to reject the null hypothesis. We suggested that, while there is no single best significance level, 1% and 5% are commonly used values.

We also dealt with the prospect of error in any hypothesis test. Because we have only sample (partial) information on which to base judgments about the population involved, we can be led into making either of two possible errors: Type I error involves rejecting the null hypothesis when it is true; Type II error involves accepting the null hypothesis when it is false.

Measuring the chances of error is of real statistical interest. We saw that the significance level (α) in a test measures the (maximum) probability of making a Type I mistake. We computed β to measure the Type II error probability. (In the chapter supplement, the OC curve idea was developed as a means for portraying visually the results of β computations.)

Finally, we developed a procedure for determining the appropriate sample size for any test of the mean which might be proposed.

KEY TERMS

Alpha
Alternative hypothesis
Beta
Bias of the null hypothesis
Null hypothesis
One-tailed hypothesis test

Operating characteristic (OC) curve
Significance level
Two-tailed hypothesis test
Type I error
Type II error

IN-CHAPTER EXERCISE SOLUTIONS

EXERCISE 2

$$H_0: \mu = 5000$$
$$H_1: \mu < 5000$$
$$\alpha = .10$$

The null sampling distribution will look like

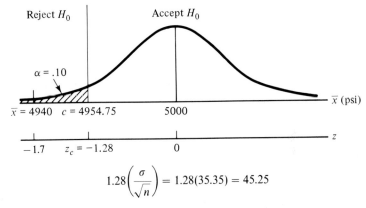

$$1.28\left(\frac{\sigma}{\sqrt{n}}\right) = 1.28(35.35) = 45.25$$

We can set the accept/reject cutoff (z_c) at 1.28 standard deviations below the mean on the z-scale or at

$$c = 5000 - 45.25 = 4954.75 \text{ psi}$$

on the original (psi) scale
On the z-scale the sample mean of 4940 is about 1.7 standard deviations below the null hypothesis mean:

$$z_{\text{sample}} = \frac{4940 - 5000}{35.5} = -1.7$$

This is clearly beyond the $z_c = 1.28$ standard deviation cutoff. We reject the null hypothesis.
    Equivalently, since the sample mean of 4940 psi falls well below the $c = 4954.75$ cutoff on the original scale, we would reject Montclair's claim (H_0).

EXERCISE 3

Basically the test looks like this:

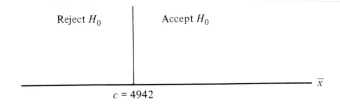

Superimposing onto the test axis the alternative sampling distribution centered on $\mu = 4920$ produces

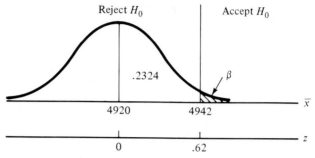

The z score for the 4942 cutoff is

$$z = \frac{4942 - 4920}{250/\sqrt{50}} = \frac{22}{35.35} = .62$$

From the normal table, the area for $z = .62$ is .2324. Therefore, the area labeled β in the figure is $.5000 - .2324 = .2676$.

EXERCISE 4

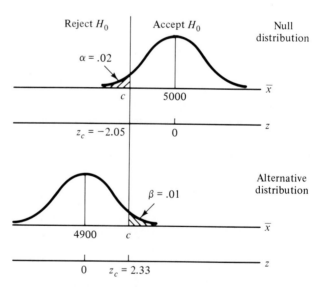

Using the null distribution centered on 5000 psi,

$$c = 5000 - 2.05\left(\frac{250}{\sqrt{n}}\right)$$

Using the alternative distribution centered on 4900 psi,

$$c = 4900 + 2.33\left(\frac{250}{\sqrt{n}}\right)$$

Setting the two c expressions equal produces

$$5000 - 2.05\left(\frac{250}{\sqrt{n}}\right) = 4900 + 2.33\left(\frac{250}{\sqrt{n}}\right)$$

Solving for n,

$$4.38\left(\frac{250}{\sqrt{n}}\right) = 100 \quad \text{or} \quad n = \left[\frac{4.38(250)}{100}\right]^2 = 120.$$

Using either of the expressions for c, we can substitute $n = 120$ and solve for the corresponding cutoff value. For example, using

$$c = 5000 - 2.05\left(\frac{250}{\sqrt{n}}\right)$$

and letting $n = 120$ produces

$$c = 5000 - 2.05\left(\frac{250}{\sqrt{120}}\right) = 5000 - 46.8 = 4953.2$$

We could just as easily have used

$$c = 4900 + 2.33\left(\frac{250}{\sqrt{120}}\right) = 4900 + 53.2 = 4953.2$$

EXERCISE 5

$$H_0: \mu = 5000$$
$$H_1: \mu \neq 5000$$
$$\alpha = .05$$

The sampling distribution would look like

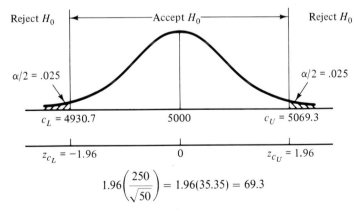

$$1.96\left(\frac{250}{\sqrt{50}}\right) = 1.96(35.35) = 69.3$$

We'll set the boundaries 1.96 standard deviations to either side of the mean:

$$c_L = 5000 - 69.3 = 4930.7 \text{ psi}$$
$$c_U = 5000 + 69.3 = 5069.3 \text{ psi}$$

Any sample mean outside the ± 1.96 standard deviation boundaries will lead us to reject Montclair's claim (H_0).

EXERCISE 6

$$H_0: \mu \le 5000$$

$$H_1: \mu > 5000$$

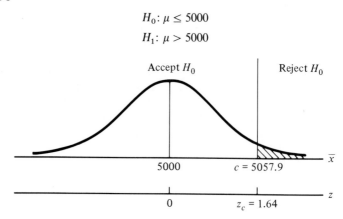

Using

$$1.64\left(\frac{250}{\sqrt{50}}\right) = 1.64(35.35) = 57.9$$

would have us set the boundary (c) at 5057.9 psi, 57.9 units above the null hypothesis mean of 5000 psi. It would take a sample mean of at least 5057.9 before we would reject the consumer group's claim.

EXERCISE 7

$$H_0: \mu \ge 5000$$

$$H_1: \mu < 5000$$

$$\alpha = .05, \quad n = 15, \quad s = 220, \quad \bar{x} = 4922$$

The null sampling distribution looks like

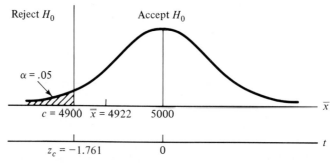

On the t scale, for $df = 15 - 1 = 14$ and a tail-end probability of .05, $t = 1.761$. Thus, any sample mean more than $t_c = 1.761$ standard deviations below the mean of 5000 will lead us to reject Montclair's claim.

Since $\sigma_{\bar{x}} = s/\sqrt{n} = 220/\sqrt{15} = 56.8$ psi, we can compute the t score for the sample result:

$$t_{sample} = \frac{4922 - 5000}{56.8} = -1.37$$

and compare it to the t_c of 1.761. Since t_{sample} is above t_c, we accept the null hypothesis. Or since 1.761 standard deviations below the mean converts to a c of

$$5000 - 1.761(56.8) = 5000 - 100 = 4900 \text{ psi}$$

on the original scale, 4922 clearly falls within the boundary. We accept the null hypothesis.

EXERCISE 8

$$H_0: \mu \geq 5000$$
$$H_1: \mu < 5000$$
$$\alpha = .05, \quad n = 50, \quad N = 500$$

The null sampling distribution for the test will look like

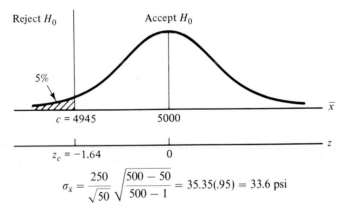

$$\sigma_{\bar{x}} = \frac{250}{\sqrt{50}} \sqrt{\frac{500 - 50}{500 - 1}} = 35.35(.95) = 33.6 \text{ psi}$$

Therefore, the cutoff (c) will be set at

$$5000 - 1.64(33.6) = 5000 - 55 = 4945 \text{ psi}$$

on the original scale.

EXERCISE S1

Basically the test looks like this:

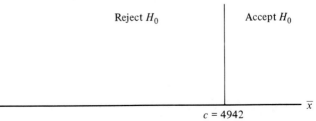

Superimposing onto the test axis the (alternative) sampling distribution centered on $\mu = 4920$ produces

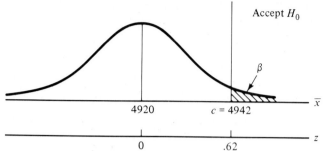

The required z score for the 4942 cutoff is

$$\frac{4942 - 4920}{250/\sqrt{50}} = \frac{22}{35.35} = .62$$

From the normal table, the area for $z = .62$ is .2324. Therefore, the area for β is $.5000 - .2324 = .2676$.

Superimposing onto the test axis the (alternative) distribution centered on $\mu = 4980$ produces

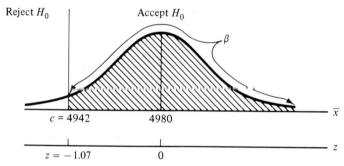

where

$$z = \frac{4942 - 4980}{35.35} = -1.07$$

From the normal table, the area for $z = 1.07$ is .3577, making the area for $\beta = .5000 + .3577 = .8577$.

EXERCISE S2

Again, the basic test looks like this:

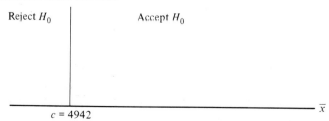

Superimposing onto the test axis the sampling distribution centered on $\mu = 5010$ produces

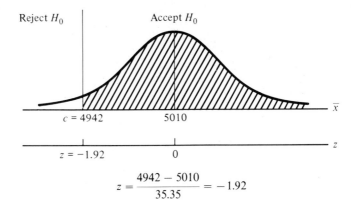

$$z = \frac{4942 - 5010}{35.35} = -1.92$$

From the normal table, for $z = 1.92$, the area is .4726.
Therefore, the area for the accept H_0 probability is equal to $.5000 + .4726 = .9726$.

EXERCISES

Assume that sampling is being done with replacement, unless otherwise indicated. If the population size is not specified, assume it's very large.

★ 1. Plexon Tire and Rubber Company claims that its Delton III tires will last an average of 50,000 miles. You take a sample of 50 of the tires to test Plexon's claim. The average life for the tires in your sample is 49,100 miles, with a sample standard deviation of 4000 miles. Is this sufficient sample evidence to reject Plexon's claim? Set up an appropriate hypothesis test, using a significance level of 10%.
   (a) Express in words and symbols the null and alternative hypotheses for the test. (Use Plexon's position as the null hypothesis.)
   (b) Show the accept/reject cutoff z_c for your test on the z-scale of the null distribution. Compute z_{sample}, the z score for the sample result, and report your conclusion.
   (c) Show c, the cutoff on the original (i.e., "average mileage") scale, along with the sample result.

  2. Golden Crest Technologies claims that the average assembly time required for its new Microtone sound synthesizer is 90 minutes (or less). You take a simple random sample of 64 recent Microtone buyers and find that the average assembly time reported by those in the sample is 110 minutes, with a sample standard deviation of 36 minutes. Is this sufficient evidence to reject Golden Crest's claim? Set up an appropriate hypothesis test, using a 5% significance level.
   (a) Express in words and symbols the null and alternative hypotheses for the test. (Use the company's claim as the null hypothesis.)
   (b) Show the accept/reject cutoff z_c for your test on the z-scale of the null distribution. Compute z_{sample}, the z score for the sample result, and report your conclusion.
   (c) Show c, the cutoff on the original (i.e., "average assembly time") scale, along with the sample result.

★ 3. Judge Lem Gough is accused of being soft on crime. He has responded to the accusation by insisting that the average sentence he has handed out to convicted bank robbers is more severe than that for any other judge in the region. In fact, he claims the average sentence he has handed down is at least 80 months. You take a simple random sample of 36 of the judge's bank robbery cases (from the total population of bank robbery cases presided over by the judge) and find that the average sentence in these cases is only 72 months (with standard deviation of 18 months).

(a) Express in words and symbols the null and alternative hypotheses for a suitable hypothesis test here. (Use the judge's position as the null hypothesis.)

(b) Show the accept/reject cutoff z_c for your test on the z-scale of the null distribution, using a significance level of 1%. Compute z_{sample}, the z score for the sample result, and report your conclusion.

(c) Show c, the cutoff on the original (i.e., "average length of sentence") scale, along with the sample result. Report your conclusion.

4. A serious tornado recently touched down in Bryce County, Kansas. Goverment policy calls for emergency support if the average damage done to the population of all farms in the county exceeds $20,000. At the moment, the government is not at all convinced that this criterion has been met. To demonstrate that average damage is more than the $20,000 figure, you take a simple random sample of 40 farms in the county and find that the average damage for the sample is $21,100 (with standard deviation of $3700). Should this be considered sufficient sample evidence to convince a reluctant government agency to provide the emergency governmental support? (That is, is this sufficient evidence to reject a $\mu \leq \$20,000$ null hypothesis?) Use a significance level of 5%.

(a) Express in words and symbols the null and alternative hypotheses for a suitable hypothesis test.

(b) Show the sampling distribution of means appropriate to the null hypothesis. Clearly identify z_c on the z-scale of the null distribution that will separate "accept H_0" from "reject H_0" sample means, using the significance level of 5%. Compute z_{sample}, the z score for the sample, and report your conclusion.

(c) Show the test on the original scale and report your conclusion.

5. A recent study reported that American grade school students spend, on average, 36.5 hours per week watching television. You randomly select a sample of 70 grade school students in the local school district to test whether this figure is accurate for your community. The sample has a mean of 34.1 hours, with standard deviation of 5.8 hours. Is this sample evidence sufficient to reject, at the 10% significance level, the 36.5-hour figure for the local school district?

★ 6. Acme Delivery claims it delivers packages anywhere in the city in an average time of 2 hours (120 minutes) or less. You randomly select 40 packages to track from pickup to delivery and find that the average delivery time for the sample is 124 minutes, with standard deviation of 18 minutes. Use this sample result to test Acme's claim at the 5% significance level. Explain your conclusion.

7. Chronic Pharmaceuticals, Inc. claims to have a new medication for the common cold, one that reduces average recovery time to 80 hours or less. Independent Testing Labs is skeptical and has selected a sample of 50 cold sufferers to whom it administers the new medication. Average recovery time for the sample is 88.4 hours with standard deviation of 32.3 hours.

(a) Using Chronic's claim as the null hypothesis, is this sample result strong enough to reject Chronic's claim at the 1% significance level?

(b) Suppose now you use Independent Testing's skeptical viewpoint to form the null hypothesis. Using again the 1% significance level, what is your conclusion?

8. Save-U-Doh convenience stores, a nationwide chain, carry the monthly *Crystal Magazine*. The company is concerned that sales of the magazine have fallen below the target quota of 60 copies (on average) per store for the month of May. You select a random sample of 50 stores in May and find that total sales in the sample are 2700, for an average sales figure of 54 copies per store. The sample also shows a standard deviation of 7 copies per store. Using a significance level of 5% and a null hypothesis which states that for the population of stores the overall (average) quota has been met, set up the appropriate hypothesis test. Based on sample results, what is your conclusion?

★ 9. It was noted in the chapter that a sample result that leads to rejecting the null hypothesis is sometimes called a *statistically significant* sample result. Now refer to problem 1 (Plexon Tires).

(a) Is the sample result (49,100 miles) statistically significant at the 10% significance level?

(b) Would this sample result be significant at the 5% level? at the 1% level?

(c) What is the minimum significance level at which this sample result would be judged statistically significant? This minimum significance level is sometimes called the "*p*-value" for the sample result and is frequently reported as part of the output whenever a computer package is used to perform a hypothesis test. (*Hint:* To produce the *p*-value, use the normal table to determine the probability of finding a value in the null distribution that is as far (or farther) from the hypothesized center as the sample result we've produced. Put more simply, determine the area in the tail of the null distribution that lies beyond the sample result.)

10. Follow the pattern of problem 9 to find the *p*-value for the sample result in Exercise 4 (Bryce County tornado).

★ 11. A large number of rivets hold the wing of an Airco 350 (a light aircraft model) to the fuselage. You are to spot check the aircraft's safety. Regulations require an average breaking strength of at least 8800 psi for the rivets. You randomly select 100 rivets and check the breaking strength for each. The average breaking strength for the sample turns out to be only 8756 psi, with standard deviation of 280 psi.

   (a) Use sample results here to test the null hypothesis that overall average breaking strength for the population of rivets satisfies the "at least 8800 psi" requirement. Use a significance level of 5%.

   (b) Disregard the sample mean reported above. Given the boundary (cutoff) you established for your hypothesis test in (a), compute the probability of a Type II error if the actual mean breaking strength for the population of rivets is only 8770.
      (i) Show the test on the original ($\bar{x}$) axis.
      (ii) Superimpose the alternative sampling distribution on the test axis.
      (iii) Show the area in the alternative distribution that lies in the accept H_0 region of the test.

   (c) Repeat your work in (b) for an actual population mean of 8720 psi.

12. Felton School of Electronics claims that the average starting salary for its recent graduates is $28,000 (or more). You contact a simple random sample of 50 of these graduates nationwide and find that the average starting salary in the sample of 50 was $27,500, with a standard deviation of $2100.

   (a) Set up a hypothesis test using Felton's claim as the null hypothesis (and 1% as the significance level).

   (b) Now ignore the specific sample mean reported here. Compute the probability that the test you have set up would lead you into a Type II error if
      (i) the actual average starting salary for the entire population of recent Felton graduates is only $27,000.
      (ii) the actual average starting salary for the entire population of recent Felton graduates is only $26,500.
   In each case, show the test on the original ($\bar{x}$) axis, superimpose the alternative sampling distribution on the test axis, and show the area in the alternative distribution that lies in the accept H_0 region of the test.

★ 13. B. Clem Soft Drinks plans to purchase advertising time on the Friday night "Dead Reckoning" TV series if it can be shown convincingly that the average age of the show's viewers is under 30. A random sample of 200 viewers is selected, and the average age for the sample turns out to be 28.3, with a standard deviation of 9.6.

   (a) Set up a hypothesis test to determine whether this is strong enough sample evidence to reject a skeptical $\mu \geq 30$ null hypothesis. Use a significance level of 5%.

   (b) Now disregard the specific sample mean reported here. How likely is it that the test you set up would lead you into making a Type II error if the actual average age for the show's population of viewers is 28?
      (i) Show the test on the original ($\bar{x}$) axis.
      (ii) Superimpose the alternative sampling distribution on the test axis.
      (iii) Show the area in the alternative distribution that lies in the accept H_0 region of the test.

14. Your firm does business with over 3800 retail outlets across the country. Your marketing department has proposed the introduction of a new product, but only if the retail outlets would be willing to order, on average, 100 or more of the new units. You randomly sample 64 of the stores and find that the average order size that can be

expected for this sample is 88 units, with a sample standard deviation of 30.

(a) Set up a hypothesis test to determine whether this is sufficient sample evidence to discourage the marketing department. (That is, is this enough sample evidence to reject a $\mu \geq 100$ null hypothesis?) Use a significance level of 5%.

(b) Now disregard the sample mean reported here. For the test you set up, compute the probability that your test would lead you into making a Type II error if the actual population mean order size is only 85 units.

★ 15. Orenco Inc. has instituted an inventory control policy which calls for severely reduced production if the average age of units in inventory exceeds 120 days. Each Friday, the company plans to take a simple random sample of items (from its population of over 15,000 items) to evaluate the current situation. You are to help them plan the sampling procedure. Specifically, you are asked to determine an appropriate sample size to use in testing a $\mu \leq 120$ days null hypothesis.

The company wants a significance level (i.e., the risk level for Type I error) of 5%. It also wants a risk no higher than 10% that—if the population of inventory items has a mean age as high as 125 days—your test would lead you to mistakenly believe that the population of items actually meets the "no more than 120 days" standard. (That is, you want β, the risk of Type II error for this case, to be 10%.) Determine the proper sample size. Assume you have a reliable estimate of the population standard deviation of 32.2 days.

16. Shed-Wate Inc. claims that with its new weight control program, average weight loss over the full span of the program is at least 28 lb. You plan to take a sample of Shed-Wate customers to test the weight-loss claim, using Shed-Wate's claim as the null hypothesis, and a significance level of 1%. You want the test to have a β value of .05 if actual average weight loss is only 26 lb. (Assume a pilot sample has produced an estimate of the population standard deviation. It's 8 lb.)

(a) How large a sample size would you recommend?

(b) Given the recommended sample size, report the cutoff value for your test.

17. Refer to problem 1 (Plexon Tires). Suppose you decide to revise your hypothesis test to control for Type II error. You want the test to have an α of .10, but you want to ensure that β is no more than .01 if the actual population average mileage is only 49,500.

(a) How large a sample would you recommend? (Use the original sample of 50 as a pilot sample.)

(b) Where should the cutoff c be set?

★ 18. If production workers are performing their jobs properly, boxes of powdered detergent produced by your company should contain, on average, 30 ounces of powder, no more and no less. You periodically select a sample of 36 boxes of detergent and measure the contents.

(a) Set up the appropriate (two-tailed) hypothesis test to process sample results, using a significance level of 5%. (Assume that the filling process has a reliable standard deviation of .3 ounce, regardless of the mean.)

(b) Suppose you select a sample and it shows an average of 29.83 ounces per box. What should you conclude?

(c) Given your answer in (b), what error could you be making? Explain.

19. Your firm has received a shipment of components which, among other things, must conform to a specification that the average diameter of components in the shipment be exactly 3.5 millimeters (mm). (That is, a "good" shipment will contain units with an average diameter of 3.5 mm.) Any variation to either side of that average is cause for concern. You plan to take a sample of 50 components to evaluate the recent shipment.

(a) If the sample mean diameter is 3.61 mm (with a standard deviation of .2 mm), what should the company conclude about the shipment as a whole? (That is, does the overall shipment conform to the specification?) Set up a two-tailed hypothesis test, using a significance level of 1%.

(b) Given your conclusion, what error could the company be making?

★ 20. Your CDX machine, if it's in proper adjustment, produces component parts that, on average, are precisely 18 mm thick, with standard deviation of .05 mm. Every hour, you take a random sample of 49 parts to measure in order to judge whether the machine continues to function properly or whether it has slipped out of adjustment and is producing components that are too thick or too thin.

(a) Set up an appropriate (two-tailed) hypothesis test, using a significance level of .02.

(b) Suppose at 11 A.M. you take a random sample of 49 units and find that the average thickness for the sample is 17.96 mm. Using the test you established above, what is your conclusion?

(c) Suppose at 1 P.M. you take a sample of 49 parts and find the average thickness in the sample is 18.07 mm. What is your conclusion?

21. You are auditing ABC Corporation's accounts receivable (5100 accounts total). ABC shows a total book value of $2,840,000 for accounts receivable. You take a simple random sample of 100 of these accounts and find the average amount receivable in this sample to be $571.23 (with standard deviation of $81.40). Use this sample information to test the hypothesis that ABC's stated book value is correct (i.e., that total accounts receivable is truly $2,840,000, no more and no less). Use a significance level of 5%. (*Hint:* If total book value is stated as $2,840,000 for the 5100 accounts, then the stated average receivable is $2,840,000/5100 or $556.86. Use this average figure to state the null hypothesis.)

★ 22. Refer to problem 1 (Plexon Tires). Assume that the sample size was only 15.

(a) Conduct the appropriate hypothesis test by computing t_{sample}, the t score for the sample result, and comparing it to t_c, the cutoff t score. Based on the results of your t test, what is your conclusion?

(b) Conduct the appropriate test by translating the cutoff value to the original scale and comparing the sample result to the cutoff? What is your conclusion?

(c) What additional population assumption is necessary here?

23. Salesman J. R. Rowe claims he spends an average of no more than $5 a day on business lunches. As company auditor, you take a random sample of 12 of J. R.'s recent lunch receipts and find that the average lunch check for the sample was $25.04, with a sample standard deviation of $8.40. Using J. R.'s $\mu \le \$5$ as the null hypothesis (and a

significance level of 1%), do the following:

(a) Conduct the appropriate hypothesis test by computing t_{sample} and comparing it to t_c. Based on the results of your t test, what is your conclusion?

(b) Conduct the appropriate test by translating the cutoff value to the original scale and comparing the sample result to the cutoff. What is your conclusion?

(c) What special population assumption is necessary in this sort of small sample case?

★ 24. As operations supervisor, you are concerned about the average time that elapses between the receipt of an order and the shipping of the order. The company prides itself on an average order-processing time of no more than 36 hours. You take a sample of 10 recent orders and find that the elapsed time from order receipt to order shipping was actually 42 hours, with a sample standard deviation of 8 hours. Determine whether this is enough sample evidence to reject a $\mu \le 36$ null hypothesis at the 5% significance level.

(a) Conduct the appropriate hypothesis test by computing t_{sample} and comparing it to t_c. Based on the results of your t test, what is your conclusion?

(b) Conduct the appropriate test by translating the cutoff value to the original scale and comparing the sample result to the cutoff. What is your conclusion?

(c) What additional population assumption is necessary here?

25. Your top salesperson claims that she makes an average of 25 sales calls per day. From her daily log you randomly select a sample of 10 days with the following sales call results:

Sample day	1	2	3	4	5	6	7	8	9	10
Sales calls	20	31	12	19	24	28	16	15	22	25

Use this data to test the hypothesis that this salesperson does average at least 25 sales calls per day. Use a significance level of 10%. (Assume the population of values represented is normally distributed.)

★ 26. From a recent production run of 300 Y-bias filaments, you select (without replacement) a random sample of 100. You find that the average operational life for the filaments in the sample is

2520 hours, with a standard deviation of 160 hours. Use these sample results to test, at the 5% significance level, the null hypothesis position that, overall, the average life for the full population of 300 filaments is no more than 2500 hours. Report your conclusion.

27. We observed in Chapter 7 that in one-tailed hypothesis tests α can be interpreted as the *maximum* probability that a particular hypothesis test will lead the user into a Type I error.

Refer to problem 26, but disregard the sample mean reported there. Show your test on the original scale and then compute the probability that the test you have set up could lead you into making a Type I error if, for the population of filaments,

(a) $\mu = 2500$ hours.

(b) $\mu = 2495$ hours.

(c) $\mu = 2490$ hours.

HYPOTHESIS TESTING FOR PROPORTIONS, MEAN DIFFERENCES, AND PROPORTION DIFFERENCES

CHAPTER OBJECTIVES

Chapter 8 should enable you to:

1. Adapt the general hypothesis testing procedure of the previous chapter to three additional cases:
   - Hypothesis tests for a population proportion
   - Hypothesis tests for the difference between two population means
   - Hypothesis tests for the difference between two population proportions

2. Describe the role of the sampling distribution in each of these cases.

3. Select and describe the appropriate significance level for each test situation.

4. Recognize the possibility of error in any hypothesis test.

5. Measure Type I and Type II error probabilities.

6. Use the t distribution and the binomial distribution as special-case sampling distributions in hypothesis testing.

Following the pattern of the Chapter 5–Chapter 6 sequence, we can readily extend our Chapter 7 hypothesis testing discussion to three additional cases:

- Hypothesis tests for a population proportion
- Hypothesis tests for the difference between two population means
- Hypothesis tests for the difference between two population proportions

If you feel comfortable with the elements of hypothesis testing that were introduced in Chapter 7, the upcoming material should make immediate sense. In fact, we'll proceed now at a slightly accelerated pace, relying more on examples and less on comprehensive explanations to develop any of the new ideas that may be required.

8.1 HYPOTHESIS TESTS FOR A POPULATION PROPORTION

To get us started, consider the following three situations:

Situation 1: A wealthy political contributor is debating whether to support your recently announced campaign for a seat on the city council. Somewhat skeptical of your chances, she tells you she will only make a significant financial contribution if you can demonstrate that you currently have the support of more than 25% of all city voters. (Conventional political wisdom says that without this minimum level of support, you have no chance of winning the general election in September.) A recent poll, using a simple random sample of 300 city voters, reports that 28% of those sampled expressed support for your candidacy. Should this be enough sample evidence to convince the prospective contributor?

Situation 2: Your firm produces units of product in fairly large batches of 5000 or more. The batches are shipped to customers throughout the region. The firm has adopted a policy which establishes an acceptable quality level of 5%; that is, any batch that contains no more than 5% defectives would be judged OK for shipment to customers. Any batch containing more than 5% defectives would be judged of inferior quality and not shipped until corrective action is taken.

Because of high testing costs (and the destructive nature of the testing process), your quality control people have instituted a sampling procedure in which a sample of 150 units from each batch will be randomly selected and tested before a decision is made to ship or not ship. The latest batch comes to your inspection station. You select and test the sample of 150 and find 12 defectives. Do you ship the batch?

Situation 3: In years past, at least 20% of the students at State U. chose to pursue public service careers upon graduation. For an editorial piece on declining student commitment to public service, the school newspaper surveyed a simple random sample of 20 currently enrolled students and found that only 1 of them expressed an interest in a career in public service. Does this sample result provide statistical evidence of a decline in student commitment?

As we'll see shortly, each of these situations provides us the opportunity to use hypothesis testing procedures as an effective framework for testing statements made about a *population proportion.*

Forming the Hypotheses

We can start with the "political contributor" example, establishing first the competing hypotheses: You either have the support of more than 25% of the voting population or you don't. To reflect the skepticism of your potential contributor, we'll go ahead and select the negative position as the null hypothesis (recall the "skeptic's" approach described in Chapter 7), and the affirmative position as the alternative. (Keep in mind that this will bias the test toward the "insufficient support" null position and put the burden of proof on you, the candidate, then, to demonstrate otherwise.) We'll show,

$H_0: \pi \le .25$ (insufficient support: no more than 25% of the voting population supports your candidacy)

$H_1: \pi > .25$ (sufficient support: more than 25% of the voting population supports your candidacy)

As always, we'll hold fast to the null until or unless there is sufficient sample evidence to the contrary.

Anticipating Possible Sample Results

Consider some possible survey results. Suppose our sample of 300 had shown a full 99% of those sampled supporting your candidacy (i.e., a $\bar{p}$ of .99). Should this be regarded as strong sample evidence to force us (and your contributor) out of believing the skeptical null and into believing the alternative? Almost without a doubt. With that sort of overwhelming sample support, it would be pretty difficult for a reasonable person to believe that overall population support is actually no more than 25%.

What about sample results of 80% or 70% or even 60%? These, too, it would seem, should qualify as the kinds of results that would lead us into rejecting the null hypothesis. Each of these sample results is simply *too unlikely* to have been produced from a population showing an overall supporter proportion of only 25% (or less).

On the other hand, a sample proportion of 26% or 27%, while greater than the null hypothesis π of 25%, would (given our predisposition to stick with the null as long as possible) probably not be enough to convince us (or your prospective contributor) to abandon the null position and embrace the alternative. We could, I think, reasonably expect these sorts of sample results to be generated by a population whose overall support proportion was actually no more than 25%.

FIGURE 8.1 **THE SAMPLING DISTRIBUTION OF PROPORTIONS**

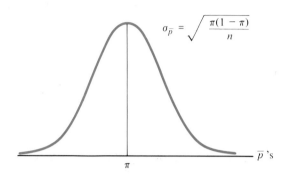

The trick, as we saw in Chapter 7, is to find a statistically defensible way to draw a clear boundary between likely and unlikely results—between the full range of accept H_0-type sample results and the full range of reject H_0-type responses.

The Sampling Distribution

Not unexpectedly, knowledge of the relevant sampling distribution—in this case, the **sampling distribution of all possible sample proportions**—plays the crucial role in differentiating likely and unlikely sample results. Familiar from our discussion in Chapter 6, we can expect this distribution to be normal (for large enough sample sizes), centered on the population proportion, π, with an easily computed standard deviation, $\sigma_{\bar{p}}$, equal to $\sqrt{\pi(1 - \pi)/n}$. (See Figure 8.1.)

The Null Sampling Distribution

If we tentatively assume (as we typically have in initiating a hypothesis testing procedure) that the null hypothesis is true, then we can center the sampling distribution on the null hypothesis population proportion (.25 for our voter illustration) and label it the *null distribution*. (See Figure 8.2.)

Choosing a Significance Level

With the null sampling distribution as reference, setting the desired **significance level** (α) will allow us to establish an appropriate accept/reject boundary for the test. (Recall that α defines exactly what we mean by sample results so unlikely under an assumption that the null hypothesis is true that, when such a sample result is produced, we have to conclude as reasonable people that the null hypothesis must be false.) We'll select a conventional α of .05 and proceed.

The Test

Having set α at 5%, finding a boundary for the test is a relatively easy matter. By placing a marker, z_c, 1.64 standard deviations above the center of the null sampl-

FIGURE 8.2   **THE π = .25 NULL SAMPLING DISTRIBUTION OF SAMPLE PROPORTIONS**

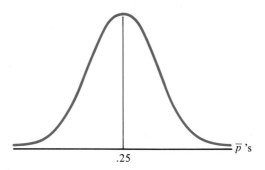

FIGURE 8.3   **SETTING THE Z-SCORE BOUNDARY ON THE NULL DISTRIBUTION OF SAMPLE PROPORTIONS**

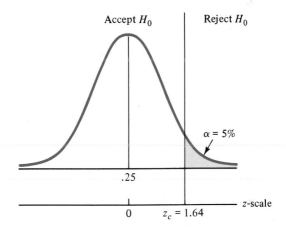

ing distribution (.25), we can effectively isolate the upper 5% of sample result possibilities, identifying sample proportions in this range as precisely the sort of "unusual" sample results that would cause us to reject the null hypothesis.

For any sample result ($\bar{p}$) above this $z_c = 1.64$ marker (i.e., beyond 1.64 standard deviations), we will conclude that you (the candidate) actually do have the support of more than 25% of the voting population. Any result below the 1.64 cutoff will lead us to conclude that you don't. (See Figure 8.3.)

> **Note:** Recall from Chapter 7 that, for a one-tailed test, choosing which tail of the sampling distribution to use in setting the boundary for the test can sometimes be confusing. A useful rule-of-thumb suggests that the direction of the alternative hypothesis inequality "arrow" will direct us immediately to the correct tail. If the arrow points right (>), then the test takes place in the right tail of the null distribution. If it points left (<), the test takes place in the left tail.

Putting Our Sample Result to the Test

Recall now that our original poll of 300 voters has reported a sample proportion of .28. With the test we've constructed, all that's left to determine is how far (in standard deviations) this reported result lies above the hypothesized null distribution center of .25. Once computed, we'll simply match this sample z value (z_{sample}) against the 1.64 standard deviations cutoff, z_c. Here,

$$z_{sample} = \frac{.28 - .25}{\sigma_{\bar{p}}}$$

To complete the calculation, we can use the $\sigma_{\bar{p}}$ expression

$$\sigma_{\bar{p}} = \sqrt{\frac{\pi(1 - \pi)}{n}}$$

to produce

$$\sigma_{\bar{p}} = \sqrt{\frac{(.25)(1 - .25)}{300}} = .025$$

the standard error of the null distribution.

> **Note:** Be sure you notice that the null hypothesis π, *not* the sample result $\bar{p}$, is used in computing the standard deviation (standard error) term for the null sampling distribution. Your instinct, based on earlier discussions, might suggest using $\bar{p}$, but remember, the null distribution represents how things would look if the null hypothesis were true (i.e., if $\pi = .25$).

Substituting the required standard error term (.025) produces the sample z score as shown:

$$z_{sample} = \frac{.28 - .25}{.025} = 1.2$$

Translation: The reported sample result ($\bar{p} = .28$) is 1.2 standard deviations above the null sampling distribution's hypothesized center of .25, well within the $z_c = 1.64$ standard deviation boundary we had established as the accept H_0/reject H_0 cutoff. (See Figure 8.4.)

Conclusion? Since z_{sample} is inside z_c, we'll accept H_0. Given the sample information presented, we don't have enough evidence to reject the null hypothesis. There simply isn't compelling statistical evidence to convince us (or, presumably, the prospective contributor) that you (the candidate) have the support of more than 25% of the voting population.

Showing the Test on the Original Axis

In Chapter 7 we established that moving the test (specifically, the boundary markers for the test) from the z-scale axis to the original scale (in this case the $\bar{p}$-scale) of the null sampling distribution may make it easier to communicate test details. Here, to set the appropriate right-hand bound (c), we'll simply multiply the 1.64 standard

FIGURE 8.4   **TESTING THE SAMPLE RESULT ON THE Z-SCALE OF THE NULL DISTRIBUTION**

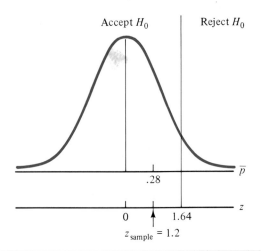

deviation cutoff, z_c, by the size of the standard deviation term (.025) and add the result to the hypothesized distribution center of .25. Accordingly,

$$c = .25 + 1.64(.025) = .291$$

sets the proper accept H_0/reject H_0 cutoff. Any sample result above 29.1% will lead us to reject the null hypothesis; any sample result at or below 29.1% will lead us to accept the null hypothesis. (See Figure 8.5.) Consistent with our earlier conclusion, our sample result of 28% falls clearly below the 29.1% boundary, well within the accept H_0 region of the test.

The Possibility of Error

As was true in the means case, any proportions hypothesis test leaves open the possibility that we might be led into making an error. In fact the same **Type I** and **Type II error** possibilities exist for every hypothesis test since they all use only sample (partial) information.

In our example, Type I error would mean concluding that you have the support of more than 25% of the voter population when you do not. Type II error would mean concluding that you do not have the support of more than 25% of the voter population when you do. (Think back to the general definitions for the two error types. Type I: Rejecting a true null hypothesis. Type II: Accepting a false null hypothesis.)

The Probability of Error

Disregard for the moment the specific sample result reported in our example (i.e., the sample $\bar{p} = .28$) and our consequent decision to accept the null hypothesis. What

FIGURE 8.5   **TESTING THE SAMPLE RESULT ON THE ORIGINAL ($\bar{p}$) SCALE OF THE NULL DISTRIBUTION**

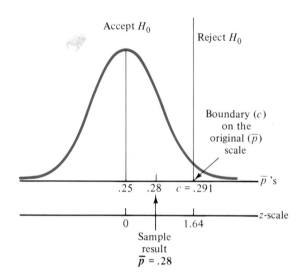

FIGURE 8.6   **COMPUTING A β VALUE FOR THE TEST**

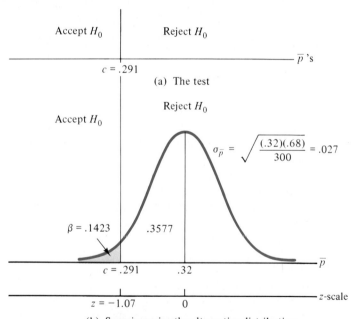

(maximum) probability of Type I error could we assign to the test we've devised? Recall that it's the value for α that routinely measures precisely this sort of risk. By choosing a significance level of .05, we have (implicitly) accepted a 5% chance that if the null hypothesis is true (with π equal to .25), we could nevertheless produce a sample result that would, under our test, lead us to mistakenly reject the null and believe that π is *greater* than .25.

What about the chances of making a Type II error? Suppose, for example, your level of voter population support is actually 32%. How likely is it that we would produce a sample result (from the corresponding alternative sampling distribution) that would lead us into incorrectly accepting the null hypothesis as stated? Figure 8.6 suggests the appropriate procedure for producing the required β value.

About the only feature deserving of special note is the computation of $\sigma_{\bar{p}}$ for the alternative distribution. As shown in the figure,

$$\sigma_{\bar{p}} = \sqrt{\frac{(.32)(.68)}{300}} = .027$$

Notice it's the alternative distribution π of .32, not the null hypothesis π of .25, that's used in the calculation. Why? Since this alternative distribution is premised on the assumption that π is actually .32, our computation of $\sigma_{\bar{p}}(.027)$ must be consistent with this assumption.

EXERCISE 1   Using the test we've devised in the voter example (with $c = .291$ and $n = 300$), compute β, the probability of Type II error, if π is actually .27.

Ans: If $\pi = .27, \beta = .7910$.

(For a more detailed solution, see the end of the chapter.)

EXERCISE 2   Apply the hypothesis testing procedure to the quality control situation described at the beginning of the chapter and restated here for convenience:

Your firm produces units of product in batches of 5000 or more. The batches are shipped to customers throughout the region. The firm has adopted a policy which establishes an acceptable quality level of 5%; that is, any batch that contains no more than 5% defectives overall would be judged OK for shipment to customers. Any batch containing more than 5% defectives would be judged of inferior quality and not shipped until corrective action is taken. Because of high testing costs (and the destructive nature of the testing process), your quality control people have installed a sampling procedure in which a sample of 150 units from each batch will be randomly selected and tested before a decision is made to ship or not ship. Use an α of 1% and show as the competing hypotheses:

$H_0: \pi \le .05$ (the batch overall meets quality standards)
$H_1: \pi > .05$ (the batch overall doesn't meet quality standards)

(a) Show the null sampling distribution, the cutoff z_c on the z-scale, and the cutoff c on the original scale.
(b) Describe the nature of Type I and Type II errors here.
(c) If the proportion of defective items in the overall batch is actually 11%, how likely is it that the test you set up in (a) could mistakenly lead you to believe that the batch is OK?
(d) If you choose a sample of 150 items from a batch and learn that the sample contains 12 defectives (i.e., $\bar{p} = 12/150 = .08$), should the batch be shipped?

Ans: (a) $z_c = 2.33$ $c = .05 + 2.33(.018) = .092$
(c) $\beta = .2451$ if $\pi = .11$
(d) Accept H_0 since $z_{sample} = 1.67 < z_c = 2.33$. (On the original scale, we would accept H_0 since $\bar{p} = .08 < c = .092$. The decision should be to ship the batch. It passes our quality control test.

(For a more detailed solution, see the end of the chapter.)

Using the FPC

As you might expect, if sampling is done without replacement, and the sample size is at least 5% of the population size, using the finite population correction factor in the usual way is in order. In such cases

$$\sigma_{\bar{p}} = \sqrt{\frac{(\pi)(1-\pi)}{n}} \sqrt{\frac{N-n}{N-1}}$$

A Small-Sample Procedure

Refer now to the student public service example outlined earlier. Here, you may recall, there was concern that student interest in public service careers has dropped—that currently less than 20% of the student population at State U. plan this kind of career. We intend to use available sample evidence (a survey of 20 randomly selected students) and a hypothesis testing framework to evaluate this proposition.

We'll begin by acknowledging the comparatively small sample size ($n = 20$) involved in the study. Given the standard large-sample criterion of 30 or more, the sample here clearly qualifies as small, too small, in fact, to allow the normal distribution to play its familiar sampling distribution role. In such a small-sample case involving a population proportion, the sampling distribution we need is actually a **binomial**. (Recall our Chapter 6 discussion.) While we sidestepped the issue of how the binomial might be used in interval estimation, we'll spend some time here seeing how it can be adapted to hypothesis testing.

Forming the Hypotheses

Although introducing the idea of a binomial sampling distribution creates a new twist in the hypothesis testing procedure, the basic method changes very little. We can set the hypotheses in the usual way, choosing as the null the past standard for student commitment, $\pi \geq 20\%$. Accordingly, we'll show

$H_0: \pi \geq .20$ (The population of current students has at least as great a commitment to public service as did their predecessors.)

$H_1: \pi < .20$ (The population of current students has a lesser commitment to public service than did their predecessors.)

As always, we'll hold fast to a belief in the null until sufficient sample evidence is produced to the contrary. (Here, we'll continue to believe that the level of student commitment hasn't diminished unless or until strong sample evidence emerges to force us out of that belief.) And what would strong and compelling evidence consist of here? A sample proportion so unexpectedly low, so unlikely under an assumption that the null hypothesis is true, that when such a result is produced we can no longer defend the null hypothesis as believable. Setting the α-value to specify the standard by which we'll judge sufficiently unlikely sample results (sufficient to reject the null) will enable us to identify an appropriate bound for the test. Here, however, we'll look to the binomial, not the normal distribution, to set the marker.

Picking the Appropriate Binomial

Turn to the binomial table in the back of the text and locate the section where $n = 20$ (to correspond to the sample size of the current example) and $p = .20$ (to correspond to the $\pi = .20$ in the null hypothesis). If the null hypothesis we've proposed is true, then this section of the binomial table describes all possible sample results that might be produced when a sample of size 20 is examined—0 affirmative responses $(x = 0)$, 1 affirmative response $(x = 1)$, 2 affirmative responses $(x = 2)$, and so on. This is precisely the distribution we'll need for our hypothesis test.

> **Note:** Notice that we're reporting our list of all possible sample results in terms of an absolute count of affirmative sample responses $(0, 1, 2, \ldots)$ rather than a relative count $(0\%$ of the sample, 5% of the sample, 10% of the sample, etc.) as we did in the large-sample case using the normal distribution. This doesn't change the fact that the binomial here truly does play the role of the relevant sampling distribution, showing the full range of sample result possibilities.

Separating Likely from Unlikely Sample Results

What sorts of sample results are likely to emerge from this binomial distribution? Checking the table (or referring to Figure 8.7), you can readily see. If, in the overall student population, the proportion of students planning a public service career is 20%, then sample results (using a sample size of 20) will likely be in the vicinity of 3 or 4 or 5 affirmative responses. That is, the most likely results will be in the immediate neighborhood of the binomial distribution mean: np (here, $n\pi$) = $20(.2) = 4$.

Notably, results below 3 are relatively unlikely. In fact, results less than 3 (i.e. 2 or fewer affirmative responses) have a rather low .206 cumulative probability $(.1369 + .0576 + .0115)$ of turning up; results below 2 (i.e. 1 or 0) have

FIGURE 8.7   **THE BINOMIAL SAMPLING DISTRIBUTION WITH π = .20**

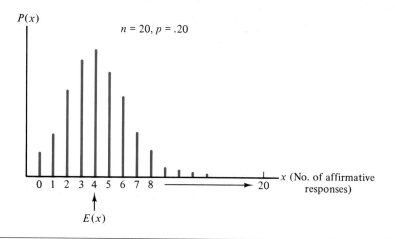

only a .0691 (cumulative) probability. (Be sure you use the table to find these probabilities.)

Suppose now we decide to set a significance level of 10% for the test, specifying that only those sample results having no more than a (cumulative) 10% chance of coming from the null distribution (in this case, those in the lower 10% tail) will cause us to reject the null hypothesis. Our job is to mark the corresponding accept/reject boundary (call it c) on the designated binomial. We'll use trial and error to see how this might be done.

Suppose we start off by setting the boundary at 3, implying that any sample result less than 3 will lead us to reject the null hypothesis and any result of 3 or more will lead us to accept it. Would this constitute a test that imposes a 10% α? The numbers say no. Using 3 as the cutoff identifies 0, 1, and 2 as sample results that would lead to rejecting the null. But, as we saw earlier, these results have a cumulative probability of 0.206—relatively low, yet clearly above the 10% significance level. Implication? Moving the cutoff lower seems in order.

We'll try moving the cutoff to 2. Will *this* boundary properly identify the left-tail results that we need (i.e., those that cumulatively have no more than a 10% chance of emerging randomly from the targeted binomial)? It sure looks like it will. Results of 0 or 1 have only a .0691 chance of selection, well within the 10% significance criterion we've set. Conclusion? To create a test using the 10% significance level, we can establish a boundary (c) at 2 and describe the hypothesis test as follows:

> In a sample of 20 randomly selected students, if the number of students who report interest in a public service career is less than 2, (i.e., 0 or 1) reject the null hypothesis and conclude that interest in public service careers has dropped below the previous norm of 20%. If the number of students who report interest in a public service career is 2 or more, accept the null hypothesis and

FIGURE 8.8   **ESTABLISHING A TEST ON THE BINOMIAL SAMPLING DISTRIBUTION WITH $\pi = .20$**

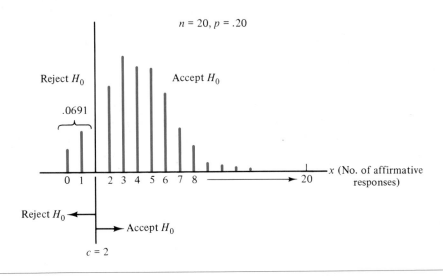

conclude that interest in public service careers has not dropped below the previous norm of 20%. (See Figure 8.8.)

Note: Be sure you're clear on our use of the cutoff c. With $c = 2$, we would accept H_0 for sample results of 2 or more, and reject H_0 for 1 or less. We will always use c as an acceptance number.

Putting the Sample Result to the Test

According to this test, our original sample result of one affirmative response in a random sample of 20 students would be judged (at the 10% significance level) sufficient evidence to reject H_0. As a consequence, we can use this 20-student study to challenge the commitment of the current student population.

EXERCISE 3   Suppose we had used a significance level of 5% in our "public service" example.

(a) Identify the appropriate boundary for the test.
(b) Determine whether our sample result of 1 in 20 would lead us to reject the null position.

Ans: (a) $c = 1$ (i.e., accept H_0 for 1 or more affirmative sample responses, reject for 0)
(b) At the 5% level, the null would not be rejected

(For a more detailed solution, see the end of the chapter.)

EXERCISE 4 Reread Exercise 2 and reconsider the quality control situation described there. Assume now that the sample size involved is 15 units. Set up a test for the $\pi \leq .05$ null hypothesis, using a significance level of 1% and report your acceptance number c. What would you conclude if the sample of 15 had 4 defectives?

Ans: (1) $c = 3$ (accept at 3, reject at 4) since $P(x > 3/n = 15, p = .05) = .0049 + .0006 = .0055$.

(2) If the number of defectives in the sample is 4, we would reject H_0.

(For a more detailed solution, see the end of the chapter.)

Note: To be technically correct, using the binomial sampling distribution is legitimate only when sampling is done with replacement. (Can you think why this should be true? It has to do with one of the binomial assumptions.) However, if the population is very large compared to the size of the sample, the binomial will work even if sampling is done without replacement.

8.2 TESTS FOR THE DIFFERENCE BETWEEN TWO POPULATION MEANS

Situation: It is frequently argued that women, on average, aren't paid the same wages as men doing comparable jobs. You want to test that proposition. Specifically, you want to see if within your own company there is statistical evidence to support this contention.

In its assembly and fabrication operations nationwide, your firm employs more than 2500 men and 2500 women who have comparable skill levels, prior training, and experience. You select a simple random sample (with replacement) of 120 men and 120 women to represent the male/female populations here. The results of the study show average monthly wages of $1680 for the men in the sample (with sample standard deviation of $182) and $1590 for the women (with standard deviation of $348). Do such results represent convincing sample evidence of a company-wide pay disparity? We can take a hypothesis testing approach to find out.

Forming the Hypotheses

The essential question is clear: Are male and female workers in the company paid the same average wage, or would we find, in looking at the entire population of men and the entire population of women working here, different average salaries? Statistically speaking, what we have is a case in which we need to test the **difference between two population means**.

In such a mean difference case, the null hypothesis is routinely selected to reflect the "no difference" position, putting the burden of proof on those contending that a difference does indeed exist. Accordingly, with

μ_1 = population average salary for all the men in the company

μ_2 = population average salary for all the women in the company we'll show

$$H_0: \mu_1 = \mu_2 \qquad \text{(there's no difference in the average salaries for men and women in the company)}$$

$$H_1: \mu_1 \neq \mu_2 \qquad \text{(there is a difference in average salaries for men and women in the company)}$$

Usefully, these hypotheses can be converted to the equivalent

$$H_0: \mu_1 - \mu_2 = 0 \qquad \text{(the difference in average salaries is 0)}$$

$$H_1: \mu_1 - \mu_2 \neq 0 \qquad \text{(the difference in average salaries is not 0)}$$

Anticipating Possible Sample Results

Suppose we had selected the two required samples of size 120 and found that the difference in the sample mean monthly salaries (call it $\bar{x}_1 - \bar{x}_2$) was actually $1,440,000. Would this sample result allow us to believe the null hypothesis as stated? That is, in light of such a dramatically large sample mean difference, could we possibly believe that the difference in the population means is 0? Clearly not! This sort of sample result would simply appear so unlikely under an assumption that the "no difference" null hypothesis is true that, given such a sample result, we could no longer believe that the null hypothesis is true. (We've been down this road before.)

And just how unlikely would such a sample result be? Knowledge of the appropriate sampling distribution (as usual) holds the key.

The Sampling Distribution

Not surprisingly, we'll need to use the sampling distribution of all possible sample mean differences. As we saw in Chapter 6, this is a distribution that looks normal (for large enough samples), is centered on the population mean difference ($\mu_1 - \mu_2$), and has a standard deviation

$$\sigma_{\bar{x}_1 - \bar{x}_2} = \sqrt{\frac{\sigma_1^2}{n_1} + \frac{\sigma_2^2}{n_2}}$$

or, when the population σ's are unknown,

$$\sqrt{\frac{s_1^2}{n_1} + \frac{s_2^2}{n_2}}$$

(See Figure 8.9.)

The Null Sampling Distribution

If the null hypothesis in our current problem is true, then the sampling distribution of sample mean differences can be properly centered on $\mu_1 - \mu_2 = 0$ and labeled the *null distribution* (see Figure 8.10).

FIGURE 8.9 **THE SAMPLING DISTRIBUTION OF SAMPLE MEAN DIFFERENCES**

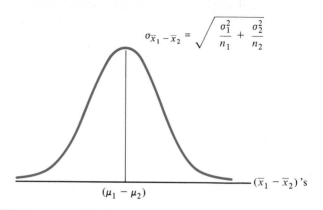

$$\sigma_{\bar{x}_1 - \bar{x}_2} = \sqrt{\frac{\sigma_1^2}{n_1} + \frac{\sigma_2^2}{n_2}}$$

$(\mu_1 - \mu_2)$

$(\bar{x}_1 - \bar{x}_2)$'s

FIGURE 8.10 **THE "NO DIFFERENCE" NULL SAMPLING DISTRIBUTION OF SAMPLE MEAN DIFFERENCES**

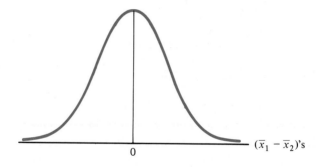

0

$(\bar{x}_1 - \bar{x}_2)$'s

Separating Likely from Unlikely Sample Results

Is this distribution likely to yield a sample mean difference value of $1,440,000? Hardly. Likely results—the sort that we would expect to produce through simple random selection—lie clearly back in the vicinity of the distribution center 0. Unlikely results appear (for this two-tailed case) in either tail. If we define precisely what we mean by unlikely results, we can establish boundaries to effectively identify these values in the upper and lower extremes. Setting α does the trick once again.

The Test

Suppose we set α, the significance level for the test, equal to a conventional 5%. Then any sample result (here, any sample mean difference) that falls in either of the 5%

FIGURE 8.11   **SETTING Z-SCORE BOUNDARIES ON THE NULL DISTRIBUTION**

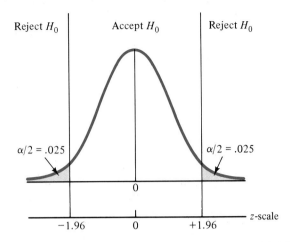

(actually 2.5%) extremes of the null sampling distribution will be judged too unlikely to allow for continued belief in the "no difference" null.

Setting specific (z_c) markers for the 5% extremes is straightforward enough. In this two-tailed test, we'll simply halve the 5% α level and establish an upper boundary, c_U, above which we should find only 2.5% of the values in the null distribution, and a lower boundary, c_L, below which we'll find a similar 2.5%. Setting markers at 1.96 standard deviations to either side of 0 will identify the proper bounds; any sample result beyond 1.96 standard deviations will lead us to reject the null hypothesis. (See Figure 8.11.)

Putting Our Sample Result to the Test

Its seems clear that we can dismiss the hypothetical $1,440,000 mean difference as lying rather far out in the right-hand reject H_0 region of the test—more than a little bit beyond the $z_c = 1.96$ standard deviation cutoff. But what about our original sample result? With a male sample of size 120 showing an average monthly salary ($\bar{x}_1$) of $1680 and a female sample of similar size showing an average monthly salary ($\bar{x}_2$) of $1590, we have a sample mean difference ($\bar{x}_1 - \bar{x}_2$) of $90. Is this sufficient to reject the proposition that there is no difference in the population averages represented? (According to our test, if this sample mean difference is beyond 1.96 standard deviations from 0, it will be judged sufficient to reject H_0.)

As usual, we need only compute a z score for the available sample result and compare it to the accept/reject cutoff z_c to reach a conclusion. In this case,

$$z_{\text{sample}} = \frac{(\bar{x}_1 - \bar{x}_2) - 0}{\sigma_{\bar{x}_1 - \bar{x}_2}}$$

where 0 represents the center of the null hypothesis sampling distribution. Omitting the 0 in the numerator, we can simply write,

$$z_{sample} = \frac{\bar{x}_1 - \bar{x}_2}{\sigma_{\bar{x}_1 - \bar{x}_2}}$$

Using the sample standard deviations, $s_1 = \$182$ and $s_2 = \$348$, to compute (estimate) the standard error term $\sigma_{\bar{x}_1 - \bar{x}_2}$ produces

$$\sigma_{\bar{x}_1 - \bar{x}_2} = \sqrt{\frac{s_1^2}{n_1} + \frac{s_2^2}{n_2}} = \sqrt{\frac{(182)^2}{120} + \frac{(348)^2}{120}} = \sqrt{1285.2} = 35.8$$

Substituting this value allows us to compute the appropriate z score for the $90 sample mean difference:

$$z_{sample} = \frac{1680 - 1590}{35.8} = 2.51$$

Conclusion? Since the sample result lies more than 1.96 standard deviations from the null hypothesis center (0)—that is, since $z_{sample} > z_c$—we'll judge the difference in sample means to be statistically significant at the 5% significance level. In other words, we'll use the sample mean difference here as sufficient evidence to reject the "no difference" null hypothesis.

Showing the Test on the Original Axis

As in the past, we may find it useful to show the test on the original scale (in this case the "average salary difference" scale) of the null distribution. All that's required is a quick translation of the $z_c = 1.96$ z-score boundaries to the equivalent average salary difference amounts. Multiplying 1.96 by the standard error term, $\sigma_{\bar{x}_1 - \bar{x}_2} = 35.8$, produces the measure we need (70.17). We can set the upper bound, c_U, $70.17 above the hypothesized center of 0, and the lower bound, c_L, $70.17 below the hypothesized center of 0. (See Figure 8.12.)

The test is now easily summarized:

Take a simple random sample of 120 employees from the male worker population and a simple random sample of 120 employees from the female worker population. Compute the difference in sample means. If the difference in sample means is greater than $70.17 (either way), reject the contention that men and women employees in the populations represented are paid the same average monthly salary.

Since our sample mean difference of $90 clearly falls in the reject H_0 region of the test, we judge the sample result statistically significant and reject the null.

The Possibility of Error

As was true in the other hypothesis testing cases we've considered, the chance that our testing procedure may lead us into making an error needs to be acknowledged.

FIGURE 8.12   **SHOWING TEST BOUNDARIES ON THE ORIGINAL SCALE OF THE NULL DISTRIBUTION**

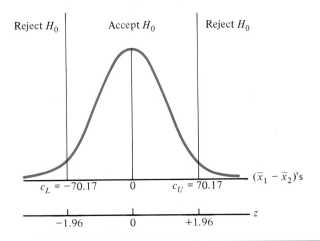

Reject H_0 Accept H_0 Reject H_0

$c_L = -70.17$ 0 $c_U = 70.17$ $(\bar{x}_1 - \bar{x}_2)$'s

-1.96 0 $+1.96$ z

Type I error, in this case, would involve mistakenly concluding that there is a difference in average salaries for men and women when there is not (i.e., rejecting a true null hypothesis). Type II error would involve believing there's no difference in average salaries when there is (i.e., accepting a false null hypothesis).

The Probability of Error

As always, α represents the (maximum) probability that our test could lead us into making a Type I error. Try the exercise below to assess your ability to compute β.

EXERCISE 5   Using the test we've established for the average monthly salary example (with $\alpha = .05$), determine the probability that this test would lead you into a Type II error if the population mean monthly salary difference for men and women $(\mu_1 - \mu_2)$ is actually $50.

Ans: If $\mu_1 - \mu_2 = 50$, $\beta = .7123$.

(For a more detailed solution, see the end of the chapter.)

The FPC

When sampling is done without replacement, and sample size is at least 5% of the population size, the FPC plays its usual role:

$$\sigma_{\bar{x}_1 - \bar{x}_2} = \sqrt{\frac{\sigma_1^2}{n_1}\left(\frac{N_1 - n_1}{N_1 - 1}\right) + \frac{\sigma_2^2}{n_2}\left(\frac{N_2 - n_2}{N_2 - 1}\right)}$$

The Small-Samples Case When σ's Are Unknown

When sample sizes are small (a combined size of less than 30) and the population standard deviations are unknown, the only adjustments that need to be made involve the introduction of the **t distribution** (with $n_1 + n_2 - 2$ degrees of freedom) to replace the normal sampling distribution. Rather than producing z scores in the testing procedure, we'll generate the corresponding t scores.

Sample standard deviations will, in these small-samples cases, be **pooled** (recall Chapter 6) to estimate the common population standard deviation that must be assumed. Accordingly, we'll use

$$s_{pooled} = \sqrt{\frac{(n_1 - 1)s_1^2 + (n_2 - 1)s_2^2}{n_1 + n_2 - 2}}$$

EXERCISE 6 In the average salary difference example, assume samples of 12 men and 12 women were selected, with the same means ($\bar{x}_1 = 1680$, $\bar{x}_2 = 1590$) and standard deviations ($s_1 = 182$, $s_2 = 348$) as before. Using a significance level of 5%, set up the appropriate hypothesis test on the t-axis of the "no difference" null sampling distribution. (You can assume that the populations represented here are normal and have equal standard deviations.)

(a) Sketch the appropriate null t distribution.
(b) Produce the cutoff t-score value, t_c, for the 5% significance level.
(c) Compute the pooled estimate of the population standard deviation.
(d) Compute the t score for the sample result $\bar{x}_1 - \bar{x}_2 = \$90$.
(e) Based on your t_{sample} computation, what is your conclusion?
(f) Now show the test on the original axis. Report the cutoff values c_L and c_U.

Ans: (b) $t_c = \pm 2.074$
(c) $s_{pooled} = 277.7$
(d) $t_{sample} = 90/113.4 = .794$ (Note: This is a t score, not a probability.)
(e) Since .794 is inside 2.074, accept the null.
(f) $2.074 \times 113.4 = 235$; so $c_L = -\$235$, $c_U = +\$235$.

(For a more detailed solution, see the end of the chapter.)

Note: You may want to change the "no difference in average salary" test to a one-tailed version in which the alternative hypothesis asserts that women are paid *less* than men for comparable jobs. This would be more consistent with commonly expressed concerns about sex discrimination in the workplace.

8.3 TESTS FOR THE DIFFERENCE BETWEEN TWO POPULATION PROPORTIONS

Last in our discussion of hypothesis testing applications is the **proportions difference** case. If you've persevered through our earlier discussions, then the method, the calculations, the twists, and the turns should be fairly instinctive.

Situation: Dazzling Premium Select, Inc., a state-of-the-art, high-tech supplier of production equipment, has recently approached you about replacing your current supplier, Dismal-But-Sincere Manufacturing. Dazzling marketing director, C. M. "Butch" Waffle, contends that the proportion of satisfied Dazzling customers far exceeds the proportion of satisfied Dismal customers. You decide to put these Dazzling claims to the test. To test for a difference in customer satisfaction, you select a random sample of 100 Dazzling customers and a sample of 120 Dismal customers, and find that 80 customers (80%) in the Dazzling sample and 90 customers (75%) in the Dismal sample reported complete satisfaction with their supplier. In light of these sample results, how does the Dazzling claim of greater customer satisfaction hold up?

Forming the Hypotheses

As we've discussed previously, when someone is making claims of superior performance, faster delivery, higher quality, and the like, it's common practice to set the null hypothesis to reflect a distinctly skeptical view of such claims. Identifying the Dazzling population as population 1 and the Dismal population as population 2, we'll consequently show

$H_0: \pi_1 \leq \pi_2$ (the proportion of satisfied customers in the Dazzling population of clients is *no greater than* the proportion of satisfied customers in the Dismal population)

$H_1: \pi_1 > \pi_2$ (the proportion of satisfied customers in the Dazzling population is greater than the proportion of satisfied customers in the Dismal population)

In choosing the "no greater than" null, we're purposely putting the burden of proof on the maker of the claims (in this case, the Dazzling people).

Adhering to the pattern of the previous section, we can translate these hypotheses to the equivalent

$H_0: \pi_1 - \pi_2 \leq 0$ (the difference in the two population satisfaction levels (proportions) is no greater than 0)

$H_1: \pi_1 - \pi_2 > 0$ (the difference in the two population satisfaction levels (proportions) is greater than 0)

The Sampling Distribution

Given the nature of the situation, the sampling distribution of all possible sample proportion differences provides the necessary framework for our test. As we've seen, this is a distribution that's normal in shape (for sufficiently large samples), centered on the population proportion difference ($\pi_1 - \pi_2$), and has a standard deviation (or standard error)

$$\sigma_{\bar{p}_1 - \bar{p}_2} = \sqrt{\frac{\pi_1(1 - \pi_1)}{n_1} + \frac{\pi_2(1 - \pi_2)}{n_2}}$$

(See Figure 8.13.)

FIGURE 8.13   **THE SAMPLING DISTRIBUTION OF SAMPLE PROPORTION DIFFERENCES**

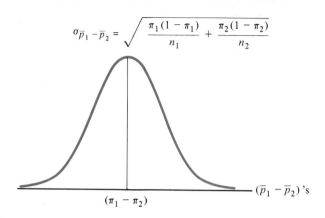

$$\sigma_{\bar{p}_1 - \bar{p}_2} = \sqrt{\frac{\pi_1(1 - \pi_1)}{n_1} + \frac{\pi_2(1 - \pi_2)}{n_2}}$$

$(\pi_1 - \pi_2)$ $(\bar{p}_1 - \bar{p}_2)$'s

FIGURE 8.14   **THE "NO DIFFERENCE" NULL SAMPLING DISTRIBUTION OF SAMPLE PROPORTION DIFFERENCES**

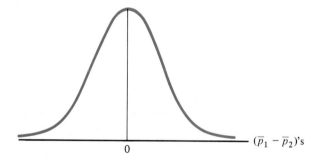

0    $(\bar{p}_1 - \bar{p}_2)$'s

The Null Sampling Distribution

If we assume that the null hypothesis is true, then the sampling distribution will be centered on 0 and designated the null distribution.

Setting the Accept/Reject Boundary

At this point, it should be apparent that we'll look to the upper tail of the null distribution to identify the sorts of sample results that would lead us to reject the null hypothesis. (Make sure this one-tailed approach makes sense to you.) To illustrate, we'll set α to 1% and proceed to draw the corresponding boundary, z_c, 2.33 standard deviations to the right of center (i.e., 2.33 standard deviations above 0). (See Figure 8.15.) Once this bound is set, any sample proportion difference that falls more than 2.33 standard deviations above 0 will be judged sufficient sample

FIGURE 8.15   **SETTING THE Z-SCORE BOUNDARY ON THE NULL DISTRIBUTION**

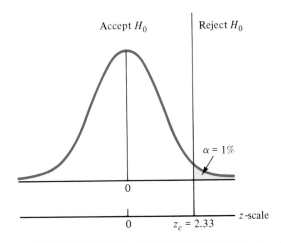

evidence to reject the null; any sample proportion inside the 2.33 standard deviation cutoff will be judged insufficient to reject the null.

Putting Our Sample Result to the Test

How does our sample proportion difference of .05 (.80 − .75) hold up under the proposed test? Simply compute the z score for .05 and compare it to the $z_c = 2.33$ standard deviation bound. Here,

$$z_{sample} = \frac{(\bar{p}_1 - \bar{p}_2) - 0}{\sigma_{\bar{p}_1 - \bar{p}_2}}$$

$$= \frac{.80 - .75}{\sigma_{\bar{p}_1 - \bar{p}_2}}$$

where 0 = center of the null distribution

$\sigma_{\bar{p}_1 - \bar{p}_2}$ = standard deviation of the null distribution

All we need is the value of the standard deviation (standard error)

$$\sigma_{\bar{p}_1 - \bar{p}_2} = \sqrt{\frac{\pi_1(1 - \pi_1)}{n_1} + \frac{\pi_2(1 - \pi_2)}{n_2}}$$

for the null distribution to complete the procedure.

Unfortunately, the required standard error computation is a bit more complicated than you might first expect. Although we did it routinely in Chapter 6, simply substituting sample proportions $\bar{p}_1$ and $\bar{p}_2$ for population proportions π_1 and π_2 in the standard error expression is, technically at least, inappropriate here. Instead, to ensure a standard error term that is perfectly consistent with the null hypothesis, we'll follow a slightly different course in providing the π_1 and π_2 estimates.

Computing the Standard Error of the Null Distribution

Since the basis for our "no difference" null distribution, the one centered on $\pi_1 - \pi_2 = 0$, is the premise that the population proportions are equal, we need to determine (or at least estimate) the one value (call it π) to which these two proportions are both equal. (In other words, by showing a null distribution centered on 0, the implication is that π_1 and π_2 are equal. The question is: equal to what?) Once determined, this common π could be used in place of the individual π_1 and π_2 values in the standard error expression to compute the standard error of the null distribution.

If this procedure is truly appropriate, the only real issue is how best to estimate the common π. (Remember, we're assuming $\pi_1 = \pi_2 = \pi$, but we don't know any of these *population* values. We only have sample results.) Not surprisingly, we'll use the available sample information to provide a reasonable guess. In fact, we can average (or pool) the sample proportions ($\bar{p}_1$ and $\bar{p}_2$) to produce a legitimate estimate of our common population proportion. Similar to the pooling of standard deviations in the small-sample mean differences case, the prescribed procedure here would have us compute a weighted average of the sample proportions according to the expression

$$\bar{p}_{\text{pooled}} = \frac{n_1\bar{p}_1 + n_2\bar{p}_2}{n_1 + n_2}$$

For our example, this means

$$\bar{p}_{\text{pooled}} = \frac{100(.80) + 120(.75)}{100 + 120} = .773$$

Note: We could have obtained the same result by combining (pooling) the number of satisfied customers in the two samples ($80 + 90 = 170$) and dividing this amount by the total of the two sample sizes ($100 + 120 = 220$). This direct approach may give a better sense of what's meant by the term *pooling*.

Substituting .773 for π_1 and π_2 in the general standard error expression produces a null distribution standard error of

$$
\begin{aligned}
\sigma_{\bar{p}_1 - \bar{p}_2} &= \sqrt{\frac{\bar{p}_{\text{pooled}}(1 - \bar{p}_{\text{pooled}})}{n_1} + \frac{\bar{p}_{\text{pooled}}(1 - \bar{p}_{\text{pooled}})}{n_2}} \\
&= \sqrt{\frac{.773(1 - .773)}{100} + \frac{.773(1 - .773)}{120}} \\
&= \sqrt{.00175 + .00146} \\
&= .057
\end{aligned}
$$

Completing the Test for Our Sample Result

We can now complete the test for our $\bar{p}_1 - \bar{p}_2 = .05$ sample result. Substituting the appropriate standard error term $\sigma_{\bar{p}_1 - \bar{p}_2} = .057$ into the z-score computation

FIGURE 8.16   **SHOWING THE BOUNDARY ON THE ORIGINAL SCALE OF THE NULL DISTRIBUTION**

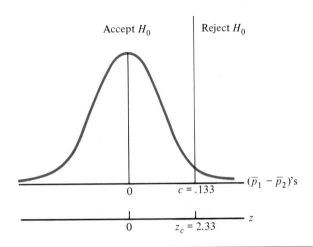

produces

$$z_{\text{sample}} = \frac{.80 - .75}{.057} = \frac{.05}{.057} = .877$$

a value well within the accept/reject cutoff of 2.33 standard deviations that had been established for the test (remember, we had selected a significance level of 1% for this one-tailed test, producing a $z_c = 2.33$).

Conclusion? Although there's clearly a sample proportion difference (.05), this difference would not be judged statistically significant. That is, the sample difference is not sufficient to establish a difference in the *population* proportions represented here. Based on these results, we could reasonably hold to a belief in the null hypothesis, dismissing the sample difference as simply due to routine (random) variation in the null sampling distribution. Put simply, the Dazzling claim isn't substantiated by sample results.

Showing the Test on the Original Axis

In the usual fashion, we can translate the test to the original axis simply by multiplying the z-score cutoff ($z_c = 2.33$) by the size of the standard error (.057) and adding the result ($2.33 \times .057 = .133$) to the null hypothesis center (0). Setting the boundary 2.33 standard deviations above 0 consequently translates into an accept/reject cutoff (c) of .133 on the "difference in sample proportion of satisfied customers" axis appropriate to the problem. (See Figure 8.16.) Obviously, our sample result of .05 falls well within the accept H_0 region of the test. Again we see that sample results don't (statistically) establish the validity of Dazzling's claim.

Errors and FPCs

Identifying error possibilities, calculating risk levels, and the like, all work as before. Finite population correction factors will be used to reduce the standard error term under the usual conditions. Thus

$$\sigma_{\bar{p}_1 - \bar{p}_2} = \sqrt{\frac{\pi_1(1 - \pi_1)}{n_1}\left(\frac{N_1 - n_1}{N_1 - 1}\right) + \frac{\pi_2(1 - \pi_2)}{n_2}\left(\frac{N_2 - n_2}{N_2 - 1}\right)}$$

when sampling is done without replacement, and sample sizes are at least 5% of the population sizes.

Small Sample Cases

As we did in Chapter 6, we'll sidestep the complications associated with the small-sample proportions difference case, leaving this issue to another time and place. To use the approach we've described, it's recommended that the two sample sizes both be at least 100 and that in each case $np \geq 5$ and $n(1 - p) \geq 5$ to ensure the normal shape of the sampling distribution.

▄▄▄▄▄▄▄ CHAPTER SUMMARY

We sought to extend our discussion of statistical hypothesis testing to three additional cases:

- Hypothesis tests for a population proportion
- Hypothesis tests for the difference between two population means
- Hypothesis tests for the difference between two population proportions

Following the pattern established in Chapter 7, we determined that characteristics of the appropriate sampling distribution play a crucial role in reaching conclusions about a target population, using only sample results.

In each case we saw a general hypothesis testing procedure that involved drawing boundaries to separate the sorts of sample results that would allow us to believe the null hypothesis (the primary position being tested) from those sample results that would essentially force us to abandon that position. While specifics differed in each case, this common theme persisted. In a minor twist, we saw the binomial distribution used to construct tests for a population proportion when the sample size is small.

Measuring the likelihood of error was recognized as a key part of any hypothesis testing procedure. As we saw in Chapter 7, the significance level α for any test measures the (maximum) chance of a Type I error—incorrectly rejecting a true null hypothesis—and β represents the chance of making a Type II error—incorrectly accepting a false null hypothesis.

KEY TERMS

Binomial distribution in hypothesis tests
Pooled standard deviation
Sampling distribution
  of proportions
  of the difference between means
  of the difference between proportions
Significance level
t distribution in hypothesis testing
Type I error
Type II error

IN-CHAPTER EXERCISE SOLUTIONS

EXERCISE 1

The test we've devised for

$$H_0: \pi \le .25$$
$$H_1: \pi > .25$$
$$n = 300$$

is basically

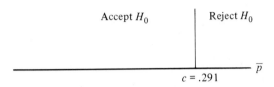

If $\pi = .27$, we can superimpose the corresponding alternative sampling distribution on the test axis:

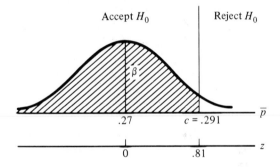

We can compute

$$\sigma_{\bar{p}} = \sqrt{\frac{(\pi)(1 - \pi)}{300}} = \sqrt{\frac{(.27)(.73)}{300}} = .026$$

and

$$z = \frac{.291 - .27}{.026} = .81$$

From the normal table, a z score of .81 corresponds to a probability of .2910. The area representing β in the figure is therefore $.5000 + .2910 = .7910$.

EXERCISE 2

$$H_0: \pi \le .05 \quad \text{(batch is OK)}$$
$$H_1: \pi > .05 \quad \text{(batch is not OK)}$$

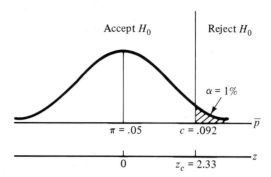

(a) $\sigma_{\bar{p}} = \sqrt{\dfrac{.05(.95)}{150}} = .018$

so $c = .05 + 2.33(.018) = .092$.

(b) Type I: concluding that the batch is not OK when it is OK.
Type II: concluding that the batch is OK when it isn't.

(c) Superimposing the alternative sampling distribution (centered on .11) on the original test axis

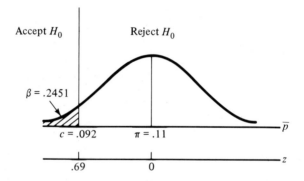

To find the area in the alternative distribution that lies in the accept H_0 region of the test, first compute

$$\sigma_{\bar{p}} = \sqrt{\dfrac{.11(.89)}{150}} = .026$$

then

$$z = \dfrac{.11 - .092}{.026} = .69$$

From the normal table, the area for $z = .69$ is .2549, so β must be $.5 - .2549 = .2451$.

(d) $z_{\text{sample}} = (.08 - .05)/.018 = 1.67$, compared to the z_c of 2.33.

EXERCISE 3

(a) For a significance level of 5%, we want to find a c such that the probability of the null binomial sampling distribution ($n = 20$, $p = .20$) yielding a value less than that c is no more than .05. Using the binomial table, you should find setting $c = 1$ accomplishes this goal (i.e., $P(x < 1/n = 20, p = .20) = P(x = 0/n = 20, p = .20) = .0115$). Visually,

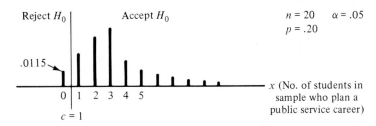

(b) Since, according to the test we've set up, any sample result of 1 or more will lead us to accept the null (i.e., no change in student commitment), the sample result of 1 in 20 is not a significant challenge to the null position (at the 5% significance level).

EXERCISE 4

$$H_0: \pi \le .05$$
$$H_1: \pi > .05$$

The null hypothesis sampling distribution is a binomial with $n = 15$, p (or π) $= .05$.

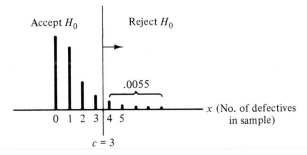

With a significance level of .01 for this one-tailed test, we want to isolate the upper 1% of the values. Using the appropriate binomial table, you should be able to see that $P(x > 3/n = 15, p = .05)$ or $P(x \ge 4/n = 15, p = .05)$ is equal to $.0049 + .0006 = .0055$, a value just below the targeted significance level of .01. Had we chosen $c = 2$, $P(x > 2/n = 15, p = .05) = .0307 + .0049 + .0006 = .0362$, a value that exceeds the targeted significance level of .01. Consequently, we use a c of 3.

EXERCISE 5

The test we've established for

$$H_0: \mu_1 - \mu_2 = 0$$
$$H_1: \mu_1 - \mu_2 \ne 0$$

looks like

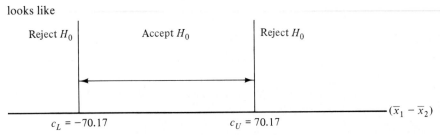

Superimposing the alternative distribution, with $\mu_1 - \mu_2 = +50$,

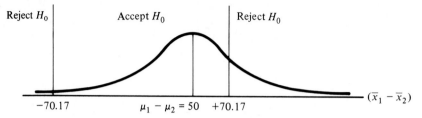

we need to establish the area in this alternative distribution that lies in the accept H_0 region of the test:

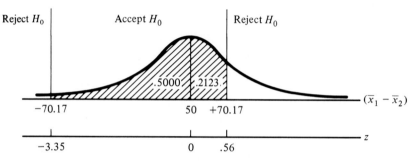

For the upper bound,

$$z = \frac{70.17 - 50}{35.8} = .56$$

which produces from the normal table a probability of .2123. Adding this to .5 yields a β value of 0.7123.

Note: Technically, we should subtract from our result the area in the alternative distribution that falls below -70.17. However, this area is so close to 0, that it would have no substantive effect on our calculation.

EXERCISE 6

(a) $H_0: \mu_1 - \mu_2 = 0$
     $H_1: \mu_1 - \mu_2 \neq 0$

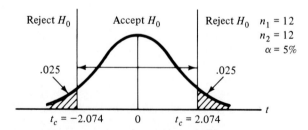

(b) Using the t table for a tail-end area of .025 and $df = 12 + 12 - 2 = 22$ produces $t_c = 2.074$.

(c) $s_{pooled} = \sqrt{\dfrac{(11)(182)^2 + (11)(348)^2}{12 + 12 - 2}}$

$= 277.7$

(d) Using the pooled standard deviations to compute the standard error of the mean differences sampling distribution produces

$$\sigma_{\bar{x}_1 - \bar{x}_2} = \sqrt{\dfrac{(277.7)^2}{12} + \dfrac{(277.7)^2}{12}}$$

$$= 113.4$$

So, $t_{sample} = 90/113.4 = .794$ (This is a t score, not a probability.)

(e) Since .794 is inside 2.074 (i.e., $t_{sample} < t_c$), accept the null.

(f) $2.074 \times 113.4 = 235$; so lower bound $c_L = -\$235$ and upper bound $c_U = +\$235$.

The test looks like

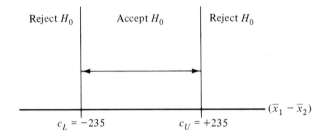

EXERCISES

Assume sampling is being done with replacement unless otherwise noted. If the population size is not specified, assume it's very large.

(Exercises 1 through 14 involve hypothesis tests for a population proportion.)

★ 1. In years past, 40% of the workers in your industry were union members. It is believed that this percentage has dropped substantially. You plan to take a simple random sample of 100 current workers to evaluate the situation. Specifically, you plan to test the hypotheses

$H_0: \pi \geq .4$ (at least 40% of all workers are union members)

$H_1: \pi < .4$ (less than 40% of all workers are union members)

using a significance level of 1%.

(a) Show the null sampling distribution.

(b) Show the 1% boundary for your test on the z-scale.

(c) Suppose the sample proportion turns out to be .26. Compute the z score for this sample result and compare it to the z-score cutoff for your test. Based on this comparison, should you reject the null hypothesis? Explain.

(d) Show the test on the original scale of the null sampling distribution.

2. A recent newspaper article suggested that (at least) 20% of the new construction projects in the city have been financed by foreign investors. You take a random sample of 150 current projects and find that 22 of them were financed by foreign investors. Is this enough evidence to counter the newspaper article's contention? Use a significance level of 5%.

(a) Show the null sampling distribution.

(b) Identify the accept/reject cutoff (boundary) for your test on both the z-scale and the original ($\bar{p}$) scale.

(c) Compute the z score for the sample result and compare it to the z-score cutoff for the test. Is your one sample result (14 out of 150) sufficient to challenge the null hypothesis?

(d) Show your test on the original scale.

★ 3. Historically, no less than 15% of the student population at Old State Tech have been minority students. There is growing concern that this percentage has fallen in recent years. You take a simple random sample of 150 students on campus and find that only 14 students in the sample could be categorized as minority students. Does this sample result represent sufficient sample evidence to challenge the historical figure of 15%? Set up an appropriate hypothesis test, using the historical position (the "no change" position) $\pi \geq .15$ as the null hypothesis. Set the significance level at 1%.
   (a) Show the null sampling distribution.
   (b) Identify the accept/reject cutoff (boundary) for your test on both the z-scale and the original ($\bar{p}$) scale.
   (c) Compute the z score for the sample result and compare it to the z-score cutoff for the test. Is your one sample result (14 out of 150) sufficient to challenge the null hypothesis?
   (d) Show your test on the original scale.

4. A recent *DayTime* magazine editorial stated that no more than 20% of the American electorate would support a Constitutional amendment restricting the states' right to control abortions. You believe the percentage is higher. In a recent survey of 500 randomly selected voters, 120 expressed support for such an amendment. Is this sufficient sample evidence to refute *DayTime's* assertion? Set up an appropriate hypothesis test, using *DayTime's* position as the null hypothesis. Test at the 5% significance level.

5. The company that produces the scratch-and-win tickets for The Filthy-Rich Instant Win Super Lottery has assured you that exactly 10% of the tickets are winners (i.e., will pay the customer at least a dollar). As an inspector for The Filthy-Rich Instant Win Super Lottery, Inc., you want to be sure this is precisely the case: too high a percentage of winners would result in excessive payoffs; too few winners might bring charges of consumer fraud and deception. You select a random sample of 200 tickets from a large order of tickets recently delivered by the ticket supplier and find that 28 of the tickets are winners. Set up a hypothesis test to determine whether this is sufficient sample evidence to reject the supplier's claim that overall the order contains 10% winners. Use a significance level of 1%.

★ 6. If more than 20% of your 250 retail outlets plan to reorder your company's new video games parabolic enhancer, you will radically increase production for the upcoming month. You contact 50 randomly selected outlets (using sampling without replacement) and find that 14 of the outlets plan to reorder. Set up a statistical hypothesis test to determine whether this is sufficient sample evidence to reject a skeptical $\pi \leq .20$ null hypothesis. (π represents the proportion all 250 outlets who plan to reorder.) Use a significance level of 5%.

7. Blisset Dental Supply has a client list of 540 dentists throughout the state. A recent business magazine article has asserted that at least 30% of all dentists have had declining revenues during the past five years. Using sampling without replacement, you survey 100 of Blisset's clients and find that only 18% of those surveyed report declining revenues. Is this sufficient sample evidence to reject the $\pi \geq .30$ for the overall population of Blisset clients? Use a significance level of 1%.

★ 8. Your cousin Howard claims that he can predict with uncanny accuracy the outcome of a coin toss. He has you toss a coin 100 times and correctly predicts the outcome of the toss in 61 cases. Set up an appropriate hypothesis test to help determine whether Howard can predict coin toss outcomes any better than a "normal" person. (*Hint:* Use as your null hypothesis $\pi \leq .5$, a position that says Howard is just randomly guessing each outcome (just like an ordinary citizen), and asserting that his chance of a successful guess on any toss is only 50%. Use a significance level of 5%.
   (a) State the null and alternative hypotheses.
   (b) Show the null sampling distribution.
   (c) Identify the accept/reject cutoff (boundary) for your test on the z-scale and compute the z score for the sample result. What do you think of Howard's performance?
   (d) Show your test on the original scale and report your conclusion.

9. Recently 80,000 signatures have been collected for a proposed ballot measure in the upcoming state election. The secretary of state requires validation of the signatures to ensure against fraud, duplication, incorrect addresses, and so on. In all, the law requires that more than 60,000

valid signatures be collected before any measure will qualify for inclusion on the ballot.

Rather than requiring validation of all the signatures submitted, standard procedure involves taking a simple random sample of the submitted signatures and validating only the signatures in the sample. For the current measure, a sample of 1000 (of the 80,000) signatures is selected, and 78% of the signatures in the sample prove valid. Should the measure be included on the ballot?

Set up the appropriate hypothesis test, using the skeptical "not-enough-signatures" position as the null hypothesis and a significance level of 5%. (*Hint:* What minimum percentage of the 80,000 signatures would have to be valid in order to meet the 60,000+ valid-signature requirement? *Answer:* more than 75%, since $(.75)(80,000) = 60,000$. Use the skeptical null hypothesis, $H_0: \pi \leq 0.75$.)

(a) State the null and alternative hypotheses in words and in symbols.

(b) Show the null sampling distribution.

(c) Identify the accept/reject cutoff (boundary) for your test on the z-scale and compute the z score for the sample result. What is your conclusion?

(d) Show the cutoff (c) on the original scale.

(e) Describe the nature of Type I and Type II errors.

★ 10. Refer to Exercise 1 (union membership). Assume that a sample result has not yet been produced. Compute β, the probability that your test could lead you into a Type II error, if the population proportion π is only .25.

(a) Show the accept/reject boundary c on the original ($\bar p$) scale for the test you devised in Exercise 1.

(b) Show the alternative distribution superimposed on the test axis.

(c) Show the area in the alternative distribution that corresponds to β.

11. It has been argued that at least 35% of all U.S. households have a head of household under the age of 30. You plan to survey 300 randomly selected households to test this proposition.

(a) Set up an appropriate hypothesis testing framework, using a significance level of 5%.

(b) How likely is it that the test you set up could lead you into making a Type II error if the actual proportion of households with a head of household under 30 is only 26%?

(i) Show the accept/reject boundary c on the original ($\bar p$) scale for the test you devised in part (a).

(ii) Show the alternative distribution superimposed on the test axis.

(iii) Show the area in the alternative distribution that corresponds to β.

★ 12. In controlling your productive process, you perform periodic inspections. As a matter of policy, if the process is producing at a defectives rate of no more than 6%, it is considered in control and should be allowed to continue unadjusted. If, however, the process is producing at a rate above 6%, it will be considered out of control and shut down for proper readjustment. At 2 P.M. you select a sample of 100 units recently produced and find nine defectives. Set up a hypothesis test to be used in processing sample results. Use a significance level of 5% and choose as your null hypothesis the "in control" position.

(a) State the null and alternative hypotheses.

(b) Show the null sampling distribution.

(c) Identify the accept/reject cutoff (boundary) for your test on the z-scale and compute the z score for the sample result. What is your conclusion?

(d) Show your test on the original scale.

(e) Describe the nature of Type I and Type II errors for this situation.

(f) Suppose at 3 P.M. the process is actually producing at a defectives rate of 14% (this is not a sample result, but rather it's the overall population defectives rate currently prevailing). How likely is it that the test you have set up could mistakenly lead you to believe that the process is in control, thus leading you into making a Type II error?

13. Refer to Exercise 8 (cousin Howard). Assume the experiment involved 10 tosses instead of 100, and that Howard correctly predicted 7 outcomes. What is your conclusion? (*Hint:* Use the binomial distribution rather than the normal approximation as the appropriate sampling distribution.)

14. You are about to take an exam consisting of 20 multiple choice questions. Each question will have four possible answers.

(a) How many questions would you expect to

answer correctly if you don't study at all and simply guess each answer. (*Hint:* You would have a 1-in-4 (25%) chance of being correct on each individual question.)

(b) How likely is it that you would answer more than 7 of the 20 questions correctly if you were guessing each answer? (Use the binomial distribution rather than the normal approximation.)

(c) How many correct answers would you have to produce to convince the instructor that you were not guessing? (Suggestion: Set up an appropriate hypothesis test, using a "just guessing" null and a significance level of 1%.)

(Exercises 15 through 27 deal with hypothesis tests for the difference between two population means.)

★ 15. Given the hypotheses

$$H_0: \mu_1 = \mu_2$$
$$H_1: \mu_1 \neq \mu_2$$

and sample results $\bar{x}_1 = 230$, $s_1 = 40$ ($n_1 = 200$) and $\bar{x}_2 = 221$, $s_2 = 32$ ($n_2 = 150$), set up an appropriate hypothesis test. Use a significance level of 5%,

(a) Show the null sampling distribution.

(b) Show the boundaries for your test on the z-scale.

(c) Compute the z score for the sample result and compare it to the z-score cutoff. What is your conclusion?

(d) Show the test on the original scale.

16. There is some question about the comparative performance of two products—one supplied by Xeron Corp., the other by Yellman Industries—routinely purchased by your firm and used as seemingly interchangeable components in the firm's manufacturing operation. Each product, because of normal variability in product quality, occasionally creates the need to shut down the manufacturing operation to make adjustments and correct problems.

You take a sample of 100 days during which Xeron units were being used in the manufacturing operation and a sample of 100 days during which Yellman units were being used. Average downtime for the Xeron sample was 53 minutes per day (with standard deviation of

18 minutes). Average downtime for the Yellman sample was 48 minutes (with standard deviation of 14 minutes). Use these results to test the null hypothesis of "equal average daily downtime" for the two populations represented. Use a significance level of 5%.

(a) Show the null sampling distribution.

(b) Show the boundaries for your test on the z-scale.

(c) Compute the z score for the sample result and compare it to the z-score cutoff. What is your conclusion?

(d) Show the test on the original scale.

★ 17. Acme Hair Products, Inc., a manufacturer of low-priced hair care products (as well as frozen dinners and VCRs), claims that its hairspray lasts longer (on average) than the industry leader's higher-priced product. You take a sample of 60 applications of Acme's hairspray and 60 applications of the leading hairspray. On average, the Acme applications held for 12.7 hours, with a standard deviation of 1.4 hours. On average, the leading hairspray lasted 11.8 hours, with a standard deviation of 1.8 hours. Set up a (one-tailed) hypothesis test for Acme's claim. Use a skeptical "no longer" null hypothesis, putting the burden of proof on Acme to make its case. Use a significance level of 2%.

18. Regal Crown Tire Company claims that its new low-cost Velvet Ride tires last, on average, longer than the leading premium tire. They report that in a test involving 200 randomly selected Velvet Ride tires and 200 premium tires, the Velvet ride tires lasted an average of 46,514 miles (with standard deviation of 3250 miles). The tires in the premium sample lasted an average of 45,854 miles (with a standard deviation of 2412 miles). Are these sample results sufficient to establish Regal's claim? (That is, are they sufficient to reject a skeptical "no better than" null hypothesis?) Construct a one-tailed hypothesis test, using a 5% significance level.

★ 19. Bose/Plotkin Laboratories claim to have produced a new drug to treat and cure the common cold. They show that for a sample treatment group of 100 randomly selected cold sufferers given the new drug, average recovery time was 79.3 hours (with a sample standard deviation of 19.2 hours). This, they argue, compared favorably to the average recovery time for a control group of 100 cold sufferers who took only a placebo. For the control group, average recovery time was 98.4 hours, with a sample standard deviation of 37.2 hours. Is this sufficient evidence to reject a skeptical "no faster" null hypothesis at the 5% significance level? Show the test.

20. Refer to Exercise 19 (cold remedy). Suppose the sample size was 12 for the test group and the control group. Reconstruct the hypothesis test and report your conclusion. What additional population assumptions would have to be made in this small-samples case?

★ 21. Two different physical conditioning programs are being tested to determine which is more effective in reducing dangerously high pulse rates. Ten randomly selected heart patients are selected for program A, and another 10 are selected for program B. After four months, program A participants showed an average pulse rate reduction of 15.2 beats per minute (with a standard deviation of 4.9 beats). Program B participants showed an average reduction in pulse rate of 12.5 beats (with a standard deviation of 5.4 beats). Set up a hypothesis test to determine whether there is a statistically significant difference in the two programs. Use a significance level of 1%. What additional population assumptions would have to be made in this small-samples case?

22. There are 400 residents living in the government-subsidized Bellmore Apartments, and 320 residents living in the nearby Justin-Peller condominium complex. You take a simple random sample of 9 Bellmore residents and a simple random sample of 12 Justin-Peller residents. The average number of tenant complaints for residents in the Bellmore sample is 13.7, with a standard deviation of 1.1. The average number of complaints for residents in the Justin-Peller sample is 12.9, with a standard deviation of .8.

Set up a hypothesis test, using a significance level of 1%, to test the null hypothesis that there is no difference in the average number of complaints from residents of the two buildings. (Assume that the two "number of complaints" populations are approximately normally distributed and have equal standard deviations.)

★ 23. A sample of six students from the Madison Grade School (total enrollment 660) is randomly selected and given a rigorous physical test. The

results

Student	1	2	3	4	5	6
Test score	522	480	610	552	390	560

A sample of five students from the Jefferson School (total size 500) is similarly selected and tested, with the following results:

Student	1	2	3	4	5
Test score	600	580	410	350	430

Use these results to test the statement that "there is no difference in average test scores for the two populations (the two schools) represented." Use a significance level of 5%. What assumption(s) must be made about the populations in this small-sample case?

24. Repeat Exercise 23, but add a sixth score, 510, to the Jefferson sample.

★ 25. Refer to Exercise 18 (Regal tires). Ignore the sample mean difference reported there. Suppose the difference in mileage for the two populations represented is actually 1000 miles ($\mu_1 - \mu_2 = 1000$). How likely is it that the test you have set up would lead you into making a Type II error?
   (a) Show the accept/reject boundary on the original scale for the test you devised in Exercise 18.
   (b) Show the alternative distribution superimposed on the test axis.
   (c) Show the area in the alternative distribution that corresponds to β.

26. You want to test the hypothesis that there is no difference in the average amount of campaign spending by Democratic versus Republican candidates in the most recent state primary elections. (It is often assumed that Republicans spend more than Democrats.) You take a sample of 50 Democratic candidates and 50 Republican candidates, with these results: Democrat sample average spending is $2.23 million, with a standard deviation of $.68 million; Republican sample average spending is $2.67 million, with a sample standard deviation of $.58 million.
   (a) Set up the test, using a significance level of 5%.
   (b) Ignore the sample means reported here. If the population spending difference is actually $.2 million, how likely is it that your test could lead you into making a Type II error?

★ 27. Your company fleet has 110 Ford Jayhawks and 130 Chevrolet Keystones. You take a sample (without replacement) of 36 of the Fords and 49 of the Chevrolets to assess whether there is a difference in the average time the two different types of trucks have spent in the garage for repairs. Looking at the maintenance records for the trucks in your sample, you see that the average downtime for the Fords during the past year was 19 days, with a sample standard deviation of 6 days. For the Chevy sample, average downtime was 17 days, with a standard deviation of 4 days. Set up an appropriate hypothesis test here, using a significance level of 5%. (Don't forget the FPC.)

(Exercises 28 through 35 deal with hypothesis tests for the difference between two population proportions.)

28. Given the hypotheses

$$H_0: \pi_1 \leq \pi_2$$
$$H_1: \pi_1 > \pi_2$$

and sample results $\bar{p}_1 = .82$ ($n_1 = 200$) and $\bar{p}_2 = .78$ ($n_2 = 300$), set up an appropriate hypothesis test. Use a significance level of 5%.
(a) Show the null sampling distribution.
(b) Show the boundary for your test on the z-scale.
(c) Show the pooled estimate of the common population proportion π.
(d) Show the standard error of the null sampling distribution.
(e) Show the z score for the sample result.
(f) What is your conclusion?
(g) Show the test on the original scale.

★ 29. It is argued that there is a difference in the rate of single-parent households in rural areas versus urban areas of the state. On a simple random sample of 1000 rural households and a simple random sample of 1000 urban households, 120 of the rural sample and 160 of the urban sample were single-parent households. Test the null position that there is "no difference" in the single-parent household percentage for the two populations represented. Use a significance level of 2%.
(a) Show the null sampling distribution.
(b) Show the boundaries for your test on the z-scale.

(c) Show the pooled estimate of the population proportion.
(d) Show the standard error for the null sampling distribution.
(e) What is the z score for the sample result?
(f) What is your conclusion?
(g) Show the test on the original scale.

30. There is concern within your firm that a serious difference exists between management and labor in their level of support for the introduction of new Japanese-style production methods. You take a simple random sample of 100 managers and a simple random sample of 100 rank-and-file production workers. Thirty percent of the management sample and 24% of the production-worker sample report opposition to the new methods. Using the "no difference" null hypothesis, set up an appropriate hypothesis test. Use a significance level of 5%. Based on your test, what is your conclusion?

★ 31. Turbo Lightning Deliveries claims that its record of on-time package delivery is superior to that of its nearest competitor, Ittle/Waite Transport. Selecting a random sample of 120 Turbo deliveries and 120 Ittle deliveries, you find that 86 of the Turbo deliveries were on time, while 70 of the Ittle deliveries were likewise on time.
(a) Can Turbo use this sample evidence to reject a hypothesis that its on-time rate is no better than Ittle's? Use a significance level of 5%.
(b) What two populations are represented?

32. In a recent survey of new car buyers, 116 of 150 randomly selected buyers of German-made cars expressed complete satisfaction with their purchase, and 104 of 150 randomly selected buyers of American-made cars expressed complete satisfaction. Are these sample results sufficient to reject a "no difference" null hypothesis at the 10% significance level?

★ 33. A recent poll of 1000 randomly selected East Germans and 1000 randomly selected West Germans showed the following results: 63% of the East German sample and 55% of the West German sample "unconditionally" supported the unification. Is this sufficient sample evidence to reject the proposition that there is no difference in the attitude of the two populations on this question? Use a 5% significance level.

34. A study of the career opportunities for its engineering and business school students has

been undertaken at Hapmore University. A sample of 100 graduating senior engineering students (from a total of 450 graduating senior Hapmore engineers) and a sample of 100 graduating business school seniors (from a total of 610 Hapmore graduating business school seniors) was randomly selected (without replacement). Each student was asked whether he or she had been successful in finding a satisfactory job. Seventy-two of the engineers and 84 of the business school seniors responded yes. Can this sample result be used to reject an argument that claims there is no difference in the percentage of graduating business majors and graduating engineering majors at Hapmore who have found satisfactory jobs? Use a significance level of 1%. Show the details of your test.

★ 35. You take a simple random sample (without replacement) of 120 items from a recent shipment of 450 HGD high-density retrobore filaments made by Spaceage Electronics. From a similar shipment of 560 filaments made by Gradient Technologies, you take a simple random sample of 130 items. The Spaceage sample shows 23 defective items, while the Gradient sample shows 18 defectives. Can this sample evidence be used to reject a claim that the defective rates for the two shipments are equal at the 5% significance level? Explain.

REGRESSION ANALYSIS

CHAPTER OBJECTIVES

Chapter 9 should enable you to:

1. Describe the general nature of regression analysis.

2. Differentiate simple versus multiple regression; linear versus nonlinear regression.

3. Produce a regression line by using the least-squares criterion.

4. Calculate and interpret measures of goodness-of-fit for any regression line.

5. Understand the sampling side of regression analysis.

6. Use sampling theory to build confidence interval estimates of the population slope and intercept terms.

7. Develop a hypothesis test to determine whether the data effectively establishes a linear relationship between variables.

8. Use a regression line for statistical estimation.

We turn now to regression analysis, a core statistical tool that has application in nearly all business-related disciplines (and in a number of nonbusiness areas as well). As a basic analytic procedure, regression analysis has been applied to financial planning, marketing research, economic forecasting, cost estimation, production scheduling, and a host of other common business-based situations.

9.1 THE NATURE OF REGRESSION ANALYSIS

In simplest terms,

> Regression analysis represents an effort to identify a relationship between two or more variables.

More technically,

> Regression analysis seeks to produce a mathematical function which effectively relates two or more variables (or factors), so that the value of one variable might be accurately predicted from a given value for the other(s).

To illustrate this application:

- We might suspect that a relationship exists between the amount of money company XYZ spends annually on advertising and the company's level of annual sales. Regression analysis might be used to help identify the nature and the extent of the suspected relationship.
- Or we might suspect that there's a relationship between the number of sales calls that our sales people make in a week and the number of confirmed sales reported. We could use regression analysis to help clearly define the apparent connection.
- On another level, we might believe there's a connection between an individual's daily exercise schedule and his or her weekly weight gain (or loss). We could turn to regression analysis in order to fill in the details of the suspected relationship.

In each case, the pattern is similar: A relationship between (or among) certain variables is suspected; regression analysis can be used to confirm the connection and to define the specifics of the relationship.

9.2 CATEGORIZING REGRESSION ANALYSIS VARIATIONS

Regression analysis comes in a variety of forms. One classification scheme distinguishes between simple and multiple regression; another draws a distinction between linear and nonlinear approaches.

Simple versus Multiple Regression

In **simple regression**, only *two* factors (i.e., two variables) are involved in the suspected relationship. (Each of the examples mentioned above would involve this kind of regression.) In fact, simple regression is sometimes referred to as *bivariate* regression to emphasize the fact that only two variables are involved.

One of the two factors in simple regression is typically labeled the **dependent variable**; the other, the **independent variable**. (Symbolically, y is most often used to designate the dependent variable, while x normally represents the independent variable.) To distinguish the two, think of the dependent variable as the "influenced" factor, and the independent variable as the factor doing the "influencing." In most applications, which variable is which will be apparent. (For example, in the advertising expenditures/sales level situation, advertising would be treated as the independent variable, since it's the factor that's influencing sales, or, at least, so we think.)

In simple regression, then, we believe that there's a relationship between an independent variable (x) and a dependent variable (y). Symbolically, we might show our suspicion in the functional form

$$y = f(x)$$

where x = independent variable
$\quad y$ = dependent variable
$\quad f$ = connecting rule or function

In **multiple regression**, suspected relationships involve a dependent variable (y) that appears to be simultaneously related to *more than one* independent variable (x_1, x_2, etc.). For example, while advertising expenditures may be a key factor in influencing company sales, we might easily produce a long list of other possible influences—things like price, general economic conditions, competitor advertising, and the like. In multiple regression, we would typically try to sort through the possibilities in order to identify the most significant factors, and then show how they collectively relate to company sales.

> **Note:** We'll avoid using the word "cause" in regression. While regression analysis is used to investigate relationships, it cannot be used to establish causality. For example, we couldn't use regression analysis to establish that company advertising *causes* company sales—only that the two factors are related.

In general, a suspected multiple regression relationship might be expressed as

$$y = f(x_1, x_2, \ldots, x_n)$$

where y = dependent variable
$\quad x_i$ = each of the independent variables
$\quad f$ = connecting rule or function

Linear versus Nonlinear Regression

We can also draw a distinction between **linear** and **nonlinear** (sometimes called **curvilinear**) **regression**. In *linear* regression, the suspected relationship between variables is believed to follow the standard linear (i.e., straight line) pattern

$$y = a + bx$$

for the bivariate case or, for the multivariate case,

$$y = a + b_1x_1 + b_2x_2 + \cdots + b_nx_n$$

where a = constant intercept term in the linear relationship
$\quad b$ or b_i = constant slope term(s)

Here, not only do we suspect a relationship, but we're willing to specify that the relationship involves one of the simplest of mathematical links—a linear function that contains no squared terms, no cubed terms, no logarithms or exponents—nothing fancy at all.

In *non*linear regression, the form of the relationship between the variables may be considerably more complex. Here, for example, quadratic expressions like

$$y = 40 + 2x + 10x^2$$

or logarithmic functions like

$$y = 1.3 \log x$$

might be used to describe the apparent variable linkage.

The Base Case

Simple linear regression can be used as a kind of "base case" illustration of general regression techniques. In one way or another, all the other forms can be portrayed as variations on the simple linear theme. Consequently, we'll devote most of our chapter to the simple linear approach, leaving for another time a more comprehensive treatment of multiple and nonlinear extensions.

9.3 THE BASIC REGRESSION PROCEDURE

Situation: Reconsider one of our earlier examples: You suspect that a relationship may exist between firm XYZ's annual advertising expenditures and the aggregate level of company sales. To take things one step further, you suspect that a *linear* relationship exists. Our intent is to use regression analysis to pursue your suspicion.

How should we get started? First of all, we'll need data. In order to carry out any regression procedure, we need to have a relevant data base of sufficient size and scope. When data are available from existing records, our job is obviously made easier. In the absence of appropriate historical data, however, we may have to devise an experiment in order to generate necessary figures.

To keep things simple, we'll assume that suitable data are, in fact, currently available for our analysis—four years of records from XYZ's corporate files:

x Advertising expenditures ($ millions)	y Annual sales ($ millions)
2	6
3	8
4	10
5	9

It appears from the table that when advertising expenditures were $2 million, total sales amounted to $6 million; when advertising expenditures were $3 million, sales receipts were $8 million; and so on. Notice that we've labeled the advertising expenditures column "x" and the sales column "y," identifying (as we had earlier) advertising as the independent variable and sales as the dependent variable.

> **Note:** We're reporting here just four years of sales-advertising figures. Most regression applications would require substantially more data. With only four observations, any conclusions we might reach will be rather tenuous. On the other hand, dealing with only four data points will allow us to keep our computational effort within reasonable limits.

Question: Given the data presented here, should we be encouraged in our belief that there's a linear connection between sales and advertising, or should we now be discouraged from pursuing that suspicion? While a quick scan of the table may be enough to decide, a graphical display of the data is often more useful. From the data provided, we can construct a simple two-dimensional plot of the sales-advertising coordinates (as shown in Figure 9.1). This sort of picture is commonly referred to as a **scatter diagram** (or **scattergram**).

Again, the question: should we be encouraged that the data support our initial suspicion of a linear connection, or does the sketch we've produced serve to discourage us from pursuing the idea any further? Even with the graph, it seems that things still aren't perfectly clear. If only the data had been a little bit different, support for our belief might have seemed a lot stronger. For example, had the last point been (5,12) instead of (5,9), the plot of the points would have looked like Figure 9.2. In this case, it's possible to connect all of the points with a single straight line, strongly suggesting that our linear relationship "hypothesis" is right on the mark. (See Figure 9.3.)

However, back in our original case, things don't appear quite as compelling. Look again at the graph in Figure 9.1. It's clearly *not* possible here to produce a single straight line that would pick up all four of the points. But should this in itself be grounds for dismissing our initial belief? The short answer is "no" (or at least

FIGURE 9.1

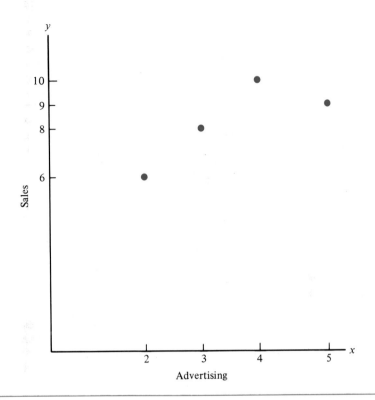

"probably not"). While it may not be possible to find a perfect linear fit for the data
we have, we might still be able to establish a case for a kind of core linear connection.
Even though no single straight line will connect all the points, we might still look for
what amounts to a pretty good fit in order to advance our case to the next level of
analysis.

> **Note:** Had the data looked like that shown in Figure 9.4, we would have been in serious
> trouble. Here, our linear hypothesis would almost certainly be abandoned.

Fitting a Line to the Data

At this point, regression analysis takes on a definite curve-fitting tone. Conceding
that no perfect linear fit to the data is possible, our job now is to find the best one we
can. In graphical terms, this means we'll want to sketch the straight line that seems to
best describe the pattern that appears in the data.

Any ideas? What seems immediately obvious is that a variety of best-fitting
lines might be proposed. For example, the line drawn in Figure 9.5, in one sense at
least, could be designated the "best" line we could find. Here we've drawn a line that

FIGURE 9.2

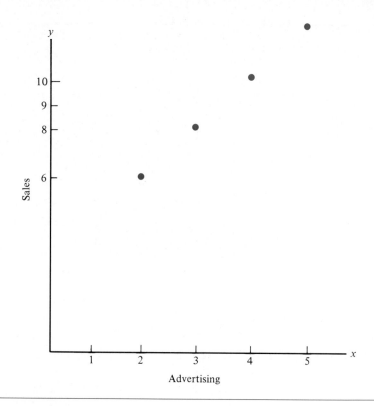

connects three of the four points in the graph. (No other straight line would link more than two.) If we define as the best-fitting line the line that maximizes the number of data points connected, then the line in Figure 9.5 is clearly the winner. If we change our definition of "best," however, then other lines may emerge as superior fits. Settling on an appropriate curve-fitting criterion is obviously crucial to finding the best-fitting line.

The Least-Squares Criterion

Standard regression analysis uses the so-called **least-squares criterion** to define the overall best-fitting line. Figure 9.6 and the discussion that follows should give you a sense of the term.

We're showing in the figure a *candidate* best-fitting line. Notice the vertical distances (or deviations) that we've labeled d_i. Not surprisingly, these d_i terms play a significant role in judging the goodness or badness of fit for the candidate line. Logically, a good-fitting line should produce small d_i distances.

FIGURE 9.3

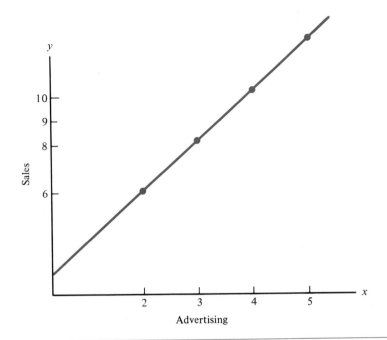

FIGURE 9.4 A SCATTER DIAGRAM SHOWING NO INDICATION OF A CONNECTION BETWEEN *x* AND *y*

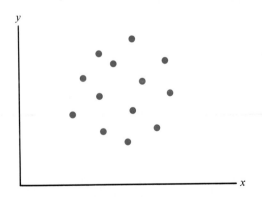

FIGURE 9.5   **A POSSIBLE BEST-FITTING LINE**

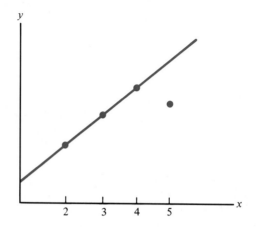

FIGURE 9.6   **FITTING THE LEAST-SQUARES LINE**

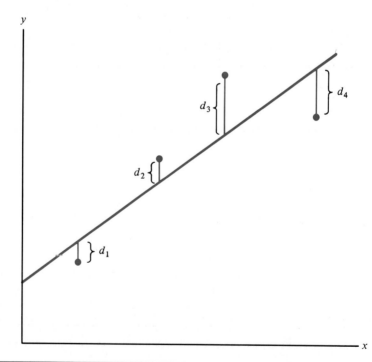

The least-squares criterion actually focuses on the *square* of the d_i values, establishing as the best-fitting line the line that

minimizes the sum of the squared vertical distances of points from the line (i.e., the sum of the d_i^2).

While other criteria might seem just as appealing, the statistical properties of the least-squares criterion make it the criterion of choice in standard regression. (We'll shortly see some of the advantages of using the least-squares approach.)

Identifying the Least-Squares Line

Implementing a least-squares procedure (i.e., finding the least-squares line) could conceivably follow one of two or three different routes. One approach, for example, would involve sketching all possible candidate lines and then computing in each case the sum of the squared deviations. Unfortunately, while this sort of search would ultimately identify the line that satisfies the least-squares objective, we could find ourselves searching for a very long time.

Luckily, a more efficient procedure is available. By reducing the job of curve fitting to a calculus problem, we can quickly identify the intercept (a) and the slope (b) for the best-fitting line. (The minimization task inherent in the least-squares criterion lends itself to a calculus-based procedure. Differential calculus, you may recall, can be used to find minimums or maximums for certain continuous functions.) As you might expect, we won't be concerned with the details of this calculus approach, only with the results.

Specifically, to find the a and b values for the line that will effectively minimize the sum of the squared distances (deviations), all we need to do is make the appropriate substitutions in the following two expressions:

$$(1) \qquad b = \frac{n\sum xy - \sum x \sum y}{n\sum x^2 - (\sum x)^2}$$

where n = number of data points (or observations)

$$(2) \qquad a = \bar{y} - b\bar{x}$$

where $\bar{y}$ = average of the y values
  $\bar{x}$ = average of the x values

Although these expressions may look a little overwhelming at first, the necessary substitutions are fairly straightforward.

> **Note:** For easier reading, we've chosen not to show subscripts for the x and y values in the summations. We'll follow this convention throughout the chapter, except where confusion may result.

Substituting to Produce the Slope and Intercept Terms

To trace the necessary a and b calculations, we'll use the table format shown below, with a column for the x values, a column for the y's, a column for the xy products,

FIGURE 9.7   **DISPLAYING THE ACTUAL LEAST-SQUARES REGRESSION LINE**

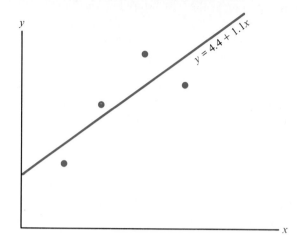

and a column for the x^2 terms. The numbers we're showing detail the computations involved in the sales-advertising example:

Advertising	Sales		
x	y	xy	x^2
2	6	12	4
3	8	24	9
4	10	40	16
5	9	45	25
$\sum x = 14$	$\sum y = 33$	$\sum xy = 121$	$\sum x^2 = 54$

Substituting in the b expression produces

$$b = \frac{4(121) - (14)(33)}{4(54) - (14)^2} = 1.1$$

By calculating the mean of the x's ($\bar{x} = 14/4 = 3.5$) and the mean of the y's ($\bar{y} = 33/4 = 8.25$) and using $b = 1.1$, we can produce the intercept term a:

$$a = 8.25 - (1.1)(3.5) = 4.4$$

Conclusion? For the sales-advertising illustration, the best-fitting line—the one that satisfies the standard least-squares regression criterion and the one that we'll now begin to label the *regression line*—has a slope of 1.1 and an intercept of 4.4. A sketch of the line is shown in Figure 9.7.

We're thus proposing a linear connection between advertising and sales in the form of a simple mathematical function

$$y = 4.4 + 1.1x$$

Such a connection would seem to suggest, for example, that an advertising budget of

FIGURE 9.8   **FITTING A LEAST-SQUARES LINE TO DATA IN WHICH NO RELATIONSHIP IS EVIDENT**

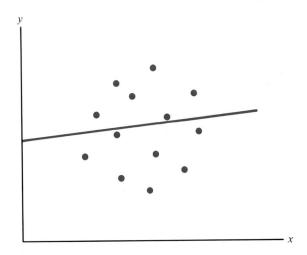

$10 million could be linked to annual sales of approximately $15.4 million. (Just substitute an x of 10 into the least squares equation, and compute the value for y.) In fact, for any given advertising budget, sales could be similarly predicted.

>   **Note:** It's important not to jump too quickly to overstated conclusions by attaching to these kinds of predictions a certainty and a precision that simply aren't justified by the data. We'll expand on this idea shortly.

9.4   PERFORMANCE MEASURES IN REGRESSION: HOW WELL DID WE DO?

Before proceeding much further, it may be appropriate to stop and evaluate how well we've done—to decide just how well the line

$$y = 4.4 + 1.1x$$

does, in fact, describe the data. (This is particularly important if we plan to make significant predictions from the line we've produced.) We know the regression line here is the best-fitting line under the least-squares criterion, but just how good a fit do we actually have? After all, even in a case that looks as shaky as the one shown in Figure 9.8, a least-squares line can be fit to the data. It turns out that a number of performance measures are available to describe goodness-of-fit—measures that might also be used to describe the apparent *strength of the relationship* between the variables involved. We'll focus on three of the most widely used measures: the standard error of estimate, the coefficient of determination (usually written r^2), and the correlation coefficient (r).

FIGURE 9.9   **DISTANCES NEEDED FOR THE STANDARD ERROR OF ESTIMATE CALCULATION**

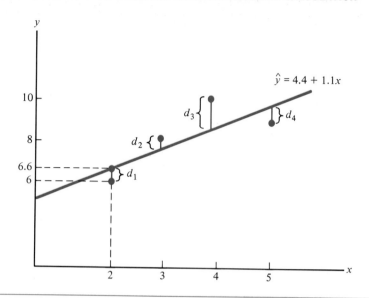

The Standard Error of Estimate

The **standard error of estimate** is essentially a standard deviation–type measure of variability in data, but in this case we'll measure the variability of the data *around the regression line*. To make the computation, we can follow the general standard deviation approach: (1) determine the distance (or deviation) of each data point from a central reference point (in this case, from the regression line), (2) square each distance, (3) sum the squared distances, (4) calculate the average squared distance, and (5) take the square root of the result.

Using Figure 9.9, we can portray the procedure in greater detail. The d_i terms shown here are at the core of our calculation. (We'll eventually refer to these d_i terms as *unexplained deviations*.) Measuring each of these terms follows the same basic pattern. Take, for example, the line segment d_1. This is simply the vertical distance of the first observation (the point (2,6)) from a corresponding point on the least-squares line lying directly above $x = 2$. To measure the size of d_1, we'll have to establish the precise y coordinate for this corresponding point on the line. How? We can simply substitute an x value of 2 into the equation for the line:

$$y = 4.4 + 1.1x = 4.4 + 1.1(2) = 6.6$$

You can see then, that an x value of 2 produces a regression line y of 6.6. Subtracting the observed y value of 6 from the regression line y of 6.6 produces exactly the d_1 distance we want. The $-.6$ result establishes that the first observation lies .6 units below the line we've proposed. (Be sure you follow along on the graph.)

Applying the same method to the second data point (3,8) produces a d_2 value of $(8 - 7.7) = .3$, leading us to conclude that the second observation lies .3 units *above* the regression line. The table below summarizes the procedure for all four of the points.

| | | | d_i |
x	y	$\hat{y}$	$(y - \hat{y})$
2	6	6.6	−.6
3	8	7.7	.3
4	10	8.8	1.2
5	9	9.9	−.9

Notice that we've used $\hat{y}$ (read y-hat) to designate the regression line values for y in order to differentiate these regression line values from the *observed* values for y shown in the original data set. In fact, from here on we'll write the regression equation as

$$\hat{y} = 4.4 + 1.1x$$

The next step in the standard error of estimate computation? Sum the squared d_i distances. The expanded table below shows the required calculations:

| | | | d_i | d_i^2 |
x	y	$\hat{y}$	$(y - \hat{y})$	$(y - \hat{y})^2$
2	6	6.6	−.6	.36
3	8	7.7	.3	.09
4	10	8.8	1.2	1.44
5	9	9.9	−.9	.81
			0	2.70

Note: As shown in the fourth column of the table, had we chosen to sum the $y - \hat{y}$ distances *without* squaring each term, the result would have been 0. This kind of symmetry is characteristic of all least-squares regression lines: the sum of the deviations around the regression line will always be 0, since the positive deviations will always offset the negative deviations. (We saw this sort of cancellation-to-zero phenomenon before—in our standard deviation discussion in Chapter 2. There, too, squaring the deviations eliminated the trivializing effect of cancellation to 0.)

As a matter of terminology, the sum of the squared d_i terms—2.70 in our example—is sometimes called the **unexplained variability** in the dependent variable. We'll discuss why shortly.

Next, averaging the sum of the squared distances produces

$$2.70/4 = .675$$

Taking the square root completes the standard error of estimate calculation:

$$\sqrt{.675} = .822 \quad \text{(which translates here to \$822,000)}$$

We can summarize the procedure with the expression below:

$$\text{S.E. of E.} = \text{STD ERROR OF EST} = \sqrt{\frac{\Sigma(y - \hat{y})^2}{n}} = .822$$

FIGURE 9.10   **A PERFECT-FIT REGRESSION LINE**

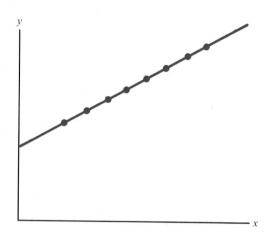

As we indicated earlier, what we have here is a measure of dispersion comparable to the standard deviation of previous chapters. It's the square root of the average squared distance of points from the regression line. As with any standard deviation measure, the larger the value, the more variability there is in the data. For the standard error of estimate, the larger the value, the more widely dispersed are the points around the regression line, and, as a consequence, the more poorly the line seems to describe (fit) the given data. At the lower extreme, a standard error of estimate value of 0 indicates a perfect fit. (See Figure 9.10.)

> **Note:** As we'll soon see, regression analysis eventually takes on an inferential tone, where samples and populations of data will be clearly distinguished. At that point, we'll differentiate between $\sigma_{y.x}$ (sigma-y-dot-x), the standard error of estimate for a *population* of points around the *population* regression line, and $s_{y.x}$ (s-y-dot-x), the standard error of estimate for a *sample* of points around a *sample* regression line. For now, though, we'll stick with our generic standard error of estimate expression, S.E. of E.

The Coefficient of Determination

While the standard error of estimate is a common measure of performance in regression, it's not the only—nor necessarily the best—way to convey how well the line describes the data. One difficulty with this measure is that there's no upper limit on the value it might take. While we mentioned that its lower bound is 0, the high side of the standard error of estimate is effectively unbounded. As a consequence, it's difficult to decide just how large the standard error of estimate would have to be before we'd be discouraged in our belief that the regression line we've produced is a reasonably good descriptor of the data involved. (Is 25.6 too large a standard error of estimate? 103? 12,568?)

An alternative measure, the **coefficient of determination**, offers something of a remedy for this unboundedness problem. Usually labeled r^2, the coefficient of determination essentially reports goodness-of-fit on a scale of 0 to 1. With both an upper bound and a lower bound on possible values, this measure makes it easier to judge just how good a fit we have.

In simplest terms, the coefficient of determination (r^2) can be described as the proportion of the total variability in y (the dependent variable) that can be *explained* (or accounted for) by y's relationship to x (the independent variable). In brief,

$$r^2 = \frac{\text{Explained variability in } y}{\text{Total variability in } y}$$

The necessary computations and the basic rationale are described below.

Total Variability

In our sales-advertising example, the total variability denominator in the r^2 expression should reflect the variability shown in the sales data we've collected. To effectively measure this total variability, we'll follow a familiar computational theme. Employing once again a sum-of-squared-deviations approach, we'll (1) compare each of the observed sales levels (y values) to the overall *average* sales figure ($\bar{y}$), (2) square the $y - \bar{y}$ differences, and (3) sum the squared terms. In short,

$$\text{Total variability} = \sum(y - \bar{y})^2$$

We can use the following table to display results.

x	y	$y - \bar{y}$	$(y - \bar{y})^2$
2	6	$6 - 8.25 = -2.25$	5.0625
3	8	$8 - 8.25 = -.25$	.0625
4	10	$10 - 8.25 = 1.75$	3.0625
5	9	$9 - 8.25 = .75$	.5625
	$\bar{y} = 33/4$	0	Total var $= 8.75$
	$= 8.25$		

Total variability here, then, is 8.75.

Explained Variability

Having established that there's variability in sales (and having quantified it as we have), the key question now centers on why? Why is there variability in sales? Why weren't sales always the same? Any ideas? Haven't we already offered our explanation? We think that there's variability in sales because sales are related to advertising expenditures, and since (as we saw in the data) advertising expenditures vary, we would necessarily expect sales to vary as well.

Indeed, our entire regression effort so far has been directed at explaining differences (variability) in sales by relating sales to advertising. Measuring just *how much* of the variability in sales we can explain by the sales-to-advertising connection should help us decide how well our regression analysis has performed.

FIGURE 9.11   **THE DISTANCE INVOLVED IN EXPLAINED VARIABILITY**

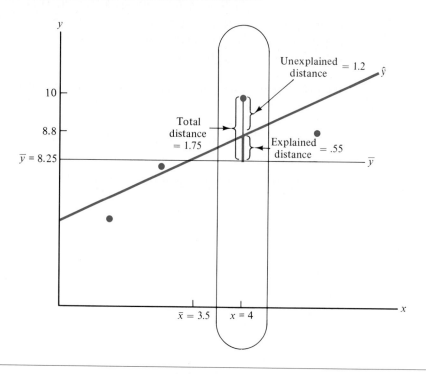

Consider again our sketch of the sales-advertising regression in Figure 9.11. To demonstrate the explained variability calculation, we've highlighted the case of $x = 4$, and drawn a horizontal line to represent the average sales level ($\bar{y}$) of $8.25 million. As shown, when x (advertising expenditures) was 4, the observed sales level was 10, a value obviously different from the 8.25 average. (Be sure to use the graph as a reference.) In fact, we've already measured this difference as part of our overall total variability calculation in the previous section, $10 - 8.25 = 1.75$.

So we're seeing that when $x = 4$, sales exceeded average sales (8.25) by a total distance of 1.75. Does this sort of difference come as much of a surprise? It shouldn't. According to our regression line, we would *expect* sales here to differ from the average because of the relationship that exists between advertising and sales. The above-average sales ($10 million)—we would argue—can be linked to the above-average level of advertising ($4 million).

But how much different from average would we expect sales to be? When $x = 4$, our regression line would predict a sales level ($\hat{y}$) of $8.8 million, leading us to expect a difference from the mean of $.55 million ($\hat{y} - \bar{y} = 8.8 - 8.25 = 0.55$). We'll call this value the *explained* part of the total distance of y from $\bar{y}$ when $x = 4$. (Note: As shown in Figure 9.11, we might refer to the remaining $10 - 8.8 = 1.2$ distance as the *unexplained* portion of the 1.75 total, the kind of d_i distance involved in our computation of the standard error of estimate. In fact, we mentioned the unexplained label before; now, hopefully, this terminology makes more sense.)

Thus (at least in the case of $x = 4$) we can account for a part of the total deviation of y (the distance of y from $\bar{y}$) by using the relationship, embodied in our regression line, that appears to exist between sales and advertising.

Note: Had our regression line been perfect, this third data point would have fallen right on the line. Thus, the total and explained distances would have been identical, implying 100% explanation. In such a case, the unexplained distance would necessarily be 0. You might take a minute to draw the corresponding picture to convince yourself.

Extending this sort of procedure to all the points will enable us to produce an aggregate explained variability term for the numerator of the r^2 expression. Focusing on the $\hat{y} - \bar{y}$ difference for each of the points, we'll once again square the individual differences (to avoid cancellation to 0) and add the results. In short,

$$\text{Explained variability} = \sum(\hat{y} - \bar{y})^2$$

The following table summarizes the appropriate work:

x	y	$\hat{y}$	$\hat{y} - \bar{y}$	$(\hat{y} - \bar{y})^2$
2	6	6.6	$6.6 - 8.25 = -1.65$	2.7225
3	8	7.7	$7.7 - 8.25 = -.55$	.3025
4	10	8.8	$8.8 - 8.25 = .55$	.3025
5	9	9.9	$9.9 - 8.25 = 1.65$	2.7225
			0	Exp var = 6.05

Putting things together,

$$r^2 = \frac{\text{Explained variability}}{\text{Total variability}}$$

$$= \frac{\sum(\hat{y} - \bar{y})^2}{\sum(y - \bar{y})^2}$$

$$= \frac{6.05}{8.75} = .691$$

Interpretation? Using the data provided, it appears that we can "explain" about 69% of the variability in sales by the sales-to-advertising connection shown in the regression relationship

$$\hat{y} = 4.4 + 1.1x$$

Not bad. Had the regression line fit the data perfectly—with all the points falling right on the line—the r^2 value would have been a perfect 1.00 (100%). On the other hand, had our line been able to explain none of the variability in sales, our r^2 result would have been 0. Given these sorts of boundaries, 69% would appear to represent a pretty good fit. We'll need to be careful, though, not to become too pleased with ourselves just yet. As we'll see shortly, there are other factors that may challenge our apparent success.

FIGURE 9.12

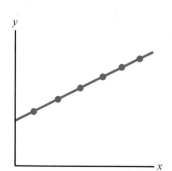
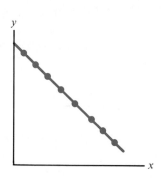

(a) Perfect positive correlation: $r^2 = 1, r = +1$ (b) Perfect negative correlation: $r^2 = 1, r = -1$

The Correlation Coefficient

The **correlation coefficient** (r) is simply the square root of r^2. In the sales/advertising example, this means

$$r = \sqrt{.691} = +.831$$

We've purposely inserted the $+$ sign to emphasize the fact that the apparent connection between sales and advertising is *positive* (or direct). That is (as our upward-sloping regression line clearly suggests), when advertising levels increase, sales levels can also be expected to increase. Had the regression line been negatively (downward) sloped, the correlation coefficient would carry a negative sign, implying an inverse connection between the two variables (i.e., when advertising levels increase, sales would *de*crease).

Given the sign possibilities and the 0-to-1 boundaries for r^2, the correlation coefficient r will always lie between -1 and $+1$. As with r^2, an r value of 0 indicates no apparent linear relationship. At the upper extreme, a "$+1$" r value implies a strong positive connection. A "-1" r value implies a strong negative link. (See Figure 9.12.)

Footnotes to the Basic Procedure

Before we proceed to the next section, a few follow-up comments may be in order.

1. As shown in Figure 9.11 and noted earlier, for any point in the data set, total deviation can be divided into explained and unexplained components, where

Total deviation = the difference between the observed y value and
the average y value ($y - \bar{y}$)

Explained deviation = the difference between the regression line y value and
the average y value ($\hat{y} - \bar{y}$)

Unexplained deviation = the difference between the observed y value and the regression line y value $(y - \hat{y})$

If we sum the squares of all the unexplained distances (i.e., $\sum(y - \hat{y})^2$) and add the result to the explained variability ($\sum(\hat{y} - \bar{y})^2$), this combined sum will always equal the total variability ($\sum(y - \bar{y})^2$). Look back at the tables in our sales-advertising example. With an explained variability of 6.05, an unexplained variability of 2.70 (from our standard error of estimate calculations), and a total variability of 8.75, the additive nature of the aggregate explained and unexplained terms is easy to see:

$$TOT = EXP + UNEXP$$
$$\sum(y - \bar{y})^2 = \sum(\hat{y} - \bar{y})^2 + \sum(y - \hat{y})^2$$
$$8.75 = 6.05 + 2.70$$

2. The value that we've labeled the *explained variability* is sometimes termed the *regression sum of squares* and labeled SSR; similarly, what we've called the *unexplained variability* is often referred to as the *residual sum of squares* or *error sum of squares* (SSE). Together they constitute the *total sum of squares* (SST). That is,

$$SST = SSR + SSE$$

Consequently, r^2 can be written as

$$r^2 = \frac{SSR}{SST}$$

or substituting SST − SSE for SSR,

$$r^2 = \frac{SST - SSE}{SST} = 1 - \frac{SSE}{SST}$$

9.5   THE INFERENCE SIDE OF REGRESSION ANALYSIS

It turns out that even when r^2 is 1 or the standard error of estimate is 0, we can't automatically conclude that there's a perfect, or even a strong, connection between the variables involved. Before we can make any defensible statements about the relationship we think we may have found, we need to recognize that there's a sampling side—a statistical inference side—to regression analysis which introduces a critical element of uncertainty into the process. (To this point, our focus has been strictly on the *descriptive* side of regression.) We'll use our sales-advertising example to demonstrate just what we mean.

For this now-familiar situation, the four data points we've collected can be viewed as a sample of four observations drawn randomly from a much larger population, the population of all observations that *might conceivably* have been selected in a random sampling of paired sales-advertising values for the firm. Figure 9.13 should give you a better idea of precisely what's involved. From the dense population of possible points shown in the scattergram, we've actually selected the four highlighted there. However, we *might* have selected a sample like any of those

FIGURE 9.13   **THE POPULATION OF POINTS FROM WHICH WE SELECTED OUR SAMPLE**

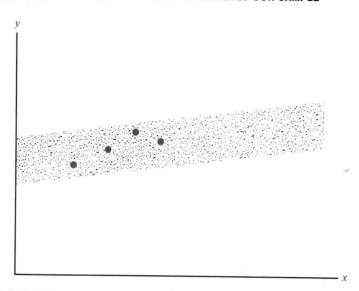

shown in Figure 9.14. And had we selected one of these other samples, the regression procedure would have produced a regression line that looked very different from the one we've taken pains to produce ($\hat{y} = 4.4 + 1.1x$).

On the inference side of regression, we'll need to deal with the capacity of any one sample-based least-squares regression line to estimate the slope and intercept terms for the line that best describes the actual "population" relationship. (From here on, we'll define the population relationship as the true relationship between x and y—the relationship that would be captured by a least-squares regression line if it were fit to the full population of data points.) What we have, then, in our regression line

$$\hat{y} = a + bx = 4.4 + 1.1x$$

is a *sample* regression line that might be used to estimate the *population* regression line

$$y = \alpha + \beta x$$

where α = intercept for the population line
        β = slope for the population line

> **Note:** Don't confuse these α (alpha) and β (beta) values with the α and β designations we used to represent the probability of Type I and Type II errors.

As you might expect, we'll need to assess just how good an estimate of the population regression line is provided by the sample regression line we've produced. In the process, we'll attempt to link the sample intercept term a to the population

FIGURE 9.14   **OTHER POSSIBLE SAMPLES FROM THE SAME POPULATION**

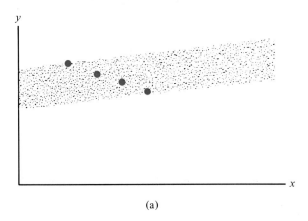

(a)

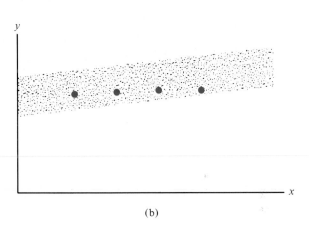

(b)

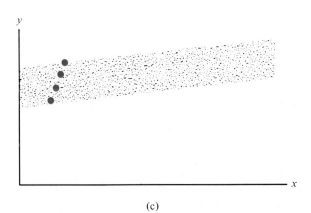

(c)

intercept term α, and the sample slope term b to the population slope term β by devising an appropriate confidence interval estimation procedure. We'll also propose a crucial hypothesis test to test the proposition that we have, in fact, uncovered evidence of a linear connection, at the *population* level, between x and y.

9.6 ESTIMATING THE INTERCEPT TERM FOR THE POPULATION RELATIONSHIP

The relationship we've produced in our sales-advertising example showed an intercept term a of $4.4 million. From our discussion above, this value can be seen now as a sample-based estimate of the real (the population) intercept term α. It's important to note that had we selected any other sample (for example, the one shown in Figure 9.15(b)), the corresponding sample intercept term would have likely been different.

Our sample intercept term (4.4), then, is just one of the many a's that we might have produced by applying the least-squares approach to a randomly selected sample of four points from the sales-advertising population. Our job now is to somehow relate this one sample a value to the overall population intercept term α. How can we create this connection? Think back to our previous four chapters. The inevitable key to linking a sample result to a population characteristic always seemed to involve a sampling distribution. The statistical link is no different here.

Note: To do what we propose in the following sections, certain fairly rigid conditions need to be met concerning the population of values from which our one sample is selected. This set of required conditions is described in the chapter supplement. For our discussion, we'll *assume* all necessary population conditions are met.

The Sampling Distribution of All Possible Sample Intercept Terms

Suppose we consider the prospect of selecting all possible samples of size 4 from our large population of points, computing in each case an a value based on least-squares regression. Producing, in this fashion, all possible a's and then showing these a's in a frequency display (a histogram) would create a familiar and perfectly predictable picture. The set of all sample a's constituting the *sampling distribution of sample intercept terms* would follow a symmetrical, bell-shaped pattern—a normal distribution for large sample cases, a t distribution for sample sizes less than 30. What's more, the center of the sampling distribution of sample a's would be exactly equal to the actual population intercept term (α). (Figure 9.16 is the type of picture we'd see.)

Building Confidence Interval Estimates of α

If the sample intercept values are indeed normally distributed and centered on α, then we can make the same kinds of observations that we made in earlier

FIGURE 9.15

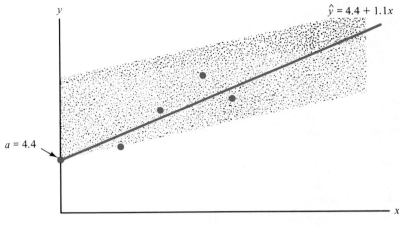

(a) The intercept term for our
sample best-fitting line

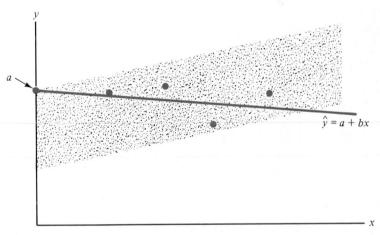

(b) The intercept term had we
chosen a different sample of
four observations

inference cases:

- 68.3% of the sample intercept terms (a's) will fall within one standard deviation of the population intercept term (α).
- 95.5% of the sample intercept terms (a's) will fall within two standard deviations of the population intercept term (α).
- And so forth.

FIGURE 9.16 **THE SAMPLING DISTRIBUTION OF ALL POSSIBLE SAMPLE INTERCEPT TERMS**

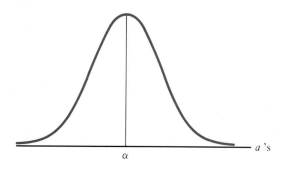

In the usual way, we can turn these kinds of statements around to create confidence interval statements that would place α (the population intercept term) in the interval

$$a \pm z \cdot (\text{std dev of the } a\text{'s distribution})$$

with z matched to a desired confidence level. All we need now is a way of computing the standard deviation of the a's distribution.

The Standard Deviation of the Sampling Distribution of a's

Although we won't try to prove it here, s_a, the standard deviation (or standard error) of the sampling distribution of a's, can be computed as

$$s_a = s_{y \cdot x} \sqrt{\frac{\sum x^2}{n \sum (x - \bar{x})^2}}$$

where n = number of observations
$\quad x$ = observed values of x
$\quad \bar{x}$ = mean value of the x's
$\quad s_{y \cdot x}$ = *sample* standard error of estimate (as explained below)

The $s_{y \cdot x}$ (read "s-y-dot-x") shown here is a slightly modified version of the generic standard error of estimate (S.E. of E.) that we discussed earlier. The modification involves changing the basic computational form

$$\text{S.E. of E.} = \sqrt{\frac{\sum (y - \hat{y})^2}{n}}$$

to

$$s_{y \cdot x} = \sqrt{\frac{\sum (y - \hat{y})^2}{n - 2}}$$

where n represents the number of observations involved.

Notice we've simply replaced the n in the original denominator with $n - 2$.

In simplest terms, $s_{y \cdot x}$ can be described as a sample-based standard error of estimate measure used to estimate the (typically unknown) population standard error of estimate (frequently labeled $\sigma_{y \cdot x}$). In essence, we're using an adjusted measure of the *sample* standard deviation of points around the *sample* regression line as an estimate of the *population* standard deviation of points around the *population* regression line. Without getting too deeply entangled in a complex explanation, we can consider the denominator adjustment ($n - 2$ instead of n) as comparable to the $n - 1$ for n adjustment we've used in cases where a sample-based standard deviation (s) is computed to estimate an unknown population standard deviation (σ). The -2 here has to do with the fact that we've had to estimate the two population line characteristics—slope (β) and intercept (α)—with sample values b and a.

Putting everything together, the basic confidence interval format used to estimate the population intercept term (α) looks like

$$a \pm z(s_a)$$

where a = sample least-squares intercept term, and

$$s_a = s_{y \cdot x} \frac{\sum x^2}{n \sum (x - \bar{x})^2}$$

$$s_{y \cdot x} = \sqrt{\frac{\sum (y - \hat{y})^2}{n - 2}}$$

In cases where sample size is less than 30, we'll replace z with t and use $n - 2$ degrees of freedom for the t-table look-up.

EXERCISE 1   Produce the 95% confidence interval estimate of the true (population) intercept term α for our sales-advertising example. Compute $s_{y \cdot x}$ using the appropriate standard error of estimate expression. (Suggestion: You may want to extend the kind of table format we've been using for computations to include a column for the x^2 and the $(x - \bar{x})^2$ terms needed to produce s_a. The following column headings should be helpful.)

x	y	$\hat{y}$	$(y - \hat{y})^2$	x^2	$(x - \bar{x})^2$

Ans: 4.4 ± 8.2 or -3.8 to 12.6

(For a more detailed solution, see the end of the chapter.)

9.7   ESTIMATING THE POPULATION SLOPE TERM

The same sort of inferential argument that we used above can be used to deal with the slope term in regression. For example, we might argue that the slope we produced in the sales-advertising regression ($b = 1.1$) is just one of the possible

FIGURE 9.17 **THE SAMPLING DISTRIBUTION OF ALL POSSIBLE SAMPLE SLOPE TERMS**

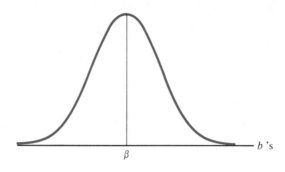

sample slope terms that we *might* have produced by selecting a sample of size 4 from the parent population. Had we taken the time to produce *all* possible samples (sticking with a sample size of 4) from the same population, we would have produced the full list of all possible sample slope terms and labeled this list the sampling distribution of b.

The Sampling Distribution of the Sample Slope b's

It should come as no surprise that (under normal conditions) the sampling distribution of all possible sample b values will follow the familiar bell-shaped pattern, conforming to the standard normal distribution when sample size is at least 30, and to a t distribution in smaller sample cases. In addition, the overall distribution mean (the center of the sampling distribution of sample slope values) is exactly equal to β, the slope of the population regression line (i.e., the slope of the line that best describes the population of points from which we select all the samples). Figure 9.17 shows the appropriate picture.

Building Confidence Interval Estimates of β

Creating confidence interval estimates of the population slope (β) based on one sample b follows recognizable lines. In fact, the standard interval form will look like:

$$b \pm z \cdot (\text{std dev of the } b\text{'s distribution})$$

where z is set according to the confidence level requirement. To compute the standard deviation (or standard error) of the b's distribution (a measure we'll label s_b), we'll use the expression

$$s_b = \frac{s_{y \cdot x}}{\sqrt{\sum(x - \bar{x})^2}}$$

where, as before,

$$s_{y.x} = \sqrt{\frac{\sum(y - \hat{y})^2}{n - 2}}$$

We can then show the confidence interval for β as

$$b \pm z(s_b)$$

or

$$b \pm z\left(\frac{s_{y.x}}{\sqrt{\sum(x - \bar{x})^2}}\right)$$

When sample size is less than 30, z is replaced by an equivalent t score with $n - 2$ degrees of freedom.

EXERCISE 2   For the sales-advertising example, construct a 95% confidence interval estimate of the true population slope (β) using our sample slope value ($b = 1.1$).

Ans: 1.1 ± 2.23 or -1.13 to $+3.33$

(For a more detailed solution, see the end of the chapter.)

Note: As a matter of terminology, the sample slope term b is sometimes called the *estimated regression coefficient* for the xy relationship. Correspondingly, s_b, the standard deviation (or standard error) of the sample b's distribution, is often labeled the **standard error of the regression coefficient**.

9.8   HYPOTHESIS TESTING IN REGRESSION ANALYSIS

In this inferential phase of regression analysis, we might, in addition to building confidence interval estimates of the slope and intercept terms, find use for constructing hypothesis tests involving either of these two values. We'll find it especially useful to test the value of the population *slope*.

In our initial discussion of the sales-advertising example, we began by *suspecting* a relationship between advertising and sales. All our work to this point has been driven by that one basic thought. Yet even though we've come quite a long way, the sales-advertising relationship remains *only a suspicion*. Our job now is to find a way to effectively test that suspicion.

The scientific test we'll devise is fairly straightforward. We'll set up a standard hypothesis test to test, in effect, these two competing positions:

H_0: There is no linear relationship between the variables

H_1: There is a linear relationship

Notice we've purposely chosen the skeptical "no relationship" position as the null, placing the burden of proof on ourselves to convincingly prove otherwise.

The Slope is the Key

To implement the test, we'll need to represent the competing positions in somewhat modified terms. Consider the following proposition concerning the population slope:

> If there's no linear connection between the two variables, then the population slope would necessarily be 0.

or, phrased in a slightly different way,

> If the population slope is not equal to 0, then there's a linear connection between the variables involved.

Figure 9.18 may serve to focus the argument. In Figure 9.18(a) the slope of the population relationship line is 0 (i.e., it's perfectly flat). Here, no matter what value of x is selected, the associated value of y will remain basically unchanged (except for some random or unexplained variability). In such a case, our best guess of y, regardless of x, is always the same (it's the average y value). x simply doesn't influence y. (Figure 9.18(b) shows a slightly more diffuse (spread out) illustration of the same basic case. As was true in 9.18(a), the slope of the population relationship line here is effectively 0, indicating that no relationship exists between the two variables shown.)

In Figure 9.18(c), the situation is markedly different. Here the slope of the line describing the population relationship is clearly not 0. Any estimate of y would have to take into account the given value of x, since here, without question, x influences y. (Figure 9.18(d) shows a more widely dispersed population in which this same kind of linear relationship, though much less precise, is clearly in evidence. Figures 9.18(e) and (f) are simply cases in which the relationship between x and y is negative.)

The crucial point? Population slope is the key to whether a linear relationship exists between the two variables involved. If the population slope is 0, we would have to conclude that no connection exists—x doesn't influence y (at least not in a linear way). If it's not equal to 0, then we've found something significant.

Back to the Test

Given our discussion of slope, can you see how the two hypotheses proposed earlier (to test for a linear relationship) might now be restated? We'll simply replace the "no

FIGURE 9.18   **CONTRASTING POPULATION SLOPE POSSIBILITIES**

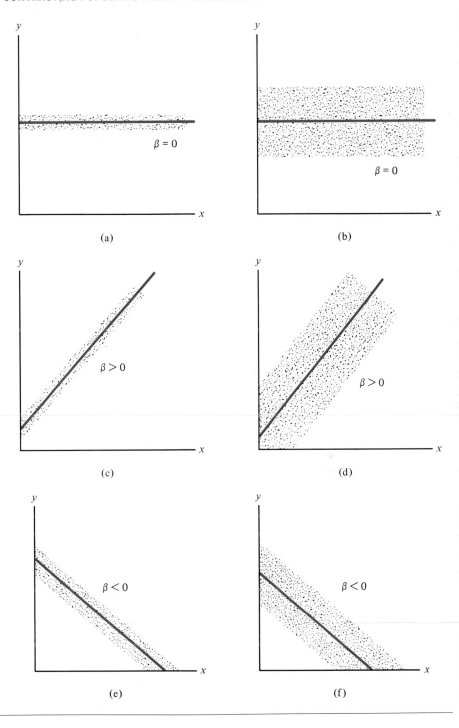

(a)

(b)

(c)

(d)

(e)

(f)

relationship" null hypothesis with

$$H_0: \text{the population slope } (\beta) \text{ is equal to } 0$$

and create as the corresponding alternative,

$$H_1: \text{the population slope } (\beta) \text{ is not equal to } 0$$

Or, more concisely,

$$H_0: \beta = 0$$
$$H_1: \beta \neq 0$$

Having established appropriate hypotheses, we'll need to decide now whether our one sample b is sufficient to reject the "no relationship" ($\beta = 0$) H_0.

> **Note:** We're showing here a nonspecific alternative hypothesis, which indicates that a two-tailed test is in order. It may sometimes be more appropriate to set up the one-tailed version of the test. For instance, in our sales-advertising example, we would likely suspect a *positive* relationship between the variables, suggesting an alternative hypothesis claiming that β is greater than 0. We'll proceed with the two-tailed version of the test and leave it to you to try the one-tailed version on your own.

The hypothesis testing procedure from here on should look pretty familiar. We'll (1) show the sampling distribution appropriate to a true null hypothesis, (2) choose a significance level to establish accept/reject boundaries, and (3) compare our one sample result to the boundaries that have been established. Figure 9.19 shows the general procedure. The sampling distribution involved is obviously the bell-shaped sampling distribution of all possible sample slope (b) values. If the null hypothesis is true (if $\beta = 0$), then this distribution of b's would be centered on 0. The standard deviation for the b's distribution, according to our previous work, would be computed as follows:

$$s_b = \frac{s_{y.x}}{\sqrt{\Sigma(x - \bar{x})^2}}$$

Given what you know about general hypothesis testing procedures, you should be able to take it from here.

EXERCISE 3   Set up a two-tailed hypothesis test to determine whether the sample data we have for the sales-advertising example establishes a linear relationship between sales and advertising. Use a significance level of 5%. (Set up the test using the appropriate t distribution with $4 - 2$ degrees of freedom.)

Ans: The t-score cutoffs are -4.303 and $+4.303$. On the original scale, the two-tailed cutoffs are -2.23 and $+2.23$. With $b = 1.1$, we cannot reject the "no relationship" H_0.

(For a more detailed solution, see the end of the chapter.)

FIGURE 9.19   **BUILDING A HYPOTHESIS TEST TO TEST $\beta = 0$**

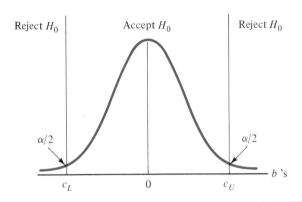

If, as in Exercise 3, we cannot reject the "no relationship" null hypothesis, it's basically back to the drawing board. In such cases, we would probably want to (1) collect more data and try again, (2) look perhaps for a nonlinear connection, or (3) try to identify another variable to explain sales. (Given that we had only four data points to work with, we would almost certainly want to enlarge the data set.)

9.9   BUILDING CONFIDENCE INTERVAL ESTIMATES OF EXPECTED POPULATION *y* VALUES

If the regression line we've produced passes the basic hypothesis test (i.e., if the sample we have leads us to reject the "no relationship" null hypothesis), then we can use the line to estimate (or predict) values for *y*. We had actually proposed this sort of prediction in an earlier discussion of regression procedures, but now the rules, as we'll see, have changed quite a bit.

In our earlier use of the least-squares line, we made predictions of *y* simply by substituting a given value of *x* in the linear equation defining the line. We've since discovered, however, that the original regression line is only a sample-based estimate of the real (population) relationship; the actual intercept term (α) for the *real* linear connection could be higher or lower than the sample intercept *a*, and the *real* slope could be greater or less than the sample-based *b*. It would seem that this sort of combined uncertainty should now be reflected in any estimate of *y* produced from our sample-based line.

In fact, since we have the means to build confidence interval estimates for each of the key population parameters (the slope and the intercept), it should come as no surprise that we can actually construct a confidence interval estimate of the expected population value of *y* for any given *x*. The expression below describes the computations involved. Using x^* to label the given value for *x*, we can say that the expected value of y (which we'll call $E(y_{x*})$) will be in the interval

$$\hat{y} \pm z(s_{y \cdot x}) \sqrt{\frac{1}{n} + \frac{(x^* - \bar{x})^2}{\sum(x - \bar{x})^2}}$$

where $E(y_{x*})$ = expected (or average) population value for y given x^*, a specific
value for x
$\hat{y}$ = sample regression line value for y
x^* = a specific value for the independent variable x

In small sample cases (with $n < 30$) t once again replaces z in the interval expression
with df equal to $n - 2$.

EXERCISE 4   For our sales-advertising example, construct a 95% confidence interval estimate of expected sales if advertising expenditures (x) are \$8 million—
i.e., estimate $E(y_8)$.

Ans: 13.2 ± 10.33 or   \$2.87 to \$23.53 million

(For a more complete solution, see the end of the chapter.)

9.10 BUILDING CONFIDENCE INTERVAL ESTIMATES OF INDIVIDUAL POPULATION *y* VALUES

The interval expression for $E(y_{x*})$ actually estimates the y-coordinate for points lying
along the (unknown) population regression line. If you consider that individual y's
in the population may themselves lie some distance from the population line
(remember, there's variability around the population line just as there's variability
around the sample regression line), it should seem reasonable that the $E(y_{x*})$ interval
would have to be expanded in order to estimate the value of any one of the
individual y's. In this case, we'll label the individual y being estimated as y_{x*} and
show the expanded interval expression as

$$\hat{y} \pm z(s_{y \cdot x}) \sqrt{1 + \frac{1}{n} + \frac{(x^* - \bar{x})^2}{\sum(x - \bar{x})^2}}$$

Clarifying the Difference between $E(y_{x*})$ and y_{x*}

The distinction between estimating an expected value for y and estimating an
individual value for y is a little tricky. As reflected in the notation we've used, $E(y_{x*})$
can actually be viewed as the average (or expected value) of the individual y
possibilities which would lie above and below the population regression line,
directly above any specified x (i.e., x^*). y_{x*} is the label we've attached to any one of
the individual y's. Figure 9.20 shows the contrast and the connection.

To pursue things just a bit further, suppose we had produced a regression line
to relate male heights to male weights. We might subsequently use the line to esti-

FIGURE 9.20   **THE ACTUAL POPULATION RELATIONSHIP CONNECTING x* TO E(y*)**

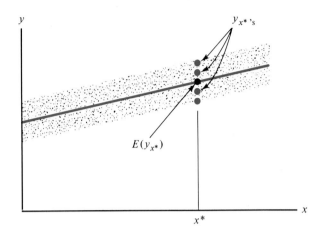

mate, for example, the *average* weight of all males who are 71 inches tall (a value we would label $E(y_{71})$), or we could use the line to estimate the actual weight of any *one* randomly selected male who is 71 inches tall (we'd label this value y_{71}). In the latter case—estimating the weight of an individual—the appropriate interval would (logically, I think) be wider than that used to estimate the overall average weight for the population of males who are 71 inches tall (since, as we've suggested in earlier discussions of sampling theory, there is typically more variability associated with individual values than with overall group averages).

Taking this contrast back to our sales-advertising situation, we could use the regression line we've produced to estimate either the *average* sales for all years in which advertising expenditures are, say, $10 million ($E(y_{10})$), or to estimate sales for any *one* particular year in which advertising is $10 million ($y_{10}$).

EXERCISE 5   Use the sales-advertising regression line to build a 95% confidence interval to estimate actual sales for next year if advertising expenditures will be $8 million. That is, estimate y_8.

Ans: 13.2 ± 11.5 or $1.7 to $24.7 million

(For a more detailed solution, see the end of the chapter.)

CHAPTER SUPPLEMENT: BASIC REGRESSION ASSUMPTIONS

To deal as we have with the inference side of regression analysis—that is, to build confidence intervals and test statistical hypotheses—several assumptions need to

FIGURE 9.21   **THE POPULATION OF POINTS SHOWN NORMALLY DISTRIBUTED AROUND THE POPULATION RELATIONSHIP LINE**

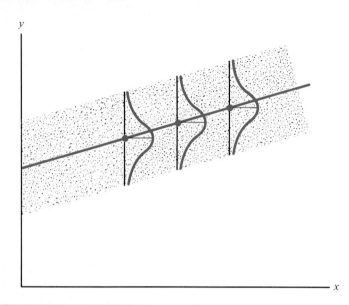

be made concerning characteristics of the population involved. These required characteristics are summarized below.

1. The population of points from which we're selecting our sample has been generated by a function of the form

$$y = \alpha + \beta x + e$$

where e (textbooks often use the Greek letter epsilon) is a normally distributed random variable. In essence, this means that we can envision a kind of "mini" normal distribution lying directly above each value of x, centered on the corresponding line value for y. Figure 9.21 indicates what's involved.

2. The standard deviation of each of these individual normal distributions is the same in each case. (This is often referred to as the condition of *homoscedasticity*, meaning equal scatter or spread.) Figure 9.22 shows the kind of condition that's required.

3. The y values generated by the core population function in (1) are independently determined at each value of x. For example, any value determined for y when $x = 2$ is not related to a value for y when $x = 1$.

At a slightly higher level of analysis, it's possible to formally test whether these population conditions are satisfied.

FIGURE 9.22

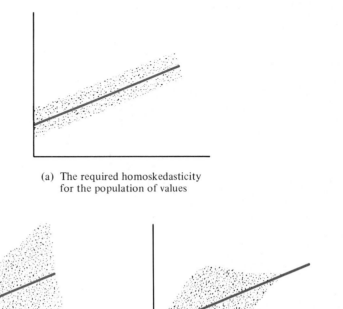

(a) The required homoskedasticity
for the population of values

(b)   (c)

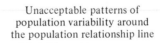

Unacceptable patterns of
population variability around
the population relationship line

CHAPTER SUMMARY

We began by describing the general nature of regression analysis. It's essentially an effort to identify a relationship between certain variables (factors). Simple regression involves just two variables (an independent variable x, which is believed to influence a dependent variable y); multiple regression involves more than two variables. Linear regression focuses on a straight-line relationship; nonlinear regression assumes a more complex connection. Our interest throughout most of the chapter was on simple linear regression.

We saw that at the heart of any regression analysis is a common curve-fitting technique, one using the least-squares criterion (minimizing the sum of squared vertical distances of points from the proposed regression line) to find the best-fitting line for any data set. Once the least-squares line is produced, a number of descriptive measures showing goodness-of-fit or strength-of-relationship can be calculated: the standard error of estimate, a measure of the standard deviation of the data points around the regression line; the coefficient of determination

(r^2), the proportion of the total variability in the dependent variable y that can be explained by the xy relationship; and the correlation coefficient (r), the signed square root of r^2.

Recognizing the sampling side of basic regression, we adapted the inference tools of previous chapters to make population estimates based on sample results. In regression, the population involved is the set of all points that we might conceivably have selected in our acquisition of data for the curve-fitting procedure. We saw that the intercept term "a" produced in least-squares regression is actually a sample result that can be used to estimate the intercept term for the population relationship. Likewise, the least-squares slope term "b" serves as the basis for building an estimate of the real (population) relationship slope.

An important hypothesis testing procedure was proposed to determine whether the sample data provide sufficient grounds to conclude that a linear link exists between the targeted variables. Should the test turn out positive—that is, should it lead us to conclude that a linear relationship exists—the regression line can be used to create predictions, in the form of confidence interval estimates, of population values for y given any specified value for x.

Finally, the essential assumptions on which the inferential side of regression analysis depends were outlined.

KEY TERMS

Coefficient of determination
Correlation coefficient
Dependent variable
Explained variability
Hypothesis tests for population slope
Interval estimates of population intercept
Interval estimates of population slope
Independent variable
Least-squares criterion

Linear regression
Multiple regression
Nonlinear regression
Scatter diagram
Simple regression
Standard error of estimate
Standard error of the slope (also known as standard error of the regression coefficient)
Unexplained variability

IN-CHAPTER EXERCISE SOLUTIONS

EXERCISE 1

Filling in the table below will provide most of the necessary values for the estimate:

x	y	$\hat{y}$	$y - \hat{y}$	$(y - \hat{y})^2$	x^2	$x - \bar{x}$	$(x - \bar{x})^2$
2	6	6.6	−.6	.36	4	−1.5	2.25
3	8	7.7	.3	.09	9	−.5	.25
4	10	8.8	1.2	1.44	16	.5	.25
5	9	9.9	−.9	.81	25	1.5	2.25
Totals: 14	33		0	2.70	54	0	5.00

$$\bar{x} = 14/4 = 3.5$$

$$s_{y.x} = \sqrt{\frac{2.70}{4 - 2}} = 1.16 \qquad s_a = s_{y.x}\sqrt{\frac{\sum x^2}{n\sum(x - \bar{x})^2}}$$

$$= 1.16\sqrt{\frac{54}{4(5.00)}} = 1.91$$

The general interval form is $a \pm t(s_a)$. Using the appropriate t score for 2 degrees of freedom and 95% confidence (and remembering that our least-squares intercept term a was 4.4), the interval is

$$4.4 \pm 4.303(1.90) \quad \text{or} \quad 4.4 \pm 8.2$$

Based on our work, we can be 95% confident that the interval -3.8 to $+12.6$ will contain the population intercept term α. Notice that according to this interval statement, the intercept term for the sales-advertising relationship could be either negative or positive; that is, the true population regression line could cross the vertical y-axis above or below the horizontal x-axis.

EXERCISE 2

Drawing on some of the values in the table in solution 1,

$$s_{y.x} = 1.16$$

$$s_b = \frac{s_{y.x}}{\sqrt{\sum(x - \bar{x})^2}} = \frac{1.16}{\sqrt{5}} = .518$$

The general interval form is $b \pm t(s_b)$. Using the appropriate t score for 2 degrees of freedom and 95% confidence (and remembering that our least squares slope term b was 1.1), the interval will look like

$$1.1 \pm 4.303(.518) \quad \text{or} \quad 1.1 \pm 2.23$$

Based on our work, we can be 95% confident that the interval -1.13 to $+3.33$ will contain the population slope term β. Notice that according to this interval statement, the population slope term here could be negative or positive; that is, the true population regression line for the sales-advertising relationship could slope either upward to the right or downward to the right. Thus, with the data we have, we really don't have a strong indication of how the variables sales and advertising may be related, or if they're related at all.

EXERCISE 3

The hypothesis test follows the general form described in the chapter:

$$H_0: \beta = 0$$

$$H_1: \beta \neq 0$$

If the null hypothesis is true, then the sampling distribution of all possible sample b's should look like this:

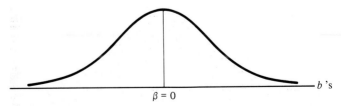

This null distribution is centered on 0. Setting t-score boundaries is pretty straightforward. For a significance level of 5%, and degrees of freedom of $4 - 2 = 2$, the t-score mark-off is

4.303. Computing the t score for the sample b on the null distribution as

$$t_{sample} = \frac{1.1 - 0}{.518} = 2.123$$

and comparing it to the cutoff t score, $t_c = 4.303$, leads us to conclude that there is not enough sample evidence to reject the skeptical null hypothesis.

Alternatively, multiplying the cutoff t score ($t_c = 4.303$) by the standard deviation s_b that we had computed in solution 2, produces a distance of

$$4.303(.518) = 2.23$$

Marking off this distance above and below the hypothesized distribution center of 0 identifies a lower bound (c_L) of -2.23 and an upper bound (c_U) of $+2.23$.

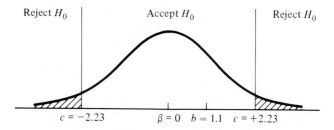

Since our one sample b is 1.1—a value which falls well inside the accept H_0 region of the null distribution—we cannot reject the "no relationship" null hypothesis. Our sample result b simply isn't sufficient evidence to establish that there is a linear connection between sales and advertising.

In either case, having determined that we cannot reject the "no relationship" null, we might now choose to collect more data and try again. (Or we might decide to look for a more complex connection or other variables to explain sales.)

EXERCISE 4

Note: Since we were unable to reject the "no relationship" null hypothesis in the hypothesis test in solution 3, we would generally not proceed with the kind of computations called for here (and in Exercise 5). However, for demonstration purposes, we've shown this next estimation step.

We again draw on information from solution 1 and use $s_{y \cdot x} = 1.16$. The general interval form is

$$\hat{y} \pm t\left(s_{y \cdot x}\sqrt{\frac{1}{n} + \frac{(x^* - \bar{x})^2}{\sum(x - \bar{x})^2}}\right)$$

Using a $\hat{y}$ of 13.2 corresponding to the given x value of 8 ($\hat{y} = 4.4 + 1.1(8) = 13.2$) and substituting appropriately, the interval becomes

$$13.2 \pm 4.303\left(1.16\sqrt{\frac{1}{4} + \frac{(8 - 3.5)^2}{5.00}}\right)$$

$$= 13.2 \pm 4.303(2.4)$$

$$= 13.2 \pm 10.33$$

We can be 95% confident that the interval 2.87 to 23.53 will contain the value for y that lies on the population regression line above $x = 8$. (This value would be labeled $E(y_8)$.) Not unexpectedly, we've produced a very wide interval, again suggesting the inadequacy of our four data points to provide much insight into the relationship between sales and advertising.

EXERCISE 5

The interval form looks a lot like that in solution 4, with one small exception:

$$\hat{y} \pm t\left(s_{y.x}\sqrt{1 + \frac{1}{n} + \frac{(x^* - \bar{x})^2}{\sum(x - \bar{x})^2}}\right)$$

Taking the necessary values from the table in solution 1, the interval becomes

$$13.2 \pm 4.303\left(1.16\sqrt{1 + \frac{1}{4} + \frac{(8 - 3.5)^2}{5.00}}\right)$$

$$= 13.2 \pm 11.5$$

We can be 95% confident that the interval 1.7 to 24.7 will contain the actual value for y (in the population of y's) that lies directly above $x = 8$. (This value would be labeled y_8.)

EXERCISES

Exercises 1 through 9 use the data provided in the following situation: As quality supervisor at Lathrop Technologies, you suspect that a linear relationship exists between the length of the QC inspection interval (i.e., the scheduled time between quality inspections) and the number of defective units produced by your assembly line operation during an 8-hour shift. You have collected the following set of five observations:

Inspection interval (min)	Defective units
50	310
80	300
90	420
110	410
120	460

★ 1. Produce the least-squares regression line for the data shown here.

★ 2. Plot the data and show the regression line you produced in Exercise 1.

★ 3. Compute the standard error of estimate for the least-squares line you produced in Exercise 1 using the generic expression

$$\text{S.E. of E.} = \text{STD ERR OF EST} = \sqrt{\frac{\sum(y - \hat{y})^2}{n}}$$

★ 4. Compute and interpret the coefficient of determination (r^2) for the regression line.

★ 5. Report the r value for the regression line (include the proper sign).

★ 6. Demonstrate the total variability = explained variability + unexplained variability relationship. Fill in values for a table with the headings shown:

x	y	$\hat{y}$	Unexplained $(y - \hat{y})^2$	Explained $(\hat{y} - \bar{y})^2$	Total $(y - \bar{y})^2$

★ 7. Switching to the inferential side of regression,
   (a) compute $s_{y.x}$ and compare it to the standard error of estimate value you produced in Exercise 3.
   (b) produce the 95% confidence interval estimate of the real population intercept term α.
   (c) produce the 95% confidence interval estimate of the real population slope term β.
   (d) construct an appropriate hypothesis test to establish whether the data support a belief that a linear relationship exists between x and y. Show the cutoff(s) on both the original axis and on the t-scale. Use a significance level of 5%. Based on test results, what is your conclusion?

★ 8. Sticking to the inferential side of things,
   (a) construct a 95% confidence interval estimate

of the expected (or average) number of defectives that would be produced by all shifts that use an inspection interval of 100 minutes (i.e., estimate $E(y_{100})$).

(b) construct a 95% confidence interval estimate of the expected (or average) number of defectives that would be produced by all shifts that use an inspection interval of 200 minutes (i.e., estimate $E(y_{200})$).

★ 9. (a) Construct a 95% confidence interval estimate of the actual number of defectives that will be produced during the next shift if an inspection interval of 100 minutes is used (i.e., estimate y_{100}).

(b) Construct a 95% confidence interval estimate of the actual number of defectives that will be produced during the next shift if an inspection interval of 200 minutes is used (i.e., estimate y_{200}).

(Be sure you are comfortable with the difference between the focus of Exercise 8 and that of Exercise 9.)

Exercises 10 through 18 use the data provided in the following situation: You suspect there is a simple linear connection between the price your firm charges for its product and the number of units sold annually. The following data are available for analysis:

Price ($/unit)	Units sold
11	150
13	140
15	100
17	110

★ 10. Produce the least-squares regression line.

11. Plot the data and show the regression line you produced in Exercise 10.

12. Compute the standard error of estimate for the least-squares line you produced in Exercise 10, using the generic expression

$$\text{S.E. of E.} = \text{STD ERR OF EST} = \sqrt{\frac{\sum(y - \hat{y})^2}{n}}$$

13. Compute and interpret the coefficient of determination (r^2) for the regression line.

14. Report the r value for the regression line (include the proper sign).

15. Demonstrate the total variability = explained variability + unexplained variability relationship. Fill in values for a table with the headings shown:

x	y	$\hat{y}$	Unexplained $(y - \hat{y})^2$	Explained $(\hat{y} - \bar{y})^2$	Total $(y - \bar{y})^2$

★ 16. Switching now to the inferential side of regression,

(a) compute $s_{y \cdot x}$ and compare it to the standard error of estimate value you produced in Exercise 12.

(b) produce the 95% confidence interval estimate of the real population intercept term α.

(c) produce the 95% confidence interval estimate of the real population slope term β.

(d) construct an appropriate hypothesis test in order to establish whether the data support a belief that a linear relationship exists between x and y. Show the cutoff(s) on the original axis and on the t-scale. Use a significance level of 5%. Based on test results, what is your conclusion?

17. Sticking to the inferential side of things,

(a) construct a 95% confidence interval estimate of the expected (or average) number of units that would be sold for all those years in which price is set equal to $5 (i.e., estimate $E(y_5)$).

(b) construct a 95% confidence interval estimate of the expected (or average) number of units that would be sold for all those years in which price is set equal to $20 (i.e., estimate $E(y_{20})$).

★ 18. (a) Construct a 95% confidence interval estimate of the actual number of units that will be sold this year if price is set equal to $5 (i.e., estimate y_5).

(b) Construct a 95% confidence interval estimate of the actual number of units that will be sold this year if price is set equal to $20 (i.e., estimate y_{20}).

Exercises 19 through 26 use the data shown in the following situation: Medical science believes that there may be a linear relationship between an individual's exercise habits and his or her serum cholesterol level. The following five observations are available:

Daily exercise (min)	Cholesterol level
20	200
40	160
50	130
80	90
10	240

19. Produce the least-squares regression line.

20. Plot the data and show the regression line you produced in Exercise 19.

21. Compute the standard error of estimate for the least-squares line you produced in Exercise 19, using the generic expression

$$\text{S.E. of E.} = \text{STD ERR OF EST} = \sqrt{\frac{\sum(y - \hat{y})^2}{n}}$$

22. Compute and interpret the coefficient of determination (r^2) for the regression line.

23. Report the r value for the regression line (include the proper sign).

24. Demonstrate the total variability = explained variability + unexplained variability relationship here. Fill in values for a table with the following headings:

			Unexplained	Explained	Total
x	y	$\hat{y}$	$(y - \hat{y})^2$	$(\hat{y} - \bar{y})^2$	$(y - \bar{y})^2$

25. Switching now to the inferential side of regression,

(a) compute $s_{y.x}$ and compare it to the standard error of estimate value you produced in Exercise 21.

(b) produce the 95% confidence interval estimate of the real population intercept term α.

(c) produce the 95% confidence interval estimate of the real population slope term β.

(d) construct an appropriate hypothesis test to establish whether the data support a belief that a linear relationship exists between x and y. Show the cutoff(s) on the original axis and on the t-scale. Use a significance level of 5%. Based on test results, what is your conclusion?

26. (a) Construct a 95% confidence interval estimate of the expected (or average) cholesterol level for all individuals who exercise 40 minutes per day (i.e., estimate $E(y_{40})$).

(b) You plan to exercise 40 minutes per day. Construct a 95% confidence interval estimate of the actual cholesterol level you can anticipate (i.e., estimate y_{40}).

27. Suppose you have done a regression analysis on 50 data points in an attempt to find a linear connection between number of classes absent and final exam score for statistics students at Old State U. The correlation coefficient turned out to be $-.60$.

(a) Discuss what the correlation coefficient

shows here.

(b) If the total variability in the 50 exam scores was 15,500, what must the explained variability have been?

(c) Using your answer to part (b), compute the unexplained variability and use it to produce the standard error of the estimate.

★ 28. Given the least squares regression line $\hat{y} = 10 + 2x$, where $x =$ units of weekly output and $y =$ total production costs, and a summary of related calculations as shown below:

$$\text{Explained variability} = \sum(\hat{y} - \bar{y})^2 = 52,000$$
$$\text{Unexplained variability} = \sum(y - \hat{y})^2 = 8000$$
$$\bar{x} = 80 \qquad \bar{y} = 170 \qquad n = 10$$
$$\sum x^2 = 7000$$
$$\sum(x - \bar{x})^2 = 20,000$$

(a) Compute $s_{y.x}$ (the sample standard error of estimate used to estimate the population standard error of estimate) for these results.

(b) Produce the 95% confidence interval estimate of the population intercept term α.

(c) Produce the 95% confidence interval estimate of the population slope term β.

(d) Construct an appropriate hypothesis test to establish whether the data support a belief that a linear relationship exists between x and y. Show the cutoffs for the test on the t-scale and the original scale. Use a two-tailed test and a significance level of 5%. Based on test results, what is your conclusion?

29. Refer to Exercise 28. Based on the information provided, construct a 95% confidence interval estimate of $E(y_x)$—the expected population value for y (i.e., the expected production cost)—when

(a) $x = 20$ (b) $x = 60$ (c) $x = 80$
(d) $x = 120$ (e) $x = 200$ (f) $x = 300$

(g) Are all the interval widths equal? Note where the interval width is a minimum. Can you make a general statement about where the interval width will always be a minimum?

30. Refer to Exercise 29. Use graph paper to plot

(a) the regression line.

(b) the upper and lower bounds for the 95% confidence intervals you've produced.

(c) Draw a contour line to connect the upper bounds you've plotted. Draw a line to connect the lower bounds as well. Notice the pattern that emerges. This is sometimes referred to as the *French horn effect*. As the given value for x gets further and further from the average x value, the interval boundaries spread farther apart, vaguely tracing the shape of a French horn. This pattern suggests extreme caution when making predictions (estimates) far out on a regression line. As you can see, the estimates produced become less and less precise.

★ 31. A recent article in *National Business* magazine reports that a strong linear connection appears to exist between regional unemployment rates and regional crime. The following data were collected from four sample regions to support this position:

Unemployment rate (%)	Nonviolent crimes per 10,000 population
10	180
8	150
12	230
6	140

(a) Produce the regression line.
(b) Compute r^2.
(c) Compute the standard error of estimate $s_{y \cdot x}$.
(d) Test the "no relationship" null hypothesis. Use a significance level of 5% and a two-tailed test.
(e) Use your regression line to produce a 95% confidence interval estimate of the expected (or average) crime level for all regions in which the unemployment rate is 11 (i.e., estimate $E(y_{11})$).
(f) Use your regression line to produce the 95% confidence interval estimate of the actual crime rate for a particular region that has an unemployment rate of 11 (i.e., estimate y_{11}).

32. Needing a simple cost estimator for current commercial construction costs in the city, you suspect that there is a clear linear relationship between the cost of the building and the amount of floor space required. You have data from five recently completed building projects:

Floor space (in 1000's of sq. ft.)	Cost ($1000's)
10	50
20	90
30	100
40	150
50	210

(a) Produce the regression line.
(b) Compute r^2.
(c) Compute the standard error of estimate $s_{y \cdot x}$.
(d) Test the "no relationship" null hypothesis. Use a significance level of 1%.
(e) Use your regression line to produce a 99% confidence interval estimate of the expected (or average) cost of all buildings in the city that would require 25,000 square feet of floor space (i.e., estimate $E(y_{25})$).
(f) Use the regression line to produce a 99% confidence interval to estimate the cost of a 25,000 square-foot building that company XYZ is about to construct in the city (i.e., estimate y_{25}).

★ 33. Below is a reproduction of output from a standard computer regression package:

		Std error	t score
Slope	6.0	1.25	4.80
Intercept	300.0		

	Sum of squares	Degrees of freedom
Regression	122,000	1
Residual	34,000	23
Total	156,000	24

(a) Explain what the Std Error value represents.
(b) Explain what regression sum of squares, residual sum of squares, and total sum of squares represent.
(c) Compute $s_{y \cdot x}$. (Note: The residual (i.e., unexplained) degrees of freedom in simple linear regression is always equal to $n - 2$. Here, then, the sample size n must be 25.)
(d) Test the "no linear relationship" hypothesis at the 5% level of significance. What is your conclusion? (*Hint:* You just need to check the t score for slope shown in the printout against the 5% cutoff t score from the t table.)
(e) Construct a 95% confidence interval estimate of the population slope term.

34. Below is part of the computer printout for a simple linear regression problem involving a study of group decision making. The independent variable (x) in the experiment was "number of second-year law students included in the group." The dependent variable (y) was "time (in minutes) to reach a consensus." Sample size was 20.

Regression relationship

$$y = 120.0 + 21.8x$$
$$(7.4)$$

Note: The value 7.4 is the standard error term for b.

Analysis of variance

		df
SSR	12000	1
SSE	18000	18
SST	30000	19

(a) Compute the coefficient of determination (r^2).
(b) Compute the correlation coefficient.
(c) Test the "no linear relationship" hypothesis at the 1% level of significance by computing the appropriate t score for the sample slope (i.e., the sample regression coefficient of x) and comparing it to the cutoff t score. Is there significant evidence of a linear connection between the two variables?

STATISTICAL
TABLES

TABLE I   BINOMIAL PROBABILITIES

n	x	.01	.05	.10	.15	.20	p .25	.30	.35	.40	.45	.50
1	0	.9900	.9500	.9000	.8500	.8000	.7500	.7000	.6500	.6000	.5500	.5000
	1	.0100	.0500	.1000	.1500	.2000	.2500	.3000	.3500	.4000	.4500	.5000
2	0	.9801	.9025	.8100	.7225	.6400	.5625	.4900	.4225	.3600	.3025	.2500
	1	.0198	.0950	.1800	.2550	.3200	.3750	.4200	.4550	.4800	.4950	.5000
	2	.0001	.0025	.0100	.0225	.0400	.0625	.0900	.1225	.1600	.2025	.2500
3	0	.9703	.8574	.7290	.6141	.5120	.4219	.3430	.2746	.2160	.1664	.1250
	1	.0294	.1354	.2430	.3251	.3840	.4219	.4410	.4436	.4320	.4084	.3750
	2	.0003	.0071	.0270	.0574	.0960	.1406	.1890	.2389	.2880	.3341	.3750
	3	.0000	.0001	.0010	.0034	.0080	.0156	.0270	.0429	.0640	.0911	.1250
4	0	.9606	.8145	.6561	.5220	.4096	.3164	.2401	.1785	.1296	.0915	.0625
	1	.0388	.1715	.2916	.3685	.4096	.4219	.4116	.3845	.3456	.2995	.2500
	2	.0006	.0135	.0486	.0975	.1536	.2109	.2646	.3105	.3456	.3675	.3750
	3	.0000	.0005	.0036	.0115	.0256	.0469	.0756	.1115	.1536	.2005	.2500
	4	.0000	.0000	.0001	.0005	.0016	.0039	.0081	.0150	.0256	.0410	.0625
5	0	.9510	.7738	.5905	.4437	.3277	.2373	.1681	.1160	.0778	.0503	.0312
	1	.0480	.2036	.3280	.3915	.4096	.3955	.3602	.3124	.2592	.2059	.1562
	2	.0010	.0214	.0729	.1382	.2048	.2637	.3087	.3364	.3456	.3369	.3125
	3	.0000	.0011	.0081	.0244	.0512	.0879	.1323	.1811	.2304	.2757	.3125
	4	.0000	.0000	.0004	.0022	.0064	.0146	.0284	.0488	.0768	.1128	.1562
	5	.0000	.0000	.0000	.0001	.0003	.0010	.0024	.0053	.0102	.0185	.0312
6	0	.9415	.7351	.5314	.3771	.2621	.1780	.1176	.0754	.0467	.0277	.0156
	1	.0571	.2321	.3543	.3993	.3932	.3560	.3025	.2437	.1866	.1359	.0938
	2	.0014	.0305	.0984	.1762	.2458	.2966	.3241	.3280	.3110	.2780	.2344
	3	.0000	.0021	.0146	.0415	.0819	.1318	.1852	.2355	.2765	.3032	.3125
	4	.0000	.0001	.0012	.0055	.0154	.0330	.0595	.0951	.1382	.1861	.2344
	5	.0000	.0000	.0001	.0004	.0015	.0044	.0102	.0205	.0369	.0609	.0938
	6	.0000	.0000	.0000	.0000	.0001	.0002	.0007	.0018	.0041	.0083	.0156
7	0	.9321	.6983	.4783	.3206	.2097	.1335	.0824	.0490	.0280	.0152	.0078
	1	.0659	.2573	.3720	.3960	.3670	.3115	.2471	.1848	.1306	.0872	.0547
	2	.0020	.0406	.1240	.2097	.2753	.3115	.3177	.2985	.2613	.2140	.1641
	3	.0000	.0036	.0230	.0617	.1147	.1730	.2269	.2679	.2903	.2918	.2734
	4	.0000	.0002	.0026	.0109	.0287	.0577	.0972	.1442	.1935	.2388	.2734
	5	.0000	.0000	.0002	.0012	.0043	.0115	.0250	.0466	.0774	.1172	.1641
	6	.0000	.0000	.0000	.0001	.0004	.0013	.0036	.0084	.0172	.0320	.0547
	7	.0000	.0000	.0000	.0000	.0000	.0001	.0002	.0006	.0016	.0037	.0078
8	0	.9227	.6634	.4305	.2725	.1678	.1001	.0576	.0319	.0168	.0084	.0039
	1	.0746	.2793	.3826	.3847	.3355	.2670	.1977	.1373	.0896	.0548	.0312
	2	.0026	.0515	.1488	.2376	.2936	.3115	.2965	.2587	.2090	.1569	.1094
	3	.0001	.0054	.0331	.0839	.1468	.2076	.2541	.2786	.2787	.2568	.2188
	4	.0000	.0004	.0046	.0185	.0459	.0865	.1361	.1875	.2322	.2627	.2734
	5	.0000	.0000	.0004	.0026	.0092	.0231	.0467	.0808	.1239	.1719	.2188
	6	.0000	.0000	.0000	.0002	.0011	.0038	.0100	.0217	.0413	.0703	.1094
	7	.0000	.0000	.0000	.0000	.0001	.0004	.0012	.0033	.0079	.0164	.0312
	8	.0000	.0000	.0000	.0000	.0000	.0000	.0001	.0002	.0007	.0017	.0039

(*continues*)

Table entries represent $P(x/n, p)$.
Example: $P(x = 3/n = 7, p = .4) = .2903$.

TABLE I   (*continued*)

n	x	.01	.05	.10	.15	.20	p .25	.30	.35	.40	.45	.50
9	0	.9135	.6302	.3874	.2316	.1342	.0751	.0404	.0207	.0101	.0046	.0020
	1	.0830	.2985	.3874	.3679	.3020	.2253	.1556	.1004	.0605	.0339	.0176
	2	.0034	.0629	.1722	.2597	.3020	.3003	.2668	.2162	.1612	.1110	.0703
	3	.0001	.0077	.0446	.1069	.1762	.2336	.2668	.2716	.2508	.2119	.1641
	4	.0000	.0006	.0074	.0283	.0661	.1168	.1715	.2194	.2508	.2600	.2461
	5	.0000	.0000	.0008	.0050	.0165	.0389	.0735	.1181	.1672	.2128	.2461
	6	.0000	.0000	.0001	.0006	.0028	.0087	.0210	.0424	.0743	.1160	.1641
	7	.0000	.0000	.0000	.0000	.0003	.0012	.0039	.0098	.0212	.0407	.0703
	8	.0000	.0000	.0000	.0000	.0000	.0001	.0004	.0013	.0035	.0083	.0176
	9	.0000	.0000	.0000	.0000	.0000	.0000	.0000	.0001	.0003	.0008	.0020
10	0	.9044	.5987	.3487	.1969	.1074	.0563	.0282	.0135	.0060	.0025	.0010
	1	.0914	.3151	.3874	.3474	.2684	.1877	.1211	.0725	.0403	.0207	.0098
	2	.0042	.0746	.1937	.2759	.3020	.2816	.2335	.1757	.1209	.0763	.0439
	3	.0001	.0105	.0574	.1298	.2013	.2503	.2668	.2522	.2150	.1665	.1172
	4	.0000	.0010	.0112	.0401	.0881	.1460	.2001	.2377	.2508	.2384	.2051
	5	.0000	.0001	.0015	.0085	.0264	.0584	.1029	.1536	.2007	.2340	.2461
	6	.0000	.0000	.0001	.0012	.0055	.0162	.0368	.0689	.1115	.1596	.2051
	7	.0000	.0000	.0000	.0001	.0008	.0031	.0090	.0212	.0425	.0746	.1172
	8	.0000	.0000	.0000	.0000	.0001	.0004	.0014	.0043	.0106	.0229	.0439
	9	.0000	.0000	.0000	.0000	.0000	.0000	.0001	.0005	.0016	.0042	.0098
	10	.0000	.0000	.0000	.0000	.0000	.0000	.0000	.0000	.0001	.0003	.0010
11	0	.8953	.5688	.3138	.1673	.0859	.0422	.0198	.0088	.0036	.0014	.0005
	1	.0995	.3293	.3835	.3248	.2362	.1549	.0932	.0518	.0266	.0125	.0054
	2	.0050	.0867	.2131	.2866	.2953	.2581	.1998	.1395	.0887	.0513	.0269
	3	.0002	.0137	.0710	.1517	.2215	.2581	.2568	.2254	.1774	.1259	.0806
	4	.0000	.0014	.0158	.0536	.1107	.1721	.2201	.2428	.2365	.2060	.1611
	5	.0000	.0001	.0025	.0132	.0388	.0803	.1321	.1830	.2207	.2360	.2256
	6	.0000	.0000	.0003	.0023	.0097	.0268	.0566	.0985	.1471	.1931	.2256
	7	.0000	.0000	.0000	.0003	.0017	.0064	.0173	.0379	.0701	.1128	.1611
	8	.0000	.0000	.0000	.0000	.0002	.0011	.0037	.0102	.0234	.0462	.0806
	9	.0000	.0000	.0000	.0000	.0000	.0001	.0005	.0018	.0052	.0126	.0269
	10	.0000	.0000	.0000	.0000	.0000	.0000	.0000	.0002	.0007	.0021	.0054
	11	.0000	.0000	.0000	.0000	.0000	.0000	.0000	.0000	.0000	.0002	.0005
12	0	.8864	.5404	.2824	.1422	.0687	.0317	.0138	.0057	.0022	.0008	.0002
	1	.1074	.3413	.3766	.3012	.2062	.1267	.0712	.0368	.0174	.0075	.0029
	2	.0060	.0988	.2301	.2924	.2835	.2323	.1678	.1088	.0639	.0339	.0161
	3	.0002	.0173	.0852	.1720	.2362	.2581	.2397	.1954	.1419	.0923	.0537
	4	.0000	.0021	.0213	.0683	.1329	.1936	.2311	.2367	.2128	.1700	.1208
	5	.0000	.0002	.0038	.0193	.0532	.1032	.1585	.2039	.2270	.2225	.1934
	6	.0000	.0000	.0005	.0040	.0155	.0401	.0792	.1281	.1766	.2124	.2256
	7	.0000	.0000	.0000	.0006	.0033	.0115	.0291	.0591	.1009	.1489	.1934
	8	.0000	.0000	.0000	.0001	.0005	.0024	.0078	.0199	.0420	.0762	.1208
	9	.0000	.0000	.0000	.0000	.0001	.0004	.0015	.0048	.0125	.0277	.0537
	10	.0000	.0000	.0000	.0000	.0000	.0000	.0002	.0008	.0025	.0068	.0161
	11	.0000	.0000	.0000	.0000	.0000	.0000	.0000	.0001	.0003	.0010	.0029
	12	.0000	.0000	.0000	.0000	.0000	.0000	.0000	.0000	.0000	.0001	.0002

(*continues*)

TABLE I (*continued*)

n	x	.01	.05	.10	.15	.20	p .25	.30	.35	.40	.45	.50
13	0	.8775	.5133	.2542	.1209	.0550	.0238	.0097	.0037	.0013	.0004	.0001
	1	.1152	.3512	.3672	.2774	.1787	.1029	.0540	.0259	.0113	.0045	.0016
	2	.0070	.1109	.2448	.2937	.2680	.2059	.1388	.0836	.0453	.0220	.0095
	3	.0003	.0214	.0997	.1900	.2457	.2517	.2181	.1651	.1107	.0660	.0349
	4	.0000	.0028	.0277	.0838	.1535	.2097	.2337	.2222	.1845	.1350	.0873
	5	.0000	.0003	.0055	.0266	.0691	.1258	.1803	.2154	.2214	.1989	.1571
	6	.0000	.0000	.0008	.0063	.0230	.0559	.1030	.1546	.1968	.2169	.2095
	7	.0000	.0000	.0001	.0011	.0058	.0186	.0442	.0833	.1312	.1775	.2095
	8	.0000	.0000	.0000	.0001	.0011	.0047	.0142	.0336	.0656	.1089	.1571
	9	.0000	.0000	.0000	.0000	.0001	.0009	.0034	.0101	.0243	.0495	.0873
	10	.0000	.0000	.0000	.0000	.0000	.0001	.0006	.0022	.0065	.0162	.0349
	11	.0000	.0000	.0000	.0000	.0000	.0000	.0001	.0003	.0012	.0036	.0095
	12	.0000	.0000	.0000	.0000	.0000	.0000	.0000	.0000	.0001	.0005	.0016
	13	.0000	.0000	.0000	.0000	.0000	.0000	.0000	.0000	.0000	.0000	.0001
14	0	.8687	.4877	.2288	.1028	.0440	.0178	.0068	.0024	.0008	.0002	.0001
	1	.1229	.3593	.3559	.2539	.1539	.0832	.0407	.0181	.0073	.0027	.0009
	2	.0081	.1229	.2570	.2912	.2501	.1802	.1134	.0634	.0317	.0141	.0056
	3	.0003	.0259	.1142	.2056	.2501	.2402	.1943	.1366	.0845	.0462	.0222
	4	.0000	.0037	.0349	.0998	.1720	.2202	.2290	.2022	.1549	.1040	.0611
	5	.0000	.0004	.0078	.0352	.0860	.1468	.1963	.2178	.2066	.1701	.1222
	6	.0000	.0000	.0013	.0093	.0322	.0734	.1262	.1759	.2066	.2088	.1833
	7	.0000	.0000	.0002	.0019	.0092	.0280	.0618	.1082	.1574	.1952	.2095
	8	.0000	.0000	.0000	.0003	.0020	.0082	.0232	.0510	.0918	.1398	.1833
	9	.0000	.0000	.0000	.0000	.0003	.0018	.0066	.0183	.0408	.0762	.1222
	10	.0000	.0000	.0000	.0000	.0000	.0003	.0014	.0049	.0136	.0312	.0611
	11	.0000	.0000	.0000	.0000	.0000	.0000	.0002	.0010	.0033	.0093	.0222
	12	.0000	.0000	.0000	.0000	.0000	.0000	.0000	.0001	.0005	.0019	.0056
	13	.0000	.0000	.0000	.0000	.0000	.0000	.0000	.0000	.0001	.0002	.0009
	14	.0000	.0000	.0000	.0000	.0000	.0000	.0000	.0000	.0000	.0000	.0001
15	0	.8601	.4633	.2059	.0874	.0352	.0134	.0047	.0016	.0005	.0001	.0000
	1	.1303	.3658	.3432	.2312	.1319	.0668	.0305	.0126	.0047	.0016	.0005
	2	.0092	.1348	.2669	.2856	.2309	.1559	.0916	.0476	.0219	.0090	.0032
	3	.0004	.0307	.1285	.2184	.2501	.2252	.1700	.1110	.0634	.0318	.0139
	4	.0000	.0049	.0428	.1156	.1876	.2252	.2186	.1792	.1268	.0780	.0417
	5	.0000	.0006	.0105	.0449	.1032	.1651	.2061	.2123	.1859	.1404	.0916
	6	.0000	.0000	.0019	.0132	.0430	.0917	.1472	.1906	.2066	.1914	.1527
	7	.0000	.0000	.0003	.0030	.0138	.0393	.0811	.1319	.1771	.2013	.1964
	8	.0000	.0000	.0000	.0005	.0035	.0131	.0348	.0710	.1181	.1647	.1964
	9	.0000	.0000	.0000	.0001	.0007	.0034	.0116	.0298	.0612	.1048	.1527
	10	.0000	.0000	.0000	.0000	.0001	.0007	.0030	.0096	.0245	.0515	.0916
	11	.0000	.0000	.0000	.0000	.0000	.0001	.0006	.0024	.0074	.0191	.0417
	12	.0000	.0000	.0000	.0000	.0000	.0000	.0001	.0004	.0016	.0052	.0139
	13	.0000	.0000	.0000	.0000	.0000	.0000	.0000	.0001	.0003	.0010	.0032
	14	.0000	.0000	.0000	.0000	.0000	.0000	.0000	.0000	.0000	.0001	.0005
	15	.0000	.0000	.0000	.0000	.0000	.0000	.0000	.0000	.0000	.0000	.0000

(*continues*)

TABLE I   (*continued*)

n	x	.01	.05	.10	.15	.20	p .25	.30	.35	.40	.45	.50
16	0	.8515	.4401	.1853	.0743	.0281	.0100	.0033	.0010	.0003	.0001	.0000
	1	.1376	.3706	.3294	.2097	.1126	.0535	.0228	.0087	.0030	.0009	.0002
	2	.0104	.1463	.2745	.2775	.2111	.1336	.0732	.0353	.0150	.0056	.0018
	3	.0005	.0359	.1423	.2285	.2463	.2079	.1465	.0888	.0468	.0215	.0085
	4	.0000	.0061	.0514	.1311	.2001	.2252	.2040	.1553	.1014	.0572	.0278
	5	.0000	.0008	.0137	.0555	.1201	.1802	.2099	.2008	.1623	.1123	.0667
	6	.0000	.0001	.0028	.0180	.0550	.1101	.1649	.1982	.1983	.1684	.1222
	7	.0000	.0000	.0004	.0045	.0197	.0524	.1010	.1524	.1889	.1969	.1746
	8	.0000	.0000	.0001	.0009	.0055	.0197	.0487	.0923	.1417	.1812	.1964
	9	.0000	.0000	.0000	.0001	.0012	.0058	.0185	.0442	.0840	.1318	.1746
	10	.0000	.0000	.0000	.0000	.0002	.0014	.0056	.0167	.0392	.0755	.1222
	11	.0000	.0000	.0000	.0000	.0000	.0002	.0013	.0049	.0142	.0337	.0667
	12	.0000	.0000	.0000	.0000	.0000	.0000	.0002	.0011	.0040	.0115	.0278
	13	.0000	.0000	.0000	.0000	.0000	.0000	.0000	.0002	.0008	.0029	.0085
	14	.0000	.0000	.0000	.0000	.0000	.0000	.0000	.0000	.0001	.0005	.0018
	15	.0000	.0000	.0000	.0000	.0000	.0000	.0000	.0000	.0000	.0001	.0002
	16	.0000	.0000	.0000	.0000	.0000	.0000	.0000	.0000	.0000	.0000	.0000
17	0	.8429	.4181	.1668	.0631	.0225	.0075	.0023	.0007	.0002	.0000	.0000
	1	.1447	.3741	.3150	.1893	.0957	.0426	.0169	.0060	.0019	.0005	.0001
	2	.0117	.1575	.2800	.2673	.1914	.1136	.0581	.0260	.0102	.0035	.0010
	3	.0006	.0415	.1556	.2359	.2393	.1893	.1245	.0701	.0341	.0144	.0052
	4	.0000	.0076	.0605	.1457	.2093	.2209	.1868	.1320	.0796	.0411	.0182
	5	.0000	.0010	.0175	.0668	.1361	.1914	.2081	.1849	.1379	.0875	.0472
	6·	.0000	.0001	.0039	.0236	.0680	.1276	.1784	.1991	.1839	.1432	.0944
	7	.0000	.0000	.0007	.0065	.0267	.0668	.1201	.1685	.1927	.1841	.1484
	8	.0000	.0000	.0001	.0014	.0084	.0279	.0644	.1134	.1606	.1883	.1855
	9	.0000	.0000	.0000	.0003	.0021	.0093	.0276	.0611	.1070	.1540	.1855
	10	.0000	.0000	.0000	.0000	.0004	.0025	.0095	.0263	.0571	.1008	.1484
	11	.0000	.0000	.0000	.0000	.0001	.0005	.0026	.0090	.0242	.0525	.0944
	12	.0000	.0000	.0000	.0000	.0000	.0001	.0006	.0024	.0081	.0215	.0472
	13	.0000	.0000	.0000	.0000	.0000	.0000	.0001	.0005	.0021	.0068	.0182
	14	.0000	.0000	.0000	.0000	.0000	.0000	.0000	.0001	.0004	.0016	.0052
	15	.0000	.0000	.0000	.0000	.0000	.0000	.0000	.0000	.0001	.0003	.0010
	16	.0000	.0000	.0000	.0000	.0000	.0000	.0000	.0000	.0000	.0000	.0001
	17	.0000	.0000	.0000	.0000	.0000	.0000	.0000	.0000	.0000	.0000	.0000
18	0	.8345	.3972	.1501	.0536	.0180	.0056	.0016	.0004	.0001	.0000	.0000
	1	.1517	.3763	.3002	.1704	.0811	.0338	.0126	.0042	.0012	.0003	.0001
	2	.0130	.1683	.2835	.2556	.1723	.0958	.0458	.0190	.0069	.0022	.0006
	3	.0007	.0473	.1680	.2406	.2297	.1704	.1046	.0547	.0246	.0095	.0031
	4	.0000	.0093	.0700	.1592	.2153	.2130	.1681	.1104	.0614	.0291	.0117
	5	.0000	.0014	.0218	.0787	.1507	.1988	.2017	.1664	.1146	.0666	.0327
	6	.0000	.0002	.0052	.0301	.0816	.1436	.1873	.1941	.1655	.1181	.0708
	7	.0000	.0000	.0010	.0091	.0350	.0820	.1376	.1792	.1892	.1657	.1214
	8	.0000	.0000	.0002	.0022	.0120	.0376	.0811	.1327	.1734	.1864	.1669
	9	.0000	.0000	.0000	.0004	.0033	.0139	.0386	.0794	.1284	.1694	.1855
	10	.0000	.0000	.0000	.0001	.0008	.0042	.0149	.0385	.0771	.1248	.1669
	11	.0000	.0000	.0000	.0000	.0001	.0010	.0046	.0151	.0374	.0742	.1214
	12	.0000	.0000	.0000	.0000	.0000	.0002	.0012	.0047	.0145	.0354	.0708
	13	.0000	.0000	.0000	.0000	.0000	.0000	.0002	.0012	.0045	.0134	.0327
	14	.0000	.0000	.0000	.0000	.0000	.0000	.0000	.0002	.0011	.0039	.0117
	15	.0000	.0000	.0000	.0000	.0000	.0000	.0000	.0000	.0002	.0009	.0031
	16	.0000	.0000	.0000	.0000	.0000	.0000	.0000	.0000	.0000	.0001	.0006
	17	.0000	.0000	.0000	.0000	.0000	.0000	.0000	.0000	.0000	.0000	.0001
	18	.0000	.0000	.0000	.0000	.0000	.0000	.0000	.0000	.0000	.0000	.0000

(*continues*)

TABLE I   (*continued*)

n	x	.01	.05	.10	.15	.20	p .25	.30	.35	.40	.45	.50
19	0	.8262	.3774	.1351	.0456	.0144	.0042	.0011	.0003	.0001	.0000	.0000
	1	.1586	.3774	.2852	.1529	.0685	.0268	.0093	.0029	.0008	.0002	.0000
	2	.0144	.1787	.2852	.2428	.1540	.0803	.0358	.0138	.0046	.0013	.0003
	3	.0008	.0533	.1796	.2428	.2182	.1517	.0869	.0422	.0175	.0062	.0018
	4	.0000	.0112	.0798	.1714	.2182	.2023	.1491	.0909	.0467	.0203	.0074
	5	.0000	.0018	.0266	.0907	.1636	.2023	.1916	.1468	.0933	.0497	.0222
	6	.0000	.0002	.0069	.0374	.0955	.1574	.1916	.1844	.1451	.0949	.0518
	7	.0000	.0000	.0014	.0122	.0443	.0974	.1525	.1844	.1797	.1443	.0961
	8	.0000	.0000	.0002	.0032	.0166	.0487	.0981	.1489	.1797	.1771	.1442
	9	.0000	.0000	.0000	.0007	.0051	.0198	.0514	.0980	.1464	.1771	.1762
	10	.0000	.0000	.0000	.0001	.0013	.0066	.0220	.0528	.0976	.1449	.1762
	11	.0000	.0000	.0000	.0000	.0003	.0018	.0077	.0233	.0532	.0970	.1442
	12	.0000	.0000	.0000	.0000	.0000	.0004	.0022	.0083	.0237	.0529	.0961
	13	.0000	.0000	.0000	.0000	.0000	.0001	.0005	.0024	.0085	.0233	.0518
	14	.0000	.0000	.0000	.0000	.0000	.0000	.0001	.0006	.0024	.0082	.0222
	15	.0000	.0000	.0000	.0000	.0000	.0000	.0000	.0001	.0005	.0022	.0074
	16	.0000	.0000	.0000	.0000	.0000	.0000	.0000	.0000	.0001	.0005	.0018
	17	.0000	.0000	.0000	.0000	.0000	.0000	.0000	.0000	.0000	.0001	.0003
	18	.0000	.0000	.0000	.0000	.0000	.0000	.0000	.0000	.0000	.0000	.0000
	19	.0000	.0000	.0000	.0000	.0000	.0000	.0000	.0000	.0000	.0000	.0000
20	0	.8179	.3585	.1216	.0388	.0115	.0032	.0008	.0002	.0000	.0000	.0000
	1	.1652	.3774	.2702	.1368	.0576	.0211	.0068	.0020	.0005	.0001	.0000
	2	.0159	.1887	.2852	.2293	.1369	.0669	.0278	.0100	.0031	.0008	.0002
	3	.0010	.0596	.1901	.2428	.2054	.1339	.0716	.0323	.0123	.0040	.0011
	4	.0000	.0133	.0898	.1821	.2182	.1897	.1304	.0738	.0350	.0139	.0046
	5	.0000	.0022	.0319	.1028	.1746	.2023	.1789	.1272	.0746	.0365	.0148
	6	.0000	.0003	.0089	.0454	.1091	.1686	.1916	.1712	.1244	.0746	.0370
	7	.0000	.0000	.0020	.0160	.0545	.1124	.1643	.1844	.1659	.1221	.0739
	8	.0000	.0000	.0004	.0046	.0222	.0609	.1144	.1614	.1797	.1623	.1201
	9	.0000	.0000	.0001	.0011	.0074	.0271	.0654	.1158	.1597	.1771	.1602
	10	.0000	.0000	.0000	.0002	.0020	.0099	.0308	.0686	.1171	.1593	.1762
	11	.0000	.0000	.0000	.0000	.0005	.0030	.0120	.0336	.0710	.1185	.1602
	12	.0000	.0000	.0000	.0000	.0001	.0008	.0039	.0136	.0355	.0727	.1201
	13	.0000	.0000	.0000	.0000	.0000	.0002	.0010	.0045	.0146	.0366	.0739
	14	.0000	.0000	.0000	.0000	.0000	.0000	.0002	.0012	.0049	.0150	.0370
	15	.0000	.0000	.0000	.0000	.0000	.0000	.0000	.0003	.0013	.0049	.0148
	16	.0000	.0000	.0000	.0000	.0000	.0000	.0000	.0000	.0003	.0013	.0046
	17	.0000	.0000	.0000	.0000	.0000	.0000	.0000	.0000	.0000	.0002	.0011
	18	.0000	.0000	.0000	.0000	.0000	.0000	.0000	.0000	.0000	.0000	.0002
	19	.0000	.0000	.0000	.0000	.0000	.0000	.0000	.0000	.0000	.0000	.0000
	20	.0000	.0000	.0000	.0000	.0000	.0000	.0000	.0000	.0000	.0000	.0000
25	0	.7778	.2774	.0718	.0172	.0038	.0008	.0001	.0000	.0000	.0000	.0000
	1	.1964	.3650	.1994	.0759	.0236	.0063	.0014	.0003	.0000	.0000	.0000
	2	.0238	.2305	.2659	.1607	.0708	.0251	.0074	.0018	.0004	.0001	.0000
	3	.0018	.0930	.2265	.2174	.1358	.0641	.0243	.0076	.0019	.0004	.0001
	4	.0001	.0269	.1384	.2110	.1867	.1175	.0572	.0224	.0071	.0018	.0004
	5	.0000	.0060	.0646	.1564	.1960	.1645	.1030	.0506	.0199	.0063	.0016
	6	.0000	.0010	.0239	.0920	.1633	.1828	.1472	.0908	.0442	.0172	.0053
	7	.0000	.0001	.0072	.0441	.1108	.1654	.1712	.1327	.0800	.0381	.0143
	8	.0000	.0000	.0018	.0175	.0623	.1241	.1651	.1607	.1200	.0701	.0322
	9	.0000	.0000	.0004	.0058	.0294	.0781	.1336	.1635	.1511	.1084	.0609

(continues)

TABLE I   (*continued*)

n	x	.01	.05	.10	.15	.20	p .25	.30	.35	.40	.45	.50
	10	.0000	.0000	.0001	.0016	.0118	.0417	.0916	.1409	.1612	.1419	.0974
	11	.0000	.0000	.0000	.0004	.0040	.0189	.0536	.1034	.1465	.1583	.1328
	12	.0000	.0000	.0000	.0001	.0012	.0074	.0268	.0650	.1140	.1511	.1550
	13	.0000	.0000	.0000	.0000	.0003	.0025	.0115	.0350	.0760	.1236	.1550
	14	.0000	.0000	.0000	.0000	.0001	.0007	.0042	.0161	.0434	.0867	.1328
	15	.0000	.0000	.0000	.0000	.0000	.0002	.0013	.0064	.0212	.0520	.0974
	16	.0000	.0000	.0000	.0000	.0000	.0000	.0004	.0021	.0088	.0266	.0609
	17	.0000	.0000	.0000	.0000	.0000	.0000	.0001	.0006	.0031	.0115	.0322
	18	.0000	.0000	.0000	.0000	.0000	.0000	.0000	.0001	.0009	.0042	.0143
	19	.0000	.0000	.0000	.0000	.0000	.0000	.0000	.0000	.0002	.0013	.0053
	20	.0000	.0000	.0000	.0000	.0000	.0000	.0000	.0000	.0000	.0003	.0016
	21	.0000	.0000	.0000	.0000	.0000	.0000	.0000	.0000	.0000	.0001	.0004
	22	.0000	.0000	.0000	.0000	.0000	.0000	.0000	.0000	.0000	.0000	.0001
30	0	.7397	.2146	.0424	.0076	.0012	.0002	.0000	.0000	.0000	.0000	.0000
	1	.2242	.3389	.1413	.0404	.0093	.0018	.0003	.0000	.0000	.0000	.0000
	2	.0328	.2586	.2277	.1034	.0337	.0086	.0018	.0003	.0000	.0000	.0000
	3	.0031	.1270	.2361	.1703	.0785	.0269	.0072	.0015	.0003	.0000	.0000
	4	.0002	.0451	.1771	.2028	.1325	.0604	.0208	.0056	.0012	.0002	.0000
	5	.0000	.0124	.1023	.1861	.1723	.1047	.0464	.0157	.0041	.0008	.0001
	6	.0000	.0027	.0474	.1368	.1795	.1455	.0829	.0353	.0115	.0029	.0006
	7	.0000	.0005	.0180	.0828	.1538	.1662	.1219	.0652	.0263	.0081	.0019
	8	.0000	.0001	.0058	.0420	.1106	.1593	.1501	.1009	.0505	.0191	.0055
	9	.0000	.0000	.0016	.0181	.0676	.1298	.1573	.1328	.0823	.0382	.0133
	10	.0000	.0000	.0004	.0067	.0355	.0909	.1416	.1502	.1152	.0656	.0280
	11	.0000	.0000	.0001	.0022	.0161	.0551	.1103	.1471	.1396	.0976	.0509
	12	.0000	.0000	.0000	.0006	.0064	.0291	.0749	.1254	.1474	.1265	.0806
	13	.0000	.0000	.0000	.0001	.0022	.0134	.0444	.0935	.1360	.1433	.1115
	14	.0000	.0000	.0000	.0000	.0007	.0054	.0231	.0611	.1101	.1424	.1354
	15	.0000	.0000	.0000	.0000	.0002	.0019	.0106	.0351	.0783	.1242	.1445
	16	.0000	.0000	.0000	.0000	.0000	.0006	.0042	.0177	.0489	.0953	.1354
	17	.0000	.0000	.0000	.0000	.0000	.0002	.0015	.0079	.0269	.0642	.1115
	18	.0000	.0000	.0000	.0000	.0000	.0000	.0005	.0031	.0129	.0379	.0806
	19	.0000	.0000	.0000	.0000	.0000	.0000	.0001	.0010	.0054	.0196	.0509
	20	.0000	.0000	.0000	.0000	.0000	.0000	.0000	.0003	.0020	.0088	.0280
	21	.0000	.0000	.0000	.0000	.0000	.0000	.0000	.0001	.0006	.0034	.0133
	22	.0000	.0000	.0000	.0000	.0000	.0000	.0000	.0000	.0002	.0012	.0055
	23	.0000	.0000	.0000	.0000	.0000	.0000	.0000	.0000	.0000	.0003	.0019
	24	.0000	.0000	.0000	.0000	.0000	.0000	.0000	.0000	.0000	.0001	.0006
	25	.0000	.0000	.0000	.0000	.0000	.0000	.0000	.0000	.0000	.0000	.0001

TABLE II   **NORMAL CURVE AREAS**

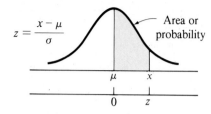

$$z = \frac{x - \mu}{\sigma}$$

z	.00	.01	.02	.03	.04	.05	.06	.07	.08	.09
.0	.0000	.0040	.0080	.0120	.0160	.0199	.0239	.0279	.0319	.0359
.1	.0398	.0438	.0478	.0517	.0557	.0596	.0636	.0675	.0714	.0753
.2	.0793	.0832	.0871	.0910	.0948	.0987	.1026	.1064	.1103	.1141
.3	.1179	.1217	.1255	.1293	.1331	.1368	.1406	.1443	.1480	.1517
.4	.1554	.1591	.1628	.1664	.1700	.1736	.1772	.1808	.1844	.1879
.5	.1915	.1950	.1985	.2019	.2054	.2088	.2123	.2157	.2190	.2224
.6	.2257	.2291	.2324	.2357	.2389	.2422	.2454	.2486	.2518	.2549
.7	.2580	.2612	.2642	.2673	.2704	.2734	.2764	.2794	.2823	.2852
.8	.2881	.2910	.2939	.2967	.2995	.3023	.3051	.3078	.3106	.3133
.9	.3159	.3186	.3212	.3238	.3264	.3289	.3315	.3340	.3365	.3389
1.0	.3413	.3438	.3461	.3485	.3508	.3531	.3554	.3577	.3599	.3621
1.1	.3643	.3665	.3686	.3708	.3729	.3749	.3770	.3790	.3810	.3830
1.2	.3849	.3869	.3888	.3907	.3925	.3944	.3962	.3980	.3997	.4015
1.3	.4032	.4049	.4066	.4082	.4099	.4115	.4131	.4147	.4162	.4177
1.4	.4192	.4207	.4222	.4236	.4251	.4265	.4279	.4292	.4306	.4319
1.5	.4332	.4345	.4357	.4370	.4382	.4394	.4406	.4418	.4429	.4441
1.6	.4452	.4463	.4474	.4484	.4495	.4505	.4515	.4525	.4535	.4545
1.7	.4554	.4564	.4573	.4582	.4591	.4599	.4608	.4616	.4625	.4633
1.8	.4641	.4649	.4656	.4664	.4671	.4678	.4686	.4693	.4699	.4706
1.9	.4713	.4719	.4726	.4732	.4738	.4744	.4750	4756	.4761	.4767
2.0	.4772	.4778	.4783	.4788	.4793	.4798	.4803	.4808	.4812	.4817
2.1	.4821	.4826	.4830	.4834	.4838	.4842	.4846	.4850	.4854	.4857
2.2	.4861	.4864	.4868	.4871	.4875	.4878	.4881	.4884	.4887	.4890
2.3	.4893	.4896	.4898	.4901	.4904	.4906	.4909	.4911	.4913	.4916
2.4	.4918	.4920	.4922	.4925	.4927	.4929	.4931	.4932	.4934	.4936
2.5	.4938	.4940	.4941	.4943	.4945	.4946	.4948	.4949	.4951	.4952
2.6	.4953	.4955	.4956	.4957	.4959	.4960	.4961	.4962	.4963	.4964
2.7	.4965	.4966	.4967	.4968	.4969	.4970	.4971	.4972	.4973	.4974
2.8	.4974	.4975	.4976	.4977	.4977	.4978	.4979	.4979	.4980	.4981
2.9	.4981	.4982	.4982	.4983	.4984	.4984	.4985	.4985	.4986	.4986
3.0	.4986	.4987	.4987	.4988	.4988	.4989	.4989	.4989	.4990	.4990

Example: The area (probability) between the mean and $z = 1.24$ is equal to .3925.

TABLE III **POISSON PROBABILITIES**

x	0.1	0.2	0.3	0.4	λ 0.5	0.6	0.7	0.8	0.9	1.0
0	.9048	.8187	.7408	.6703	.6065	.5488	.4966	.4493	.4066	.3679
1	.0905	.1637	.2222	.2681	.3033	.3293	.3476	.3595	.3659	.3679
2	.0045	.0164	.0333	.0536 ·	.0758	.0988	.1217	.1438	.1647	.1839
3	.0002	.0011	.0033	.0072	.0126	.0198	.0284	.0383	.0494	.0613
4	.0000	.0001	.0002	.0007	.0016	.0030	.0050	.0077	.0111	.0153
5	.0000	.0000	.0000	.0001	.0002	.0004	.0007	.0012	.0020	.0031
6	.0000	.0000	.0000	.0000	.0000	.0000	.0001	.0002	.0003	.0005
7	.0000	.0000	.0000	.0000	.0000	.0000	.0000	.0000	.0000	.0001

x	1.1	1.2	1.3	1.4	λ 1.5	1.6	1.7	1.8	1.9	2.0
0	.3329	.3012	.2725	.2466	.2231	.2019	.1827	.1653	.1496	.1353
1	.3662	.3614	.3543	.3452	.3347	.3230	.3106	.2975	.2842	.2707
2	.2014	.2169	.2303	.2417	.2510	.2584	.2640	.2678	.2700	.2707
3	.0738	.0867	.0998	.1128	.1255	.1378	.1496	.1607	.1710	.1804
4	.0203	.0260	.0324	.0395	.0471	.0551	.0636	.0723	.0812	.0902
5	.0045	.0062	.0084	.0111	.0141	.0176	.0216	.0260	.0309	.0361
6	.0008	.0012	.0018	.0026	.0035	.0047	.0061	.0078	.0098	.0120
7	.0001	.0002	.0003	.0005	.0008	.0011	.0015	.0020	.0027	.0034
8	.0000	.0000	.0001	.0001	.0001	.0002	.0003	.0005	.0006	.0009
9	.0000	.0000	.0000	.0000	.0000	.0000	.0001	.0001	.0001	.0002

x	2.1	2.2	2.3	2.4	λ 2.5	2.6	2.7	2.8	2.9	3.0
0	.1225	.1108	.1003	.0907	.0821	.0743	.0672	.0608	.0550	.0498
1	.2572	.2438	.2306	.2177	.2052	.1931	.1815	.1703	.1596	.1494
2	.2700	.2681	.2652	.2613	.2565	.2510	.2450	.2384	.2314	.2240
3	.1890	.1966	.2033	.2090	.2138	.2176	.2205	.2225	.2237	.2240
4	.0992	.1082	.1169	.1254	.1336	.1414	.1488	.1557	.1622	.1680
5	.0417	.0476	.0538	.0602	.0668	.0735	.0804	.0872	.0940	.1008
6	.0146	.0174	.0206	.0241	.0278	.0319	.0362	.0407	.0455	.0504
7	.0044	.0055	.0068	.0083	.0099	.0118	.0139	.0163	.0188	.0216
8	.0011	.0015	.0019	.0025	.0031	.0038	.0047	.0057	.0068	.0081
9	.0003	.0004	.0005	.0007	.0009	.0011	.0014	.0018	.0022	.0027
10	.0001	.0001	.0001	.0002	.0002	.0003	.0004	.0005	.0006	.0008
11	.0000	.0000	.0000	.0000	.0000	.0001	.0001	.0001	.0002	.0002
12	.0000	.0000	.0000	.0000	.0000	.0000	.0000	.0000	.0000	.0001

x	3.1	3.2	3.3	3.4	λ 3.5	3.6	3.7	3.8	3.9	4.0
0	.0450	.0408	.0369	.0344	.0302	.0273	.0247	.0224	.0202	.0183
1	.1397	.1304	.1217	.1135	.1057	.0984	.0915	.0850	.0789	.0733
2	.2165	.2087	.2008	.1929	.1850	.1771	.1692	.1615	.1539	.1465
3	.2237	.2226	.2209	.2186	.2158	.2125	.2087	.2046	.2001	.1954
4	.1734	.1781	.1823	.1858	.1888	.1912	.1931	.1944	.1951	.1954

(*continues*)

Example: $P(x = 4/\lambda = 2.2) = .1082$.

TABLE III   (*continued*)

x	3.1	3.2	3.3	3.4	λ 3.5	3.6	3.7	3.8	3.9	4.0
5	.1075	.1140	.1203	.1264	.1322	.1377	.1429	.1477	.1522	.1563
6	.0555	.0608	.0662	.0716	.0771	.0826	.0881	.0936	.0989	.1042
7	.0246	.0278	.0312	.0348	.0385	.0425	.0466	.0508	.0551	.0595
8	.0095	.0111	.0129	.0148	.0169	.0191	.0215	.0241	.0269	.0298
9	.0033	.0040	.0047	.0056	.0066	.0076	.0089	.0102	.0116	.0132
10	.0010	.0013	.0016	.0019	.0023	.0028	.0033	.0039	.0045	.0053
11	.0003	.0004	.0005	.0006	.0007	.0009	.0011	.0013	.0016	.0019
12	.0001	.0001	.0001	.0002	.0002	.0003	.0003	.0004	.0005	.0006
13	.0000	.0000	.0000	.0000	.0001	.0001	.0001	.0001	.0002	.0002
14	.0000	.0000	.0000	.0000	.0000	.0000	.0000	.0000	.0000	.0001

x	4.1	4.2	4.3	4.4	λ 4.5	4.6	4.7	4.8	4.9	5.0
0	.0166	.0150	.0136	.0123	.0111	.0101	.0091	.0082	.0074	.0067
1	.0679	.0630	.0583	.0540	.0500	.0462	.0427	.0395	.0365	.0337
2	.1393	.1323	.1254	.1188	.1125	.1063	.1005	.0948	.0894	.0842
3	.1904	.1852	.1798	.1743	.1687	.1631	.1574	.1517	.1460	.1404
4	.1951	.1944	.1933	.1917	.1898	.1875	.1849	.1820	.1789	.1755
5	.1600	.1633	.1662	.1687	.1708	.1725	.1738	.1747	.1753	.1755
6	.1093	.1143	.1191	.1237	.1281	.1323	.1362	.1398	.1432	.1462
7	.0640	.0686	.0732	.0778	.0824	.0869	.0914	.0959	.1002	.1044
8	.0328	.0360	.0393	.0428	.0463	.0500	.0537	.0575	.0614	.0653
9	.0150	.0168	.0188	.0209	.0232	.0255	.0280	.0307	.0334	.0363
10	.0061	.0071	.0081	.0092	.0104	.0118	.0132	.0147	.0164	.0181
11	.0023	.0027	.0032	.0037	.0043	.0049	.0056	.0064	.0073	.0082
12	.0008	.0009	.0011	.0014	.0016	.0019	.0022	.0026	.0030	.0034
13	.0002	.0003	.0004	.0005	.0006	.0007	.0008	.0009	.0011	.0013
14	.0001	.0001	.0001	.0001	.0002	.0002	.0003	.0003	.0004	.0005
15	.0000	.0000	.0000	.0000	.0001	.0001	.0001	.0001	.0001	.0002

x	5.1	5.2	5.3	5.4	λ 5.5	5.6	5.7	5.8	5.9	6.0
0	.0061	.0055	.0050	.0045	.0041	.0037	.0033	.0030	.0027	.0025
1	.0311	.0287	.0265	.0244	.0225	.0207	.0191	.0176	.0162	.0149
2	.0793	.0746	.0701	.0659	.0618	.0580	.0544	.0509	.0477	.0446
3	.1348	.1293	.1239	.1185	.1133	.1082	.1033	.0985	.0938	.0892
4	.1719	.1681	.1641	.1600	.1558	.1515	.1472	.1428	.1383	.1339
5	.1753	.1748	.1740	.1728	.1714	.1697	.1678	.1656	.1632	.1606
6	.1490	.1515	.1537	.1555	.1571	.1584	.1594	.1601	.1605	.1606
7	.1086	.1125	.1163	.1200	.1234	.1267	.1298	.1326	.1353	.1377
8	.0692	.0731	.0771	.0810	.0849	.0887	.0925	.0962	.0998	.1033
9	.0392	.0423	.0454	.0486	.0519	.0552	.0586	.0620	.0654	.0688
10	.0200	.0220	.0241	.0262	.0285	.0309	.0334	.0359	.0386	.0413
11	.0093	.0104	.0116	.0129	.0143	.0157	.0173	.0190	.0207	.0225
12	.0039	.0045	.0051	.0058	.0065	.0073	.0082	.0092	.0102	.0113
13	.0015	.0018	.0021	.0024	.0028	.0032	.0036	.0041	.0046	.0052
14	.0006	.0007	.0008	.0009	.0011	.0013	.0015	.0017	.0019	.0022
15	.0002	.0002	.0003	.0003	.0004	.0005	.0006	.0007	.0008	.0009
16	.0001	.0001	.0001	.0001	.0001	.0002	.0002	.0002	.0003	.0003
17	.0000	.0000	.0000	.0000	.0000	.0001	.0001	.0001	.0001	.0001

(*continues*)

TABLE III   (*continued*)

x	6.1	6.2	6.3	6.4	λ 6.5	6.6	6.7	6.8	6.9	7.0
0	.0022	.0020	.0018	.0017	.0015	.0014	.0012	.0011	.0010	.0009
1	.0137	.0126	.0116	.0106	.0098	.0090	.0082	.0076	.0070	.0064
2	.0417	.0390	.0364	.0340	.0318	.0296	.0276	.0258	.0240	.0223
3	.0848	.0806	.0765	.0726	.0688	.0652	.0617	.0584	.0552	.0521
4	.1294	.1249	.1205	.1162	.1118	.1076	.1034	.0992	.0952	.0912
5	.1579	.1549	.1519	.1487	.1454	.1420	.1385	.1349	.1314	.1277
6	.1605	.1601	.1595	.1586	.1575	.1562	.1546	.1529	.1511	.1490
7	.1399	.1418	.1435	.1450	.1462	.1472	.1480	.1486	.1489	.1490
8	.1066	.1099	.1130	.1160	.1188	.1215	.1240	.1263	.1284	.1304
9	.0723	.0757	.0791	.0825	.0858	.0891	.0923	.0954	.0985	.1014
10	.0441	.0469	.0498	.0528	.0558	.0588	.0618	.0649	.0679	.0710
11	.0245	.0265	.0285	.0307	.0330	.0353	.0377	.0401	.0426	.0452
12	.0124	.0137	.0150	.0164	.0179	.0194	.0210	.0227	.0245	.0264
13	.0058	.0065	.0073	.0081	.0089	.0098	.0108	.0119	.0130	.0142
14	.0025	.0029	.0033	.0037	.0041	.0046	.0052	.0058	.0064	.0071
15	.0010	.0012	.0014	.0016	.0018	.0020	.0023	.0026	.0029	.0033
16	.0004	.0005	.0005	.0006	.0007	.0008	.0010	.0011	.0013	.0014
17	.0001	.0002	.0002	.0002	.0003	.0003	.0004	.0004	.0005	.0006
18	.0000	.0001	.0001	.0001	.0001	.0001	.0001	.0002	.0002	.0002
19	.0000	.0000	.0000	.0000	.0000	.0000	.0000	.0001	.0001	.0001

x	7.1	7.2	7.3	7.4	λ 7.5	7.6	7.7	7.8	7.9	8.0
0	.0008	.0007	.0007	.0006	.0006	.0005	.0005	.0004	.0004	.0003
1	.0059	.0054	.0049	.0045	.0041	.0038	.0035	.0032	.0029	.0027
2	.0208	.0194	.0180	.0167	.0156	.0145	.0134	.0125	.0116	.0107
3	.0492	.0464	.0438	.0413	.0389	.0366	.0345	.0324	.0305	.0286
4	.0874	.0836	.0799	.0764	.0729	.0696	.0663	.0632	.0602	.0573
5	.1241	.1204	.1167	.1130	.1094	.1057	.1021	.0986	.0951	.0916
6	.1468	.1445	.1420	.1394	.1367	.1339	.1311	.1282	.1252	.1221
7	.1489	.1486	.1481	.1474	.1465	.1454	.1442	.1428	.1413	.1396
8	.1321	.1337	.1351	.1363	.1373	.1382	.1388	.1392	.1395	.1396
9	.1042	.1070	.1096	.1121	.1144	.1167	.1187	.1207	.1224	.1241
10	.0740	.0770	.0800	.0829	.0858	.0887	.0914	.0941	.0967	.0993
11	.0478	.0504	.0531	.0558	.0585	.0613	.0640	.0667	.0695	.0722
12	.0283	.0303	.0323	.0344	.0366	.0388	.0411	.0434	.0457	.0481
13	.0154	.0168	.0181	.0196	.0211	.0227	.0243	.0260	.0278	.0296
14	.0078	.0086	.0095	.0104	.0113	.0123	.0134	.0145	.0157	.0169
15	.0037	.0041	.0046	.0051	.0057	.0062	.0069	.0075	.0083	.0090
16	.0016	.0019	.0021	.0024	.0026	.0030	.0033	.0037	.0041	.0045
17	.0007	.0008	.0009	.0010	.0012	.0013	.0015	.0017	.0019	.0021
18	.0003	.0003	.0004	.0004	.0005	.0006	.0006	.0007	.0008	.0009
19	.0001	.0001	.0001	.0002	.0002	.0002	.0003	.0003	.0003	.0004
20	.0000	.0000	.0001	.0001	.0001	.0001	.0001	.0001	.0001	.0002
21	.0000	.0000	.0000	.0000	.0000	.0000	.0000	.0000	.0001	.0001

(*continues*)

TABLE III *(continued)*

x	8.1	8.2	8.3	8.4	λ 8.5	8.6	8.7	8.8	8.9	9.0
0	.0003	.0003	.0002	.0002	.0002	.0002	.0002	.0002	.0001	.0001
1	.0025	.0023	.0021	.0019	.0017	.0016	.0014	.0013	.0012	.0011
2	.0100	.0092	.0086	.0079	.0074	.0068	.0063	.0058	.0054	.0050
3	.0269	.0252	.0237	.0222	.0208	.0195	.0183	.0171	.0160	.0150
4	.0544	.0517	.0491	.0466	.0443	.0420	.0398	.0377	.0357	.0337
5	.0882	.0849	.0816	.0784	.0752	.0722	.0692	.0663	.0635	.0607
6	.1191	.1160	.1128	.1097	.1066	.1034	.1003	.0972	.0941	.0911
7	.1378	.1358	.1338	.1317	.1294	.1271	.1247	.1222	.1197	.1171
8	.1395	.1392	.1388	.1382	.1375	.1366	.1356	.1344	.1332	.1318
9	.1256	.1269	.1280	.1290	.1299	.1306	.1311	.1315	.1317	.1318
10	.1017	.1040	.1063	.1084	.1104	.1123	.1140	.1157	.1172	.1186
11	.0749	.0776	.0802	.0828	.0853	.0878	.0902	.0925	.0948	.0970
12	.0505	.0530	.0555	.0579	.0604	.0629	.0654	.0679	.0703	.0728
13	.0315	.0334	.0354	.0374	.0395	.0416	.0438	.0459	.0481	.0504
14	.0182	.0196	.0210	.0225	.0240	.0256	.0272	.0289	.0306	.0324
15	.0098	.0107	.0116	.0126	.0136	.0147	.0158	.0169	.0182	.0194
16	.0050	.0055	.0060	.0066	.0072	.0079	.0086	.0093	.0101	.0109
17	.0024	.0026	.0029	.0033	.0036	.0040	.0044	.0048	.0053	.0058
18	.0011	.0012	.0014	.0015	.0017	.0019	.0021	.0024	.0026	.0029
19	.0005	.0005	.0006	.0007	.0008	.0009	.0010	.0011	.0012	.0014
20	.0002	.0002	.0002	.0003	.0003	.0004	.0004	.0005	.0005	.0006
21	.0001	.0001	.0001	.0001	.0001	.0002	.0002	.0002	.0002	.0003
22	.0000	.0000	.0000	.0000	.0001	.0001	.0001	.0001	.0001	.0001

x	9.1	9.2	9.3	9.4	λ 9.5	9.6	9.7	9.8	9.9	10
0	.0001	.0001	.0001	.0001	.0001	.0001	.0001	.0001	.0001	.0000
1	.0010	.0009	.0009	.0008	.0007	.0007	.0006	.0005	.0005	.0005
2	.0046	.0043	.0040	.0037	.0034	.0031	.0029	.0027	.0025	.0023
3	.0140	.0131	.0123	.0115	.0107	.0100	.0093	.0087	.0081	.0076
4	.0319	.0302	.0285	.0269	.0254	.0240	.0226	.0213	.0201	.0189
5	.0581	.0555	.0530	.0506	.0483	.0460	.0439	.0418	.0398	.0378
6	.0881	.0851	.0822	.0793	.0764	.0736	.0709	.0682	.0656	.0631
7	.1145	.1118	.1091	.1064	.1037	.1010	.0982	.0955	.0928	.0901
8	.1302	.1286	.1269	.1251	.1232	.1212	.1191	.1170	.1148	.1126
9	.1317	.1315	.1311	.1306	.1300	.1293	.1284	.1274	.1263	.1251
10	.1198	.1210	.1219	.1228	.1235	.1241	.1245	.1249	.1250	.1251
11	.0991	.1012	.1031	.1049	.1067	.1083	.1098	.1112	.1125	.1137
12	.0752	.0776	.0799	.0822	.0844	.0866	.0888	.0908	.0928	.0948
13	.0526	.0549	.0572	.0594	.0617	.0640	.0662	.0685	.0707	.0729
14	.0342	.0361	.0380	.0399	.0419	.0439	.0459	.0479	.0500	.0521
15	.0208	.0221	.0235	.0250	.0265	.0281	.0297	.0313	.0330	.0347
16	.0118	.0127	.0137	.0147	.0157	.0168	.0180	.0192	.0204	.0217
17	.0063	.0069	.0075	.0081	.0088	.0095	.0103	.0111	.0119	.0128
18	.0032	.0035	.0039	.0042	.0046	.0051	.0055	.0060	.0065	.0071
19	.0015	.0017	.0019	.0021	.0023	.0026	.0028	.0031	.0034	.0037
20	.0007	.0008	.0009	.0010	.0011	.0012	.0014	.0015	.0017	.0019
21	.0003	.0003	.0004	.0004	.0005	.0006	.0006	.0007	.0008	.0009
22	.0001	.0001	.0002	.0002	.0002	.0002	.0003	.0003	.0004	.0004
23	.0000	.0001	.0001	.0001	.0001	.0001	.0001	.0001	.0002	.0002
24	.0000	.0000	.0000	.0000	.0000	.0000	.0000	.0001	.0001	.0001

TABLE IV   **VALUES OF *t***

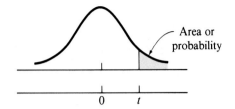

Degrees of Freedom			Right Tail Areas		
(df)	.10	.05	.025	.01	.005
1	3.078	6.314	12.706	31.821	63.657
2	1.886	2.920	4.303	6.965	9.925
3	1.638	2.353	3.182	4.541	5.841
4	1.533	2.132	2.776	3.747	4.604
5	1.476	2.015	2.571	3.365	4.032
6	1.440	1.943	2.447	3.143	3.707
7	1.415	1.895	2.365	2.998	3.499
8	1.397	1.860	2.306	2.896	3.355
9	1.383	1.833	2.262	2.821	3.250
10	1.372	1.812	2.228	2.764	3.169
11	1.363	1.796	2.201	2.718	3.106
12	1.356	1.782	2.179	2.681	3.055
13	1.350	1.771	2.160	2.650	3.012
14	1.345	1.761	2.145	2.624	2.977
15	1.341	1.753	2.131	2.602	2.947
16	1.337	1.746	2.120	2.583	2.921
17	1.333	1.740	2.110	2.567	2.898
18	1.330	1.734	2.101	2.552	2.878
19	1.328	1.729	2.093	2.539	2.861
20	1.325	1.725	2.086	2.528	2.845
21	1.323	1.721	2.080	2.518	2.831
22	1.321	1.717	2.074	2.508	2.819
23	1.319	1.714	2.069	2.500	2.807
24	1.318	1.711	2.064	2.492	2.797
25	1.316	1.708	2.060	2.485	2.787
26	1.315	1.706	2.056	2.479	2.779
27	1.314	1.703	2.052	2.473	2.771
28	1.313	1.701	2.048	2.467	2.763
29	1.311	1.699	2.045	2.462	2.756
30	1.310	1.697	2.042	2.457	2.750
40	1.303	1.684	2.021	2.423	2.704
60	1.296	1.671	2.000	2.390	2.660
120	1.289	1.658	1.980	2.358	2.617
∞	1.282	1.645	1.960	2.326	2.576

Source: From M. Merrington, "Table of Percentage Points of the *t*-Distribution," *Biometrika*, 1941, *32*, 300. Reproduced by permission of the *Biometrika* Trustees.
Example: The *t*-value for df = 15 and a right-hand tail area of .05 is 1.753.

SOLUTIONS TO
STARRED EXERCISES

CHAPTER 2

1. a. $\bar{x} = \dfrac{\sum\limits_{i=1}^{n} x_i}{n}$

$\bar{x} = \dfrac{100 + 90 + 110 + 80 + 120 + 140 + 100 + 60}{8}$

$= \dfrac{800}{8}$

$= \mathbf{100}$

b. median (middle vale in data set) = 100

60, 80, 90, **100**   100, 110, 120, 140

$\uparrow$
med

c. mode (most frequently occurring value)

$= \mathbf{100}$ (twice)

d. range (largest value–smallest value) $= 140 - 60$

$= \mathbf{80}$

e. $\text{MAD} = \dfrac{\sum\limits_{i=1}^{n} |x_i - \bar{x}|}{n}$

$= \dfrac{|60 - 100| + |80 - 100| + |90 - 100| + |100 - 100| + |100 - 100| + |110 - 100| + |120 - 100| + |140 - 100|}{8}$

$= \dfrac{40 + 20 + 10 + 0 + 0 + 10 + 20 + 40}{8} = \dfrac{140}{8} = \mathbf{17.5}$

f. $\text{VAR} = \dfrac{\sum\limits_{i=1}^{n} (x_i - \bar{x})^2}{n}$

$= \dfrac{(60 - 100)^2 + (80 - 100)^2 + (90 - 100)^2 + (100 - 100)^2 + (100 - 100)^2 + (110 - 100)^2 + (120 - 100)^2 + (140 - 100)^2}{8}$

$= \dfrac{40^2 + 20^2 + 10^2 + 0^2 + 0^2 + 10^2 + 20^2 + 40^2}{8}$

$= \dfrac{1600 + 400 + 100 + 0 + 0 + 100 + 400 + 1600}{8}$

$= \dfrac{4200}{8} = \mathbf{525}$

g. $\text{STD DEV} = \sqrt{\text{VAR}}$

$= \sqrt{525} = \mathbf{22.9}$

4. The median (100) or the mode (100) would seem to be the best measures to use to represent the results. The mean (135) doesn't really represent a "typical" value in the skewed data set. It has been severely influenced by the one extreme value, 280.

6. VAR $= \dfrac{(60^2 + 80^2 + 90^2 + 100^2 + 100^2 + 110^2 + 120^2 + 140^2) - 8(100^2)}{8}$

$= \dfrac{84200 - 80000}{8} = \dfrac{4200}{8} = \textbf{525}$

STD DEV $= \sqrt{525} = \textbf{22.9}$

8.

x	$f(x)$
10	31
15	46
20	28
25	15
30	10

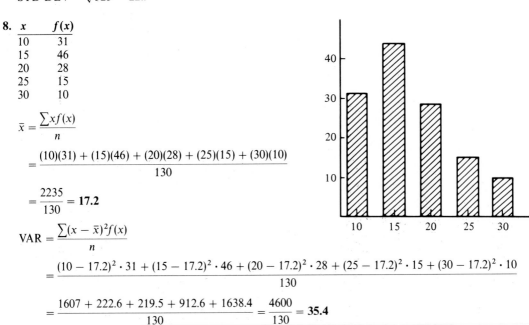

$\bar{x} = \dfrac{\sum x f(x)}{n}$

$= \dfrac{(10)(31) + (15)(46) + (20)(28) + (25)(15) + (30)(10)}{130}$

$= \dfrac{2235}{130} = \textbf{17.2}$

VAR $= \dfrac{\sum (x - \bar{x})^2 f(x)}{n}$

$= \dfrac{(10 - 17.2)^2 \cdot 31 + (15 - 17.2)^2 \cdot 46 + (20 - 17.2)^2 \cdot 28 + (25 - 17.2)^2 \cdot 15 + (30 - 17.2)^2 \cdot 10}{130}$

$= \dfrac{1607 + 222.6 + 219.5 + 912.6 + 1638.4}{130} = \dfrac{4600}{130} = \textbf{35.4}$

STD DEV $= \sqrt{35.4} = \textbf{5.9}$

10.

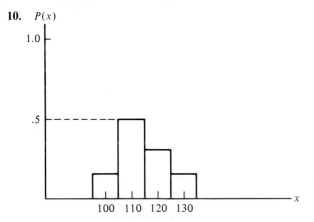

$\bar{x} = 100(.1) + 110(.5) + 120(.3) + 130(.1) = 10 + 55 + 36 + 13 = \textbf{114}$

VAR$(x) = (100 - 114)^2(.1) + (110 - 114)^2(.5) + (120 - 114)^2(.3) + (130 - 114)^2(.1)$

$= 19.6 + 8 + 10.8 + 25.6 = \textbf{64}$

STD DEV $= \sqrt{64} = \textbf{8}$

12. a.

x	f(x)	f(output ≤ x)
40	12	12
42	8	20
44	6	26
46	12	38
48	2	40

c.

x	f(x)	f(output ≥ x)
40	12	40
42	8	28
44	6	20
46	12	14
48	2	2

b. CUM $f(x)$

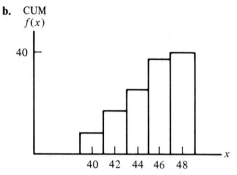

d. CUM $f(x)$

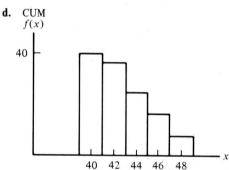

15. a. $f(x)$

Number of workers

b. $\bar{x} = \dfrac{50(20) + 150(50) + 250(30) + 350(20) + 450(15) + 550(10) + 650(5)}{150} = \dfrac{38,500}{150} = \mathbf{256.7}$

$\text{VAR}(x) = \dfrac{\begin{array}{c}(50 - 256.7)^2 20 + (150 - 256.7)^2 50 + (250 - 256.7)^2 30 + (350 - 256.7)^2 20 \\ + (450 - 256.7)^2 15 + (550 - 256.7)^2 10 + (650 - 256.7)^2 5\end{array}}{150}$

$= \dfrac{3,793,333}{150} = \mathbf{25,288.9}$

$\text{STD DEV}(x) = \sqrt{25,288.9} = \mathbf{159}$

17. a.

Interval	Midpoint	Frequency
0–under 2	1	15
2–under 4	3	7
4–under 6	5	19
6–under 8	7	13
8–under 10	9	6
		60

b. $f(x)$

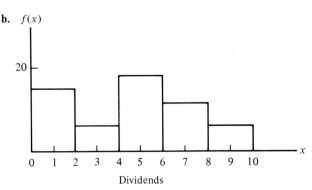

Dividends

c.
$$\bar{x} = \frac{1(15) + 3(7) + 5(19) + 7(13) + 9(6)}{60} = \frac{276}{60} = 4.6$$

$$VAR(x) = \frac{(1 - 4.6)^2 15 + (3 - 4.6)^2 7 + (5 - 4.6)^2 19 + (7 - 4.6)^2 13 + (9 - 4.6)^2 6}{60}$$

$$= \frac{406.4}{60} = 6.77$$

$$STD\ DEV(x) = \sqrt{6.77} = 2.6$$

20. Stock A:

$$\bar{x} = 23$$
$$STD\ DEV = 7.72$$
$$COEFF\ OF\ VAR = \frac{7.72}{23} = .336$$

Stock B:

$$\bar{x} = 123$$
$$STD\ DEV = 7.72$$
$$COEFF\ OF\ VAR = \frac{7.72}{123} = .063$$

Using the coeff. of variation, there appears to be considerably less variability in stock B prices. (Standard deviation measures show equal variability.)

22.

Year	Begin amt.	End amt.	Ratio	Rate of return
1	10000	10600	1.06	.06
2	10600	12720	1.20	.20
3	12720	14882	1.17	.17
4	14882	20091	1.35	.35

a. (1) Arithmetic mean: $\dfrac{.06 + .20 + .17 + .35}{4} = .195$

(2) Geometric mean: $\sqrt[4]{(1.06)(1.20)(1.17)(1.35)} = [(1.06)(1.20)(1.17)(1.35)]^{1/4}$

$$= 1.1906 \quad \text{for an "average" rate of .1906 per year.}$$

b. (1) Using the arithmetic mean rate of .195,
$10,000(1.195)(1.195)(1.195)(1.195) = **\$20,393**$

(2) Using the geometric mean rate of .1906,
$10,000(1.1906)(1.1906)(1.1906)(1.1906) = **\$20,093**$

Since the actual ending amount is $20,091, the geometric mean is clearly more accurate as a measure of average rate of return.

24. Weighted average $= \dfrac{150(\$20) + 95(\$30) + 10(\$50)}{255} = \dfrac{6350}{255} = **\$24.90**$

26. $\dfrac{800(80) + 1500(92) + 1200(78)}{3500} = \dfrac{295600}{3500} = **84.46** \text{ octane}$

27. 86% of those who took the exam scored the same as or below Climber on the verbal portion of the exam (i.e., he is in the top 14%).

35% of those who took the exam scored the same as or below Climber on the math portion (i.e., he is in the lower 35%).

It would seem that Climber needs to work on his math skills.

CHAPTER 3

2. a. No   **b.** No   **c.** Yes   **d.** Yes   **e.** Yes   **f.** No

Note: All but (f) could actually be argued either way. We would need to see some numbers before any firm conclusions could be drawn.

4. a. $P(A) = .6$ $P(B) = .8$ $P(A \cap B) = .5$

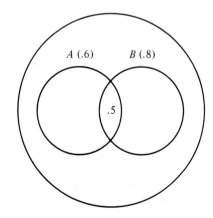

A (.6) B (.8)

.5

b. $P(A \cup B) = P(A) + P(B) - P(A \cap B)$

$$= .6 + .8 - .5 = .9$$

c. $P(A/B) = \dfrac{P(A \cap B)}{P(B)} = \dfrac{.5}{.8} = .625$

d. $P(\text{neither}) = 1 - P(A \cup B) = 1 - .9 = .1$

e. No, since $P(A/B) \neq P(A)$; that is, $.625 \neq .6$

6. $P(N) = .4$ $P(T) = .7$ $P(N \cap T) = .3$

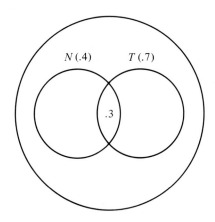

a. $P(N \cup T) = P(N) + P(T) - P(N \cap T)$
$= .4 + .7 - .3 = .8$

b. $P(\text{neither}) = 1 - P(N \cup T) = 1 - .8 = .2$

c. $P(T \cap N')$:   Look at the T circle in the Venn diagram to find what's in T but *not* in N: $(.7 - .3) = .4$. We're focusing here, then, on the "half-moon" area in T that doesn't include any of N. $(P(T) - P(N \cap T)) = .7 - .3 = .4$

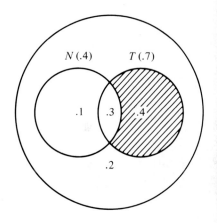

d. $P(N/T) = \dfrac{P(N \cap T)}{P(T)} = \dfrac{.3}{.7} = .428$

e. No, since $P(N/T) \neq P(N)$
$.428 \neq .4$

f. No, since $P(N \cap T) \neq 0$

9. PR = Primary fails
B = Backup fails

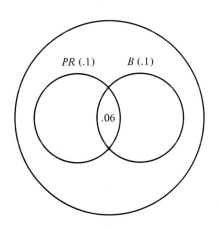

$P(PR) = .1$
$P(B) = .1$
$P(PR \cap B) = .06$

a. $P(\text{at least one will function properly}) = 1 - P(\text{both fail})$

$= 1 - P(PR \cap B)$

$= 1 - .06 = .94$

b. $P(B/PR) = \dfrac{P(B \cap PR)}{P(PR)} = \dfrac{.06}{.10} = .60$

c. No, since

$P(B/PR) \neq P(B)$

that is, $.60 \neq .10$

11. $P(A) = .6 \qquad P(A') = .4$

$P(B) = .2 \qquad P(B') = .8$

Assuming independence allows us to compute $P(A \cap B) = (.6)(.2) = .12$
Using a Venn diagram,

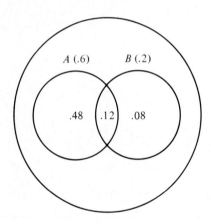

$P(\text{Job}) = P(A \cup B) = P(A) + P(B) - P(A \cap B)$

$= .6 + .2 - .12 = .68$

Using a probability tree,

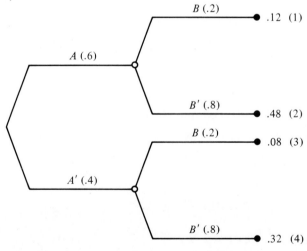

$P(\text{Job})$, from node (1), (2), and (3) is $.12 + .48 + .08 = .68$

12. 6—NY
    4—Medford

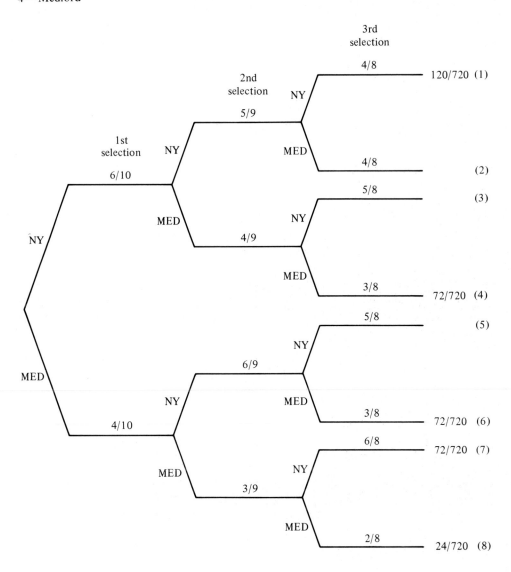

a. $P(3 \text{ NY}) = (6/10)(5/9)(4/8) = 120/720$ (node 1)

b. $P(1 \text{ NY, } 2 \text{ MED}) = (6/10)(4/9)(3/8)$ (node 4)
$\phantom{P(1 \text{ NY, } 2 \text{ MED}) =} + (4/10)(6/9)(3/8)$ (node 6)
$\phantom{P(1 \text{ NY, } 2 \text{ MED}) =} + (4/10)(3/9)(6/8)$ (node 7)
$\phantom{P(1 \text{ NY, } 2 \text{ MED}) =} \overline{ 216/720 = .30}$

c. $P(0 \text{ NY}) = P(3 \text{ MED}) = (4/10)(3/9)(2/8)$ (node 8)
$\phantom{P(0 \text{ NY})} = 24/720 = .033$

15. $P(D) = .7$

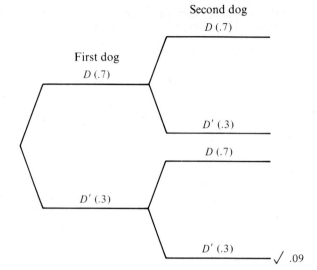

a. $P(\text{undetected}) = P(D' \cap D') = (.3)(.3) = .09$

b. (1) with 3 dogs:
   $P(\text{undetected}) = P(D' \cap D' \cap D') = (.3)(.3)(.3) = .027$

   (2) Retraining so that $P(D) = .9$:
   $P(\text{undet'ed}) = P(D' \cap D') = (.1)(.1) = .01$

Retraining produces a lower probability of nondetection.

17.

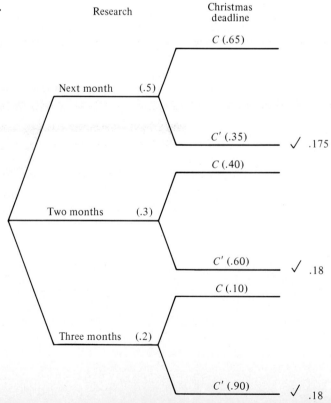

P(not on mkt by Christmas) $= .175 + .18 + .18 = .535$

19. $D =$ item tests "defective"

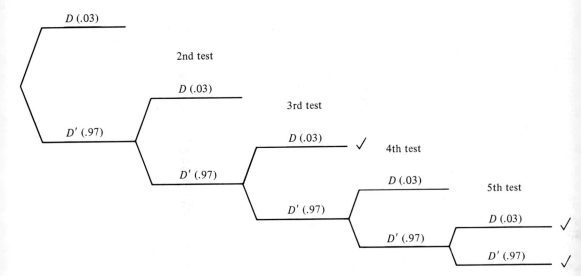

1st test

D (.03)

2nd test

D (.03)

D' (.97)

3rd test

D (.03) ✓

D' (.97)

4th test

D (.03)

D' (.97)

5th test

D (.03) ✓

D' (.97)

D' (.97) ✓

a. P(test exactly 3) $= P(D')P(D')P(D)$

$\phantom{P(\text{test exactly 3})} = (.97)(.97)(.03)$

$\phantom{P(\text{test exactly 3})} = .028$

b. P(accept) $= P$(all 5 not defective)

$\phantom{P(\text{accept})} = (.97)(.97)(.97)(.97)(.97)$

$\phantom{P(\text{accept})} = .859$

c. P(at least 3 tests) $= P$(exactly 3) $+ P$(exactly 4) $+ P$(exactly 5)

$\phantom{P(\text{at least 3 tests})} = (.97)(.97)(.03) + (.97)(.97)(.97)(.03) + (.97)(.97)(.97)(.97)$

$\phantom{P(\text{at least 3 tests})} = .028 + .027 + .885$

$\phantom{P(\text{at least 3 tests})} = .94$

22.

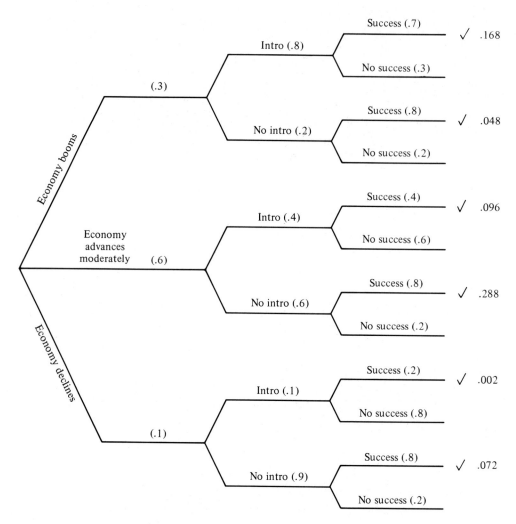

P(success) = .168 + .048 + .096 + .288 + .002 + .072

= **.674**

Since P(success) is greater than the required 60%, proceed with plans to introduce.

25.

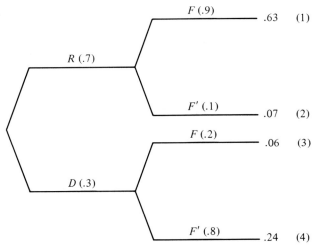

R = Republican
D = Democrat
F = Favors deployment

a. From nodes (1) and (3) $P(F) = .63 + .06 = .69$

b. $P(D/F) = \dfrac{P(D \cap F)}{P(F)} = \dfrac{.06}{.69} = .087$

c. $P(R/F') = \dfrac{P(R \cap F')}{P(F')} = \dfrac{.07}{.31} = .226$

28.

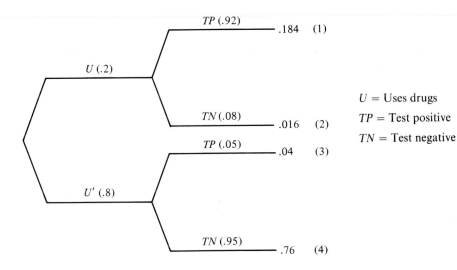

U = Uses drugs
TP = Test positive
TN = Test negative

a. $P(TP) = .184 + .04 = .224$ (nodes (1) & (3))

b. $P(U/TP) = \dfrac{P(U \cap TP)}{P(TP)} = \dfrac{.184}{.224} = .821$

30.

	A	A'	
P	.2	.1	.3
P'	.2	.5	.7
	.4	.6	1.0

A = Saw ad

P = Purchased product

a. $P(A \cup P) = .2 + .1 + .2 = .5$ or $P(A \cup P) = P(A) + P(P) - P(A \cap P)$

$$= .4 + .3 - .2 = .5$$

b. $P(P) = .3$ **c.** $P(A/P) = \dfrac{P(A \cap P)}{P(P)} = \dfrac{.2}{.3} = .667$ **d.** $P(P/A) = \dfrac{P(P \cap A)}{P(A)} = \dfrac{.2}{.4} = .5$

e. No, since $P(P/A) \neq P(P)$, that is, $.5 \neq .3$

32.

	B	B'	
A	.5	.1	.6
A'	.3	.1	.4
	.8	.2	1.0

a. $P(A \cup B) = P(A) + P(B) - P(A \cap B)$

$$= .6 + .8 - .5 = .9$$

b. $P(\text{exactly 1 sale}) = P(A \cap B') + P(A' \cap B)$

$$= .1 + .3 = .4$$

c. $P(B/A') = \dfrac{P(B \cap A')}{P(A')} = \dfrac{.3}{.4} = .75$ **d.** $P(A'/B') = \dfrac{P(A' \cap B')}{P(B')} = \dfrac{.1}{.2} = .5$

e.

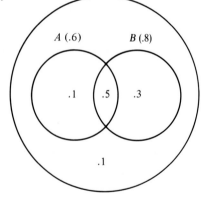

34. $P(N) = .4$ $P(T) = .7$ $P(N \cap T) = .3$

	T	T'	
N	.3	.1	.4
N'	.4	.2	.6
	.7	.3	1.0

a. $P(T' \cap N') = .2$ **b.** $P(T \cap N') = .4$ **c.** $P(N/T') = \dfrac{P(N \cap T')}{P(T')} = \dfrac{.1}{.3} = .333$

d. $P(T'/N') = \dfrac{P(T' \cap N')}{P(N')} = \dfrac{.2}{.6} = .333$ **e.**

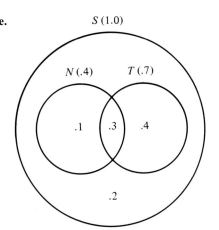

35. From problem 26:

	Mojne	Grove	
Defect	.028	.021	.049
Not defect	.672	.279	.951
	.70	.30	1.00

39. G = gang member D = drug related crime
$P(G) = .65$ $P(G/D) = .86$ $P(D) = .70$

a. $P(D \cap G) = P(D) \cdot P(G/D)$ **b.** $P(D/G) = \dfrac{P(D \cap G)}{P(G)} = \dfrac{.602}{.65} = \mathbf{.926}$

$\qquad = (.7)(.86) = \mathbf{.602}$

c.

	D	D'	
G	**.602**	.048	.65
G'	.098	.252	.35
	.70	.30	1.00

$P(D/G') = \dfrac{P(D \cap G')}{P(G')} = \dfrac{.098}{.35}$

$\qquad = \mathbf{.28}$

41. a. $_{12}C_5 = \dfrac{12!}{(12-5)!5!} = 792$ **b.** $(_4C_2)(_8C_3) = \left[\dfrac{4!}{(4-2)!2!}\right]\left[\dfrac{8!}{(8-3)!3!}\right]$

$\qquad\qquad\qquad\qquad\qquad\qquad\qquad = (6)(56)$

$\qquad\qquad\qquad\qquad\qquad\qquad\qquad = \mathbf{336}$

CHAPTER 4

1.

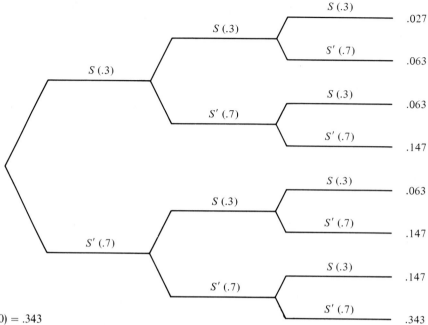

$P(x = 0) = .343$
$P(x = 1) = .147 + .147 + .147 = .441$
$P(x = 2) = .063 + .063 + .063 = .189$
$P(x = 3) = .027$

x	$P(x)$
0	.343
1	.441
2	.189
3	.027
	1.000

2. $P(x)$

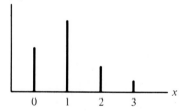

$$
\begin{aligned}
E(x) = 0(.343) &= 0 \\
1(.441) &= .441 \\
2(.189) &= .378 \\
+\,3(.027) &= .081 \\
\hline
&= .900
\end{aligned}
$$

That is,

$E(x) = .9$

$$VAR(x) = (0 - .9)^2(.343) = .27783$$
$$(1 - .9)^2(.441) = .00441$$
$$(2 - .9)^2(.189) = .22869$$
$$(3 - .9)^2(.027) = .11907$$
$$= .63$$

so

$STD\ DEV(x) = \sqrt{.63} = \mathbf{.794}$

3. $P(x = 0/n = 3, p = .3) = \dfrac{3!}{3!0!}(.3)^0(.7)^3$

$$= 1(1)(.343) = \mathbf{.343}$$

$P(x = 1/n = 3, p = .3) = \dfrac{3!}{2!1!}(.3)^1(.7)^2$

$$= 3(.3)(.49) = \mathbf{.441}$$

$P(x = 2/n = 3, p = .3) = \dfrac{3!}{1!2!}(.3)^2(.7)^1$

$$= 3(.09)(.7) = \mathbf{.189}$$

$P(x = 3/n = 3, p = .3) = \dfrac{3!}{0!3!}(.3)^3(.7)^0$

$$= 1(.027)1 = \mathbf{.027}$$

$E(x) = np = 3(.3) = .9$

$VAR(x) = np(1 - p) = 3(.3)(.7) = .63$

$STD\ DEV(x) = \sqrt{np(1 - p)} = \sqrt{.63} = \mathbf{.794}$

8. a. $P(x = 4/n = 10, p = .3) = \mathbf{.2001}$

   b. $P(x = 8/n = 20, p = .6) = P(x = 12/n = 20, p = .4) = \mathbf{.0355}$

   c. $P(x \le 12/n = 15, p = .7) = P(x \ge 3/n = 15, p = .3)$

$$= .1700 + .2186 + .2061 + .1472 + .0811 + .0348 + .0116$$
$$+ .0030 + .0006 + .0001$$
$$= \mathbf{.8731}$$

   d. $P(7 \le x \le 13/n = 20, p = .5) = .0739 + .1201 + .1602 + .1762 + .1602 + .1201 + .0739$
$$= \mathbf{.8846}$$

   e. $P(14 \le x \le 18/n = 30, p = .6) = P(12 \le x \le 16/n = 30, p = .4)$
$$= .1474 + .1360 + .1101 + .0783 + .0489$$
$$= \mathbf{.5207}$$

10. a. $P(x \ge 20/n = 30, p = .4) = .002 + .0006 + .0002 = \mathbf{.0028}$

b. $P(x \le 12/n = 30, p = .4) = .1474 + .1396 + .1152 + .0823 + .0505 + .0263$

$$+ .0115 + .0041 + .0012 + .0003 = \textbf{.5784}$$

c. $P(15 \le x \le 21/n = 30, p = .4) = .0783 + .0489 + .0269 + .0129 + .0054 + .0020 + .0006 = \textbf{.175}$

d. "Fewer than 8" means 7 or less:

$P(x \le 7/n = 30, p = .4) = .0263 + .0115 + .0041 + .0012 + .0003 = \textbf{.0434}$

e. $E(x) = np = 30(.4) = 12$

$\text{VAR}(x) = np(1 - p) = 7.2$

$\text{STD DEV}(x) = \sqrt{7.2} = 2.68.$

Note: If 60% don't know, then 40% do know; thus our use of $p = .4$.

12. a. $P(x \ge 12/n = 20, p = .5) = .1201 + .0739 + .0370 + .0148 + .0046 + .0011 + .0002$

$$= \textbf{.2517}$$

b. $P(x \ge 16/n = 20, p = .5) = .0046 + .0011 + .0002 = \textbf{.0059}$

c. $P(x \ge 12/n = 20, p = .25) = .0008 + .0002 = \textbf{.001}$

Note 1: Using $p = .5$ reflects the 50/50 chance of guessing right on each true/false question.
Note 2: 60% of 20 = 12, the required minimum number correct to pass the test.

14. a. $P(x \ge 3/n = 20, p = .10) = 1 - p(x \le 2/n = 20, p = .10)$

$$= 1 - [.1216 + .2702 + .2852) = \textbf{.323}$$

b. $P(x \le 1/n = 20, p = .10) = .1216 + .2702 = \textbf{.3918}$

c. $P(1 \le x \le 4/n = 20, p = .10) = .2702 + .2852 + .1901 + .0898$

$$= \textbf{.8353}$$

16. a. $\sigma = 10$

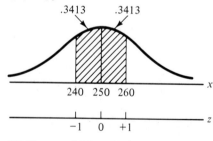

$$z = \frac{260 - 250}{10} = 1$$

$$z = \frac{240 - 250}{10} -1$$

$P(240 \le x \le 260) = .3413 + .3413 = \textbf{.6826}$

b.

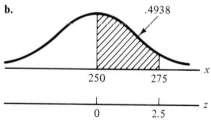

$$z = \frac{275 - 250}{10} = 2.5$$

$P(250 \le x \le 275) = .4938$

c.

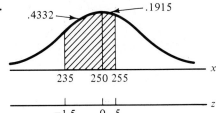

$$z = \frac{235 - 250}{10} = -1.5$$

$$z = \frac{255 - 250}{10} = .5$$

$P(235 \le x \le 255) = .4332 + .1915 = .6247$

d.

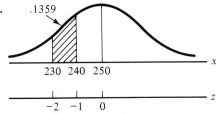

$$z = \frac{230 - 250}{10} = -2$$

$$z = \frac{240 - 250}{10} = -1$$

$P(230 \le x \le 240) = .4772 - .3413 = .1359$

e.

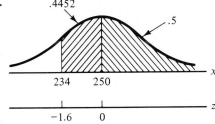

$$z = \frac{234 - 250}{10} = -1.6$$

$P(x \ge 234) = .4452 + .5 = .9452$

f.

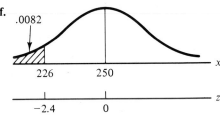

$$z = \frac{226 - 250}{10} = -2.4$$

$P(x \le 226) = .5 - .4918 = \mathbf{.0082}$

18. a.

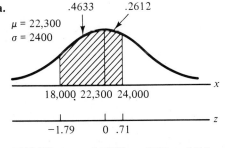

$\mu = 22,300$
$\sigma = 2400$

$$z = \frac{18,000 - 22,300}{2400} = -1.79$$

$$z = \frac{24,000 - 22,300}{2400} = .71$$

$P(18,000 \le x \le 24,000) = .4633 + .2612 = \mathbf{.7245}$

b.

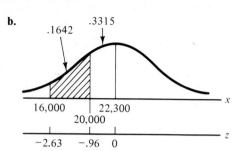

$$z = \frac{16,000 - 22,300}{2400} = -2.63$$

$$z = \frac{20,000 - 22,300}{2400} = -.96$$

$P(16,000 \leq x \leq 20,000) = .4957 - .3315 = \mathbf{.1642}$

c.

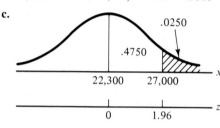

$$z = \frac{27,000 - 22,300}{2400} = 1.96$$

$P(x \geq 27,000) = .5 - .4750 = .0250$

d.

$$z = \frac{17,000 - 22,300}{2400} = -2.21$$

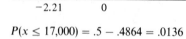

$P(x \leq 17,000) = .5 - .4864 = .0136$

e.

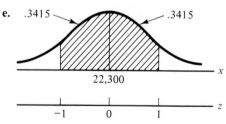

Set boundaries at $+1$ and -1 std. deviations:

$\$22,300 \pm 1(\$2400)$ or $\$19,900$ to $\$24,700$

19. a.

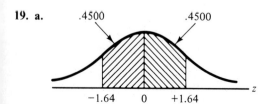

$z = \pm 1.64$

b.

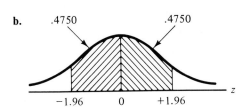

.4750   .4750 $z = \pm 1.96$

-1.96 0 $+1.96$

c.

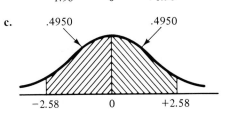

.4950   .4950 $z = \pm 2.58$

-2.58 0 $+2.58$

21. a.

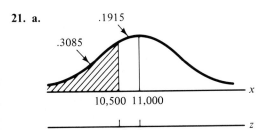

.1915 $\mu = 11,000$

.3085 $\sigma = 1000$

10,500 11,000

$-.5$ 0

You will have at least 1500 unsold units if demand is less than or equal to $12,000 - 1500 = 10,500$ units. Using this in the normal demand distribution:

$$z = \frac{10,500 - 11,000}{1000} = -.5$$

From the table, the area for $z = .5$ is .1915. $\therefore$ $P(x \le 10,500) = .5 - .1915 = .3085$

b. A shortage of at least 500 units will occur if demand is at least $12,000 + 500 = 12,500$ units.

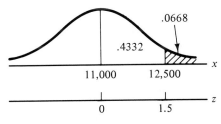

.0668 $z = \frac{12,500 - 11,000}{1000} = 1.5$

.4332

11,000   12,500

0   1.5

For $z = 1.5$, area is .4332, so

$$P(x \ge 12,500) = .5 - .4332 = .0668$$

c. No shortage will occur if demand is 12,000 units or less.

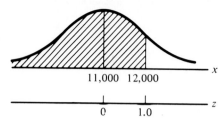

$$z = \frac{12,000 - 11,000}{1000} = 1.0$$

For $z = 1.0$, area is .3413.

$$P(x \le 12,000) = .5 + .3413 = \mathbf{.8413}$$

d.

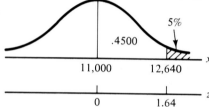

Produce at a level 1.64 standard deviations above 11,000: $11,000 + 1.64(1000) = 11,000 + 1640$ |

$$= 12,640 \text{ units}$$

23. a.

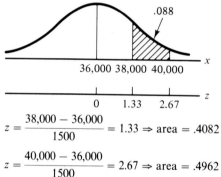

$$z = \frac{38,000 - 36,000}{1500} = 1.33 \Rightarrow \text{area} = .4082$$

$$z = \frac{40,000 - 36,000}{1500} = 2.67 \Rightarrow \text{area} = .4962$$

$$P(38,000 \le x \le 40,000) = .4962 - .4082 = .088$$

b.

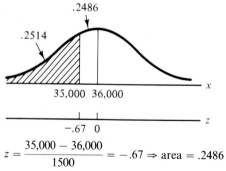

$$z = \frac{35,000 - 36,000}{1500} = -.67 \Rightarrow \text{area} = .2486$$

$$P(x \le 35,000) = .5 - .2486 = .2514$$

c.

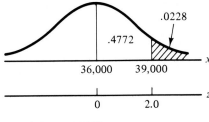

$$z = \frac{39,000 - 36,000}{1500} = 2.0$$

area $= .4772$

$P(x \geq 39,000) = .5 - .4772 = \mathbf{.0228}$

d.

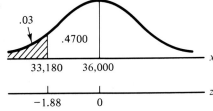

Set warranty at $36,000 - 1.88(1500) = 36,000 - 2,820$

$$= \mathbf{33,180} \text{ miles}$$

24. a. $P(x = 2/\lambda = 3) = \dfrac{\lambda^x e^{-\lambda}}{x!} = \dfrac{3^2(2.718)^{-3}}{2!}$ **b.** $P(x = 4/\lambda = 1) = \dfrac{1^4(2.718)^{-1}}{4!} = \dfrac{1}{2.718(24)}$

$$= \frac{9}{(2.718)^3 2!} = \frac{9}{40.158} = \mathbf{.224}$$

$$= \mathbf{.0153}$$

c. $P(x < 3/\lambda = 6) = P(x \leq 2/\lambda = 6) = P(x = 0/\lambda = 6) + P(x = 1/\lambda = 6) + P(x = 2/\lambda = 6)$

$$P(x = 0/\lambda = 6) = \frac{6^0(2.718)^{-6}}{0!} = \frac{1}{(2.718)^6} = \frac{1}{403.2} = \mathbf{.0025}$$

$$P(x = 1/\lambda = 6) = \frac{6^1(2.718)^{-6}}{1!} = \frac{6}{403.2} = \mathbf{.0149}$$

$$P(x = 2/\lambda = 6) = \frac{6^2(2.718)^{-6}}{2!} = \frac{36}{2(403.2)} = \mathbf{.0446}$$

$P(x < 3) = .0025 + .0149 + .0446 = \mathbf{.062}$

26. $P(x)$

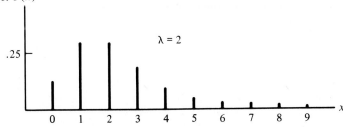

$$E(x) = 0(.1353) + 1(.2707) + 2(.2707) + 3(.1804) + 4(.0902) + 5(.0361) + 6(.0120) + 7(.0034)$$
$$+ 8(.0009) + 9(.0002) = 1.9994 \approx 2 = \lambda$$
$$VAR(x) = (0 - 2)^2(.1353) + (1 - 2)^2(.2707) + (2 - 2)^2(.2707)$$
$$+ (3 - 2)^2(.1804) + (4 - 2)^2(.0902) + (5 - 2)^2(.0361)$$
$$+ (6 - 2)^2(.0120) + (7 - 2)^2(.0034) + (8 - 2)^2(.0009)$$
$$+ (9 - 2)^2(.0002) = 1.9972 \approx 2$$
$$STD\ DEV(x) = \sqrt{VAR} = \sqrt{2}$$

31. a. $P(x \geq 11/\lambda = 8) = .0722 + .0481 + .0296 + .0169 + .0090 + .0045 + .0021 + .0009 + .0004$
$$+ .0002 + .0001 = \textbf{.184}$$

b. Should have a crew that can handle up to (and including) **13** malfunctions since
$P(x \geq 14/\lambda = 8) = .0169 + .0090 + .0045 + .0021 + .0009 + .0004 + .0002 + .0001 = \textbf{.0341}$
If they could only handle 12,
$P(x \geq 13/\lambda = 8) = .0637$, which is too high.

33. a. $P(x = 3/n = 200, p = .02) \approx P(x = 3/\lambda = 4) = \textbf{.1954}$

b. $P(x = 1/n = 50, p = .01) \approx P(x = 1/\lambda = .5) = \textbf{.3033}$

c. $P(x = 0/n = 1000, p = .002) \approx P(x = 0/\lambda = 2) = \textbf{.1353}$

d. $P(x \leq 2/n = 100, p = .01) \approx P(x \leq 2/\lambda = 1) = .3679 + .3679 + .1839 = \textbf{.9197}$

34. a. $P(20 \leq x \leq 25/n = 50, p = .5) \approx P(19.5 \leq x \leq 25.5/\mu = 25, \sigma = 3.54) = .4951$

Areas for $z = .14$, area $= .0557$

for $z = 1.55$, area $= \dfrac{.4394}{.4951}$

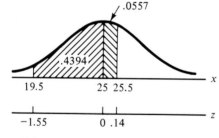

$\mu = np = 25$

$\sigma = \sqrt{np(1 - p)} = 3.54$

$z = \dfrac{25.5 - 25}{3.54} = .14$

$z = \dfrac{19.5 - 25}{3.54} = -1.55$

b. $P(x \geq 48/n = 100, p = .4) \approx P(x \geq 47.5/\mu = 40, \sigma = 4.9)$

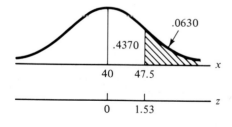

$\mu = np = 100(.4) = 40$

$\sigma = \sqrt{np(1 - p)} = 4.9$

$$z = \frac{47.5 - 40}{4.9} = 1.53 \qquad \text{Area: .4370,}$$

so the tail-end probability is $.5 - .4370 = \mathbf{.0630}$

c. $P(x \le 85/n = 1000, p = .1) \approx P(x \le 85.5/\mu = 100, \sigma = 9.49) = .0630$

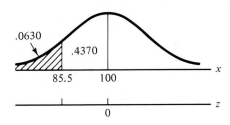

$\mu = np = 100$

$\sigma = \sqrt{np(1 - p)} = 9.49$

$z = \dfrac{85.5 - 100}{9.49} = 1.53$

Area: .4370, so tail-end area is $.5 - .4370 = .0630$

d. $P(x \ge 230/n = 5000, p = .05) \approx .9082$

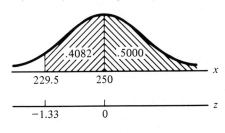

$\mu = np = 250$

$\sigma = \sqrt{np(1 - p)} = 15.41$

$$z = \frac{229.5 - 250}{15.41} = -1.33 \qquad \text{Area: .4082}$$

Total area: $.4082 + .5000 = \mathbf{.9082}$

36. a. $P(x \ge 5/\lambda = .4) = e^{-\lambda t} = e^{-.4(5)} = \dfrac{1}{2.718^2}$

$$= \mathbf{.1353}$$

b. The answers are identical, which indicates the close relationship between the two distributions.

CHAPTER 5

1. $\bar{x} \pm z(\sigma/\sqrt{n})$ or $\bar{x} \pm z(s/\sqrt{n})$

$1040 \pm 1.96[83/\sqrt{100}]$

1040 ± 16.27

1023.73 to 1056.27

Interpretation: There's a 95% probability that the interval will contain the overall mean score for the population of Milar students who took the exam.

2. $\bar{x} \pm z(\sigma/\sqrt{n})$ or $\bar{x} \pm z(s/\sqrt{n})$

a. $13.6 \pm 1.96(5.2/\sqrt{49})$

$13.6 \pm 1.96(.74)$

$13.6 \pm 1.45 \Rightarrow 12.15$ yr to 15.05 yr

b. 99%: $13.6 \pm 2.58(5.2/\sqrt{49})$

$13.6 \pm 2.58(.74) \Rightarrow 13.6 \pm 1.91$

11.69 yr to 15.51 yr

80%: $13.6 \pm 1.28(.74) \Rightarrow 13.6 \pm .95$

12.65 yr to 14.55 yr

c.

Confid.	STD error	Sampling error
80%	$s/\sqrt{n} = .74$ yr	$z(s/\sqrt{n}) = .95$ yr
95%	$s/\sqrt{n} = .74$ yr	$z(s/\sqrt{n}) = 1.45$ yr
99%	$s/\sqrt{n} = .74$ yr	$z(s/\sqrt{n}) = 1.91$ yr

5. In this small sample case, with σ unknown and s used to estimate σ, the t distribution replaces the normal distribution. $(df = 12 - 1 = 11)$

$\bar{x} \pm t(s/\sqrt{n})$

$1040 \pm 2.201[83/\sqrt{12}]$

$1040 \pm 2.201[23.96]$

$1040 \pm 52.74 \Rightarrow 987.26$ to 1092.74

Assumption: The population of values must be normal.

8. Compute the sample mean: $\bar{x} = 8240/10 = 824$.

Compute the sample STD. DEV.: $s = \sqrt{\dfrac{\Sigma(x - \bar{x})^2}{n - 1}} = \sqrt{\dfrac{(840 - 824)^2 + \cdots + (850 - 824)^2}{10 - 1}} = 33.4$

$\bar{x} \pm t(s/\sqrt{n})$

$824 \pm 1.833(33.4/\sqrt{10})$

$824 \pm 1.833(10.56)$

$824 \pm 19.4 \Rightarrow$ **804.6 to 843.4**

10. $\bar{x} \pm z(\sigma/\sqrt{n})$ or $\bar{x} \pm z(s/\sqrt{n})$

a. $6500 \pm 1.96(1600/\sqrt{200})$

$6500 \pm 1.96(113.14)$

$6500 \pm 221.75 \Rightarrow \$6278.25$ to 6721.75

b. Sampling without replacement, with the sample size more than 5% of the population size, use the FPC $\sqrt{(N - n)/(N - 1)}$.

$6500 \pm 1.96\left[(113.14)\left(\sqrt{\dfrac{2000 - 200}{2000 - 1}}\right)\right]$

$6500 \pm 1.96[(113.14)(.95)]$

$6500 \pm 1.96[107.48]$

$6500 \pm 210.67 \Rightarrow \$6289.33$ to 6710.67

The FPC has reduced the standard error term (and the sampling error term) by about 5% (FPC $= .95$).

11. If the interval estimate of the *average* is $\$6500 \pm \$210.67$, then the interval estimate of the *total* is

$$[2000]\$6500 \pm [2000]\$210.67 = \$13,000,000 \pm \$421,340$$

$$\Rightarrow \$12,578,660 \text{ to } \$13,421,340$$

In general, the interval estimate for a total will look like

$$N\bar{x} \pm N\left[z\frac{\sigma}{\sqrt{n}}\right]$$

14. a. Technically the FPC should be used. However, it will have very little material effect on results:

$$FPC = \sqrt{\frac{1600 - 16}{1600 - 1}} = .995$$

Recommendation: As mentioned in the chapter, unless the sample size is at least 5% of the population size, don't bother with the FPC.

b. The t-distribution is appropriate since the sample standard deviation is being used to estimate the population standard deviation.

c. With a small sample ($n < 30$), the assumption of a normal population is required.

d. We will have to assume that the sample standard deviation ($s = 5.3$) was computed using this adjustment. Don't use $n - 1$ again in computing the standard error:

$$\sigma_{\bar{x}} \approx s/\sqrt{n}$$

In sum, the interval should be

$$16.2 \pm 2.131(5.3/\sqrt{16})$$

17. set $z(\sigma/\sqrt{n}) = 10$

$$1.96(83/\sqrt{n}) = 10$$

$$\sqrt{n} = \frac{1.96(83)}{10} \Rightarrow n = \left[\frac{1.96(83)}{10}\right]^2$$

$$n = 265$$

20. Compute the pilot sample standard deviation s (using a sample mean of 62):

$$s = \sqrt{\frac{\sum(x - \bar{x})^2}{n - 1}} = 12.77$$

Now set $z(s/\sqrt{n}) = 1$

$$1.64(12.77/\sqrt{n}) = 1$$

$$\sqrt{n} = \frac{1.64(12.77)}{1} \Rightarrow n = [1.64(12.77)]^2$$

$$n = 439$$

22. set $z(\sigma/\sqrt{n}) = 20$

$$2.58(\sigma/\sqrt{n}) = 20$$

$$\sqrt{n} = \frac{2.58(120)}{20} \Rightarrow n = \left[\frac{2.58(120)}{20}\right]^2$$

$$n = 240$$

23. Given the n of 240 from exercise 22, use the FPC-related adjustment for sample size:

$$n' = \left[\frac{240}{1 + n/N}\right] = \left[\frac{240}{1 + 240/245}\right]$$

$$= 240/1.979 = \mathbf{121}$$

25. a. $\mu = \dfrac{10 + 20 + 30}{3} = 20$

$$\sigma = \sqrt{\frac{[10 - 20]^2 + [20 - 20]^2 + [30 - 20]^2}{3}}$$

$$= \sqrt{200/3} = 8.16$$

b.

Samples	Values	Sample means ($\bar{x}$)
A, A	10, 10	10
A, B	10, 20	15
A, C	10, 30	20
B, A	20, 10	15
B, B	20, 20	20
B, C	20, 30	25
C, A	30, 10	20
C, B	30, 20	25
C, C	30, 30	30

c. See above for sample means.

d. The mean of the sample means:

$$\frac{10 + 15 + 20 + 15 + 20 + 25 + 20 + 25 + 30}{9} = \mathbf{20}$$

e. The standard deviation of the list of sample means:

$$\sqrt{\frac{(10 - 20)^2 + (15 - 20)^2 + (20 - 20)^2 + \cdots + (30 - 20)^2}{9}} = \sqrt{\frac{300}{9}}$$

$$= 5.77$$

f. Plotting the $\bar{x}$ distribution produces

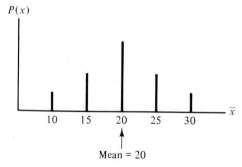

Observations:

(1) Shape: Symmetric (sort of normal)

(2) The center: 20, exactly equal to the population mean

(3) STD DEV: 5.77, equal to

$$\sigma/\sqrt{n} = 8.16/\sqrt{2} = 5.77$$

thus showing that the STD DEV of the sampling distribution of sample means is equal to the population standard deviation divided by the square root of the sample size.

CHAPTER 6

1. $.64 \pm 1.96\left(\sqrt{\dfrac{(.64)(.36)}{200}}\right) = .64 \pm 1.96(.034) = .64 \pm .067$ or   .573 to .707

Interpretation: There is a 95% probability that the interval .573 to .707 contains the proportion of all *World Review* subscribers who favor the old format.

4. $\bar{p} = \dfrac{610}{1000} = .61$

a. $.61 \pm 1.96\left(\sqrt{\dfrac{.61(.39)}{1000}}\right) = .61 \pm 1.96(.015)$

$= .61 \pm .029$ or   .581 to .639

b. STD error: $\sigma_p = \sqrt{\dfrac{\bar{p}(1-\bar{p})}{n}} = .015$

Sampling error:

$$z\sqrt{\dfrac{\bar{p}(1-\bar{p})}{n}} = 1.96(.015) = .029$$

5. Set $z\sqrt{\dfrac{\pi(1-\pi)}{n}} = .04$

Then

a. $1.96\sqrt{\dfrac{(.12)(.88)}{n}} = .04$

$n = \left[\dfrac{1.96\sqrt{(.12)(.88)}}{.04}\right]^2 = 254$

b. $1.64\sqrt{\dfrac{(.12)(.88)}{n}} = .04$

$n = \left[\dfrac{1.64\sqrt{(.12)(.88)}}{.04}\right]^2 = 178$

c. For the 95% interval,

$$n = \left[\dfrac{1.96\sqrt{.5(.5)}}{.04}\right]^2 = 600$$

For the 90% interval,

$$n = \left[\dfrac{1.64\sqrt{.5(.5)}}{.04}\right]^2 = 420$$

7. Set $1.96\sqrt{\dfrac{(.5)(.5)}{n}} = .02$

$$n = \left[\dfrac{1.96\sqrt{(.5)(.5)}}{.02}\right]^2 = 2401$$

9. a. The 95% interval for the *proportion* would be

$$.125 \pm 1.96\left(\sqrt{\frac{(.125)(.875)}{120}}\right)\left(\sqrt{\frac{500-120}{500-1}}\right)$$

$$.125 \pm .051$$

The 95% interval for the *total* would be

$$500(.125) \pm 500(.051)$$

$$62.5 \pm 25.5 \quad \text{or} \quad 37 \text{ to } 88 \text{ components}$$

b. The 90% interval for the *proportion* is

$$.125 \pm 1.64\left(\sqrt{\frac{(.125)(.875)}{120}}\right)\left(\sqrt{\frac{500-120}{500-1}}\right)$$

$$.125 \pm .043$$

The 90% interval for the *total* would be

$$500(.125) \pm 500(.043)$$

$$62.5 \pm 21.5 \quad \text{or} \quad 41 \text{ to } 84 \text{ components}$$

12. a. Set

$$z\sqrt{\frac{\pi(1-\pi)}{n}} = .04$$

$$1.96\sqrt{\frac{.5(1-.5)}{n}} = .04$$

Solving for n:

$$n = \left[\frac{1.96(\sqrt{.5(.5)})}{.04}\right]^2 = 600$$

b. $z\sqrt{\dfrac{.5(1-.5)}{600}} = .02$ produces a z score of $z = \dfrac{.02}{\sqrt{.5(1-.5)/600}} = \dfrac{.02}{.0204} = .98$

Using the normal table for $z = .98$ produces a probability of .3365, implying a confidence level of

$$2 \times .3365 = \mathbf{.6730} \quad \text{or} \quad 67.3\%$$

13. a. $\pi = 2/4 = .5$

c. See (b).

b.

Samples	$\bar{p}$	$(\bar{p} - \bar{\bar{p}})^2$
A, A	1	$(.5)^2 = .25$
A, B	.5	$0^2 = 0$
A, C	.5	$0^2 = 0$
A, D	1	$.5^2 = .25$
B, A	.5	$0^2 = 0$
B, B	0	$(-.5)^2 = .25$
B, C	0	$(-.5)^2 = .25$
B, D	.5	$0^2 = 0$
C, A	.5	$0^2 = 0$
C, B	0	$(-.5)^2 = .25$
C, C	0	$(-.5)^2 = .25$
C, D	.5	$0^2 = 0$
D, A	1	$.5^2 = .25$
D, B	.5	$0^2 = 0$
D, C	.5	$0^2 = 0$
D, D	1	$.5^2 = .25$
	8	2

$$\bar{\bar{p}} = 8/16 = .5$$

$$\sigma_{\bar{p}} = \sqrt{2/16}$$

$$= \sqrt{1/8}$$

$$= .354$$

d. Symmetric in shape (bell shaped?)

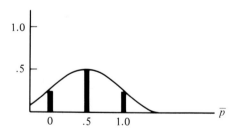

e. Average of sample $\bar{p}$'s $= \dfrac{\Sigma \bar{p}}{16} = \dfrac{8}{16} = .5 = \pi$ (We've labeled the average $\bar{p}$ as $\bar{\bar{p}}$.)

f. See (b).

$$\sigma_{\bar{p}} = \sqrt{\dfrac{\Sigma(\bar{p} - \bar{\bar{p}})^2}{16}} = \sqrt{\dfrac{2}{16}} = \sqrt{\dfrac{1}{8}} = .354$$

g. $\sigma_{\bar{p}} = \sqrt{\dfrac{\pi(1 - \pi)}{n}} = \sqrt{\dfrac{(.5)(.5)}{2}} = \sqrt{\dfrac{1}{8}} = .354$

As predicted by statistical theory, the results in (f) match the results in (g).

15. System 1   System 2

$n_1 = 50$ $n_2 = 50$

$\bar{x}_1 = 29.2$ $\bar{x}_2 = 13.1$

$s_1 = 4.5$ $s_2 = 3.5$

To estimate $\mu_1 - \mu_2$: $(\bar{x}_1 - \bar{x}_2) \pm z \sqrt{\dfrac{\sigma_1^2}{n_1} + \dfrac{\sigma_2^2}{n_2}}$

$(29.2 - 13.1) \pm 1.64 \sqrt{\dfrac{4.5^2}{50} + \dfrac{3.5^2}{50}} = 16.1 \pm 1.64\sqrt{.405 + .245}$

$= 16.1 \pm 1.64(\sqrt{.65})$

$= 16.1 \pm 1.32 \text{ days}$

or 14.78 to 17.42 days.

If we were to examine all orders processed under system 1 and all orders processed under system 2, we can be 90% confident that the interval 14.78 to 17.42 days would contain the actual difference in average processing time for these 2 populations.

17. Northeast   Southeast

$n_1 = 500$ $n_2 = 400$

$\bar{x}_1 = \$9.84$ $\bar{x}_2 = \$7.48$

$s_1 = \$1.23$ $s_2 = \$2.04$

To estimate $\mu_1 - \mu_2$: $(9.84 - 7.48) \pm 1.64 \sqrt{\dfrac{1.23^2}{500} + \dfrac{2.04^2}{400}} = \$2.36 \pm 1.64(.116)$

$= \$2.36 \pm \$.19$

or \$2.17 to \$2.55/hr difference.

19. a. Poor students   Good students

$n_1 = 1000$ $n_2 = 1400$

$\bar{x}_1 = 31.5$ $\bar{x}_2 = 22.4$

$s_1 = 6.1$ $s_2 = 4.7$

At the 99% confidence level, $\mu_1 - \mu_2$ should be in the interval

$$(31.5 - 22.4) \pm 2.58 \sqrt{\frac{6.1^2}{1000} + \frac{4.7^2}{1400}} = 9.1 \pm 2.58(.23)$$

$$9.1 \text{ hr} \pm .6 \text{ hr}$$

b. $n_1 = 10$ $\qquad\qquad$ $n_2 = 14$

$\bar{x}_1 = 31.5$ $\qquad\qquad$ $\bar{x}_2 = 22.4$

$s_1 = 6.1$ $\qquad\qquad$ $s_2 = 4.7$

Switching to the t distribution and pooling the sample standard deviations:

$$s_{pooled} = \sqrt{\frac{(10-1)(6.1)^2 + (14-1)(4.7)^2}{10 + 14 - 2}} = \sqrt{28.26} = 5.3$$

$df = 22$

$$(31.5 - 22.4) \pm t \sqrt{\frac{5.3^2}{10} + \frac{5.3^2}{14}}$$

$$9.1 \pm 2.819(2.19) \quad \text{or} \quad 9.1 \pm 6.2 \text{ hr}$$

c. (1) Equal standard deviations $\qquad$ (2) Normally distributed

22. a. Brand W $\qquad\qquad\qquad\qquad\qquad\qquad$ Brand F

$n_1 = 5$ $\qquad\qquad\qquad\qquad\qquad\qquad$ $n_2 = 5$

$\bar{x}_1 = 150$ $\qquad\qquad\qquad\qquad\qquad\qquad$ $\bar{x}_2 = 144$

$$s_1 = \sqrt{\frac{(156-150)^2 + \cdots + (150-150)^2}{5-1}} \qquad s_2 = \sqrt{\frac{(145-144)^2 + \cdots + (149-144)^2}{5-1}}$$

$$= 4.47 \qquad\qquad\qquad\qquad\qquad\qquad\qquad\qquad = 4.0$$

b. Pooling produces

$$s_{pooled} = \sqrt{\frac{(5-1)(4.47)^2 + (5-1)(4.0)^2}{5+5-2}} = 4.24$$

c. $\qquad (\bar{x}_1 - \bar{x}_2) \pm t \sqrt{\frac{s_{pooled}^2}{n_1} + \frac{s_{pooled}^2}{n_2}}$

$$= (150 - 144) \pm 2.306 \sqrt{\frac{(4.24)^2}{5} + \frac{(4.24)^2}{5}}$$

$$= 6 \pm 2.306(2.68)$$

$$= 6 \pm 6.18 \quad \text{or} \quad -.18° \text{ to } 12.18° \text{ difference}$$

24. *Population 1* $\qquad\qquad\qquad$ *Population 2*

All Republican members $\qquad$ All Democrat members

$N_1 = 102$ $\qquad\qquad\qquad\qquad$ $N_2 = 110$

$n_1 = 40$ $\qquad\qquad\qquad\qquad$ $n_2 = 40$

$\bar{x}_1 = 7800$ $\qquad\qquad\qquad\qquad$ $\bar{x}_2 = 6250$

$s_1 = 1400$ $\qquad\qquad\qquad\qquad$ $s_2 = 1100$

$$(\bar{x}_1 - \bar{x}_2) \pm z\sqrt{\frac{\sigma_1^2}{n_1}\left(\frac{N_1 - n_1}{N_1 - 1}\right) + \frac{\sigma_2^2}{n_2}\left(\frac{N_2 - n_2}{N_2 - 1}\right)}$$

$$= (7800 - 6250) \pm 1.96\left(\sqrt{\frac{1400^2}{40}\left(\frac{102 - 40}{102 - 1}\right) + \frac{1100^2}{40}\left(\frac{110 - 40}{110 - 1}\right)}\right)$$

$$= 1550 \pm 1.96\sqrt{49000(.614) + 30250(.642)}$$

$$= 1550 \pm 1.96(\sqrt{49506})$$

$$= 1550 \pm 1.96(222.5)$$

$$\$1550 \pm \$436$$

or $1114 to $1986 difference

25. Ivy bound   Others

$n_1 = 200$ $n_2 = 200$

$\bar{x}_1 = 1130$ $\bar{x}_2 = 1080$

$s_1 = 56$ $s_2 = 84$

a. $(1130 - 1080) \pm 1.96\left(\sqrt{\frac{(56)^2}{200} + \frac{(84)^2}{200}}\right) = 50 \pm 1.96\sqrt{50.96} = 50 \pm 1.96(7.14)$

$$= 50 \pm 14 \quad \text{or} \quad 36 \text{ to } 64 \text{ points difference}$$

b. The two populations might be described here as

(1) All students who have taken the ivy bound course

(2) All students who have not taken the ivy bound course

c. Set the sampling error term

$$z\sqrt{\frac{\sigma_1^2}{n_1} + \frac{\sigma_2^2}{n_2}} = 10$$

Using the s_1 and s_2 values from (a) to estimate σ_1 and σ_2, the expression becomes

$$1.96\sqrt{\frac{(56)^2}{n_1} + \frac{(84)^2}{n_2}} = 10$$

If $n_1 = n_2$, then we can define $n = n_1 = n_2$ and show

$$1.96\sqrt{\frac{(56)^2 + (84)^2}{n}} = 10$$

Solving for n:

$$n = \left[\frac{1.96\sqrt{(56)^2 + (84)^2}}{10}\right]^2$$

so

$$n = \left[\frac{1.96(100.95)}{10}\right]^2 = 392$$

Result: Take a sample of $n_1 = 392$ and $n_2 = 392$ from the respective populations.

27.   *Population 1*   *Population 2*

All men   All women

$n_1 = 800$ $n_2 = 800$

$\bar{p}_1 = .64$ $\bar{p}_2 = .43$

$\pi_1 - \pi_2$ should be in the interval

$$(\bar{p}_1 - \bar{p}_2) \pm z \sqrt{\frac{\pi_1(1 - \pi_1)}{n_1} + \frac{\pi_2(1 - \pi_2)}{n_2}}$$

or

$$(\bar{p}_1 - \bar{p}_2) \pm z \sqrt{\frac{\bar{p}_1(1 - \bar{p}_1)}{n_1} + \frac{\bar{p}_2(1 - \bar{p}_2)}{n_2}}$$

$$= (.64 - .43) \pm 1.64 \sqrt{\frac{(.64)(.36)}{800} + \frac{(.43)(.57)}{800}}$$

$$= .21 \pm 1.64(.024)$$

$$= .21 \pm .04 \quad \text{or} \quad .17 \text{ to } .25$$

If we surveyed all men and all women on the issue, there is a 90% probability that the interval 17% to 25% would contain the actual difference in the proportions in the two populations who have absolutely no interest.

29.   Alcott   Brookside

shipment   shipment

$n_1 = 150$ $n_2 = 150$

$\bar{p}_1 = \dfrac{15}{150} = .10$ $\bar{p}_2 = \dfrac{9}{150} = .06$

The interval estimate should look like

$$(\bar{p}_1 - \bar{p}_2) \pm z \sqrt{\frac{\pi_1(1 - \pi_1)}{n_1} + \frac{\pi_2(1 - \pi_2)}{n_2}}$$

or

$$(\bar{p}_1 - \bar{p}_2) \pm z \sqrt{\frac{\bar{p}_1(1 - \bar{p}_1)}{n_1} + \frac{\bar{p}_2(1 - \bar{p}_2)}{n_2}}$$

$$= (.10 - .06) \pm 1.96 \sqrt{\frac{(.1)(.9)}{150} + \frac{(.06)(.94)}{150}}$$

$$= .04 \pm 1.96(.031)$$

$$= .04 \pm .061 \quad \text{or} \quad -.021 \text{ to } .101$$

It would be difficult to say, based on these sample results, just which of the two shipments is better. (Note: we've omitted the FPCs because the sample sizes are less than 5% of the population sizes.)

32. a.

Japanese-car buyers	American-car buyers
$n_1 = 200$	$n_2 = 200$
$\bar{p} = \dfrac{142}{200} = .71$	$\bar{p} = \dfrac{120}{200} = .60$

$$(\bar{p}_1 - \bar{p}_2) \pm z \sqrt{\frac{\bar{p}_1(1 - \bar{p}_1)}{n_1} + \frac{\bar{p}_2(1 - \bar{p}_2)}{n_2}}$$

$$= (.71 - .60) \pm 1.64 \sqrt{\frac{(.71)(.29)}{200} + \frac{(.6)(.4)}{200}}$$

$$= .11 \pm 1.64(.047)$$

$$= .11 \pm .077 \quad \text{or} \quad .033 \text{ to } .187$$

b. Set

$$1.64 \sqrt{\frac{(.71)(.29)}{n_1} + \frac{(.6)(.4)}{n_2}} = .04$$

Since n_1 will equal n_2, we can set $n_1 = n_2 = n$:

$$1.64 \sqrt{\frac{(.71)(.29) + (.6)(.4)}{n}} = .04$$

so

$$n = \left[\frac{1.64(\sqrt{.71(.29) + (.6)(.4)})}{.04} \right]^2 = 750$$

thus, $n_1 = n_2 = 750$

c. Using the worst-case .5 as an estimate of π_1 and π_2:

$$1.64 \sqrt{\frac{(.5)(.5) + (.5)(.5)}{n}} = .04$$

so

$$n = \left[\frac{1.64\sqrt{.5(.5) + .5(.5)}}{.04} \right]^2 = 840$$

thus, $n_1 = n_2 = 840$

CHAPTER 7

1. a. H_0: The average life for Delton III tires will be at least 50,000 mi.

H_1: The average life for Delton III's will be lest than 50,000 mi.

or

$H_0: \mu \geq 50,000$	$n = 50$
$H_1: \mu < 50,000$	$\bar{x} = 49,100$
	$s = 4000$

b.

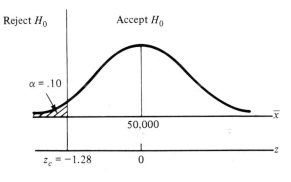

$$\sigma_{\bar{x}} = \frac{4000}{\sqrt{50}} = 566 \text{ mi}$$

$$z_{sample} = \frac{49,100 - 50,000}{566} = -1.59$$

Since -1.59 is outside the $z_c = -1.28$ boundary, reject H_0.

c.

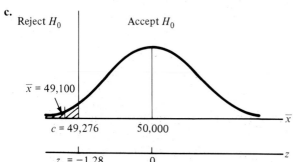

$$c = 50,000 - 1.28(\sigma_{\bar{x}})$$
$$c = 50,000 - 1.28(566) = 50,000 - 724$$
$$= 49,276 \text{ mi}$$

3. a. H_0: The judge is tough on bank robbers.

H_1: The judge is not tough on bank robbers.

or

$H_0: \mu \geq 80$ $n = 36$

$H_1: \mu < 80$ $\bar{x} = 72$

        $s = 18$

b.

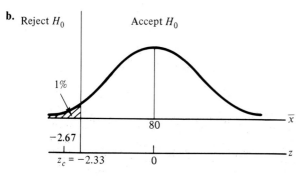

$$\sigma_{\bar{x}} = \frac{\sigma}{\sqrt{n}} = \frac{18}{\sqrt{36}} = 3 \text{ mos.}$$

Since $z_{sample} = \dfrac{\bar{x} - \mu}{\sigma_{\bar{x}}} = \dfrac{72 - 80}{3} = -2.67$, we will reject the judge's claim.

c.

Reject H_0 Accept H_0

$\bar{x} = 72$

$c = 73.01$ 80 $\bar{x}$

-2.33 0 z

$$c = 80 - 2.33\left(\frac{\sigma}{\sqrt{n}}\right) = 80 - 2.33(3)$$
$$= 80 - 6.99$$
$$= \mathbf{73.01}$$

6. H_0: Acme's claim is true.

H_1: Acme's claim is false.

or

H_0: $\mu \le 120$ $n = 40$

H_1: $\mu > 120$ $\bar{x} = 124$

$s = 18$

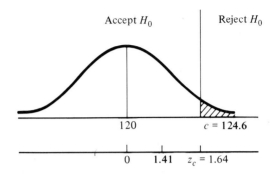

Accept H_0 Reject H_0

$$\sigma_{\bar{x}} = \frac{5}{\sqrt{n}} = \frac{18}{\sqrt{40}} = 2.84$$

$$z_{\text{sample}} = \frac{124 - 120}{2.84}$$

120                    $c = 124.6$

0   1.41 $z_c = 1.64$

$$= \textbf{1.41}$$

We cannot reject the null since $z_{\text{sample}} < z_c$.
On the original scale,

$$c = 120 + 1.64(2.84)$$

so $c = \textbf{124.6}$

Since $\bar{x}(124)$ is inside the cutoff ($c = 124.6$), we cannot reject the null.

9. a. Yes, since

$$z_{\text{sample}} = \frac{49,100 - 50,000}{566} = -1.59$$

places the sample result beyond the $z_c = -1.28$ cutoff, we reject H_0.

b. No, since $z_{\text{sample}} = -1.59$ is inside the $z_c = -1.64$ boundary (using $\alpha = 5\%$) we would *not* reject H_0.

No, since $z_{\text{sample}} = -1.59$ is inside the $z_c = -2.33$ boundary (using $\alpha = 1\%$) we would *not* reject H_0.

c. We need to find the probability that a sample mean will lie at least 1.59 standard deviations below the null hypothesis population mean (i.e., the probability that a sample mean of 49,100 or less would be selected from the null distribution).
From the normal table, $.5 - .4441 = \textbf{.0559}$, so if α were set at 5.59%, our 49,100 sample result would just barely be judged statistically significant.

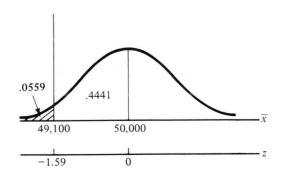

.0559

.4441

49,100   50,000

-1.59 0

11. $H_0: \mu \geq 8800$ (The rivets meet the standard.)

$H_1: \mu < 8800$ (The rivets do not meet the standard.)

a.

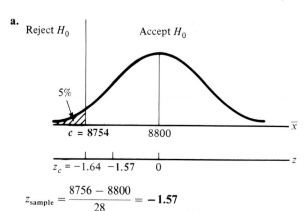

$$\sigma_{\bar{x}} = \frac{280}{\sqrt{100}} = 28 \text{ psi}$$

$$z_{\text{sample}} = \frac{8756 - 8800}{28} = -1.57$$

Conclusion: Since -1.57 is inside $z_c = -1.64$, we'll accept the null and conclude that the overall population average is at least 8800 psi.

On the original scale, $c = 8800 - 1.64(28) = 8754$ psi, so $c = \mathbf{8754}$.

b. Showing the test as

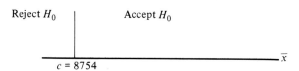

we can superimpose the alternative sampling distribution here as shown below.

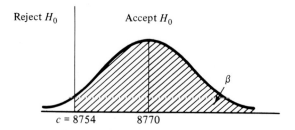

To compute β, we need only find what part of the alternative distribution falls in the accept H_0 region of the original test. To do this, use the normal table

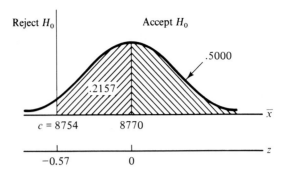

$$z = \frac{8754 - 8770}{28} = -.57$$ From the table, the area for a z of .57 = .2157.

Therefore β = .2157 + .5000 = **.7157**

c.

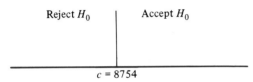

Superimposing the alternative distribution centered on $\mu = 8720$:

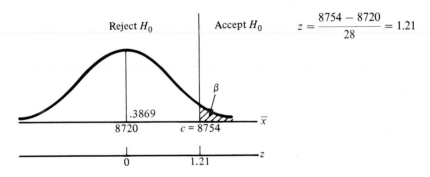

$$z = \frac{8754 - 8720}{28} = 1.21$$

From the normal table, for $z = 1.21$, the area is .3869. Thus, β is .5000 − .3869 = **.1131**.

13. $H_0: \mu \geq 30$ (The average age of viewers is at least 30.)

$H_1: \mu < 30$ (The average age of viewers is less than 30.)

a.

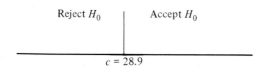

$n = 200$

$\bar{x} = 28.3$

$s = 9.6$

$$\sigma_{\bar{x}} = \frac{9.6}{\sqrt{200}} = .67$$

Since $z_{\text{sample}} = \dfrac{28.3 - 30}{.67} = -2.54$, we reject H_0.

On the original scale,

Reject H_0 | Accept H_0

$c = 28.9$

$c = 30 - 1.64(.67) = \mathbf{28.9}$

b. Superimpose the alternative distribution centered on 28.

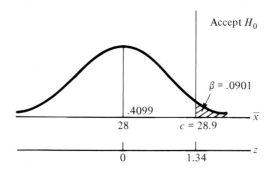

$$z = \frac{28 - 28.9}{.67} = 1.34 \quad \text{For a } z = 1.34, \text{ the area is } .4099; \text{ therefore } \beta = .5 - .4099 = .0901$$

15. $H_0: \mu \le 120$ (The average age of the inventory is 120 days or less.)

$H_1: \mu > 120$ (The average age of the inventory exceeds 120 days.)

$c = 120 + 1.64(\sigma_{\bar{x}})$ (from the null distribution)

$c = 125 - 1.28(\sigma_{\bar{x}})$ (from the alternative distribution)

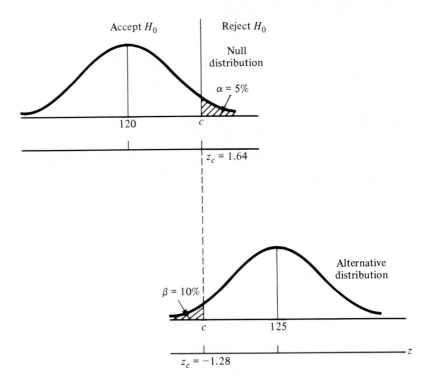

Setting $120 + 1.64\sigma_{\bar{x}} = 125 - 1.28\sigma_{\bar{x}}$, we can substitute $\sigma/\sqrt{n}$ for $\sigma_{\bar{x}}$ and show

$$120 + 1.64\sigma/\sqrt{n} = 125 - 1.28\sigma/\sqrt{n}$$

Using 32.2 as the estimate of σ, this becomes

$$120 + 1.64(32.2/\sqrt{n}) = 125 - 1.28(32.2/\sqrt{n}) \Rightarrow 2.92(32.2/\sqrt{n}) = 5$$

$$\Rightarrow \sqrt{n} = 2.92(32.2)/5 = 18.8 \quad \text{so} \quad n = 18.8^2 = \textbf{354}$$

18. a. Set up a two-tailed test for

$H_0: \mu = 30$ (The average fill-weight is 30 oz.)

$H_1: \mu \neq 30$ (The average fill-weight is not 30 oz.)

$\sigma = .3$

$n = 36$

$\sigma_{\bar{x}} = .3/\sqrt{36} = .05$

b. $z_{sample} = (29.83 - 30)/.05 = -3.4$

Therefore, we'll **reject** H_0.

c. We *could* be making a Type I error anytime we reject a null hypothesis.

20. a. Set up a two-tailed test for

$H_0: \mu = 18$ $n = 49$

$H_1: \mu \neq 18$ $\sigma = .05$

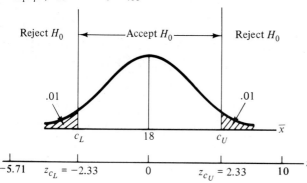

$\sigma_{\bar{x}} = .05/\sqrt{49} = .007$

b. $z_{sample} = (17.96 - 18)/.007 = -5.71$

Therefore, **reject** H_0.

c. $z_{sample} = (18.07 - 18)/.007 = 10$. Therefore, reject H_0.

22. $H_0: \mu \geq 50,000$

$H_1: \mu < 50,000$

a.

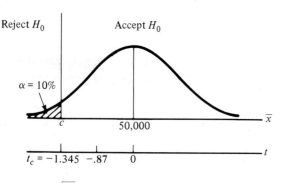

$\bar{x} = 49,100$ $s = 4000$

$n = 15$

$df = 15 - 1 = 14$

$\sigma_{\bar{x}} - 4000/\sqrt{15} = 1033$

From the t table using 14 df and a tail-end area of .10, $t_c = 1.345$.

$$t_{sample} = \frac{49,100 - 50,000}{4000/\sqrt{15}} = -\frac{900}{1033} = -.87$$

Since $-.87$ is inside the -1.345 cutoff, we will not be able to reject the null.

b. On the original scale,

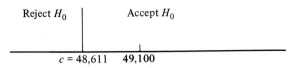

$$\bar{x} = \textbf{49,100} \qquad c = 50,000 - 1.345(1033) = 48,611 \text{ mi}$$

c. Population of values must be normal.

24. $H_0: \mu \le 36$

$H_1: \mu > 36$

a.

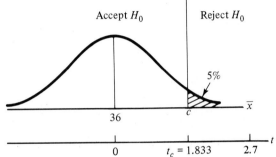

$n = 10 \qquad df = 10 - 1 = 9$

$\bar{x} = 42$

$s = 8$

$$\sigma_{\bar{x}} = 8/\sqrt{10} = 2.53$$

From the t table with 9 df and a tail-end area of .05,

$$t_c = 1.833$$

$$t_{\text{sample}} = (42 - 36)/2.53 = \textbf{2.37}$$

Since t_{sample} is outside t_c, reject H_0.

b. On the original scale,

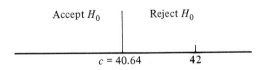

$$c = 36 + 1.833(2.53) = 40.64$$

$$\bar{x} = \textbf{42}$$

c. Population of values must be normally distributed.

26. Use the FPC

$H_0: \mu \le 2500$

$H_1: \mu > 2500$

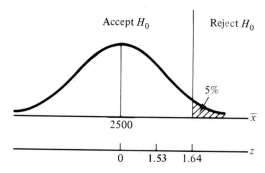

$\bar{x} = 2520$

$s = 160$

$n = 1.00$

$N = 300$

$$\sigma_{\bar{x}} = \frac{s}{\sqrt{n}} \sqrt{\frac{N-n}{N-1}} = \frac{160}{\sqrt{100}} \sqrt{\frac{300-100}{300-1}} = 16(.818) = \mathbf{13.1}$$

$$z_{\text{sample}} = \frac{2520 - 2500}{13.1} = \mathbf{1.53}$$

Since z_{sample} is inside the z_c cutoff, accept the null.

CHAPTER 8

1. a, b. $H_0: \pi \geq .40$

        $H_1: \pi < .40$

$$\sigma_{\bar{p}} = \sqrt{\frac{(.4)(.6)}{100}} = .049$$

c. $z_{\text{sample}} = \dfrac{(\bar{p} - \pi)}{\sigma_{\bar{p}}} = \dfrac{.26 - .40}{.049} = -2.86$

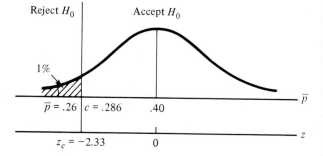

     Reject the null hypothesis.

d. On the original scale,

     $c = \pi - 2.33\sigma_{\bar{p}}$

        $= .40 - 2.33(.049) = \mathbf{.286}$

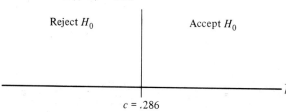

3. a, b. $H_0: \pi \geq .15$ (The proportion of minority students on campus is at least .15.)

$H_1: \pi < .15$ (The proportion of minority students on campus is less than .15.)

$\bar{p} = 14/150 = \mathbf{.093}$

$\sigma_{\bar{p}} = \sqrt{\dfrac{(.15)(.85)}{150}} = .029$

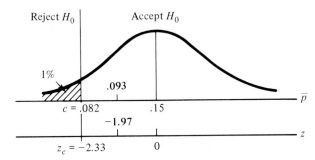

c. $z_{\text{sample}} = \dfrac{(\bar{p} - \pi)}{\sigma_{\bar{p}}} = \dfrac{.093 - .15}{.029} = \mathbf{-1.97}$

Since -1.97 is inside $z_c = -2.33$, we cannot reject H_0.

d. On the original scale,

$c = \pi - z_c \sigma_{\bar{p}}$

$= .15 - 2.33(.029)$

$= \mathbf{.082}$

6. $H_0: \pi \leq .20$

$H_1: \pi > .20$

$\bar{p} = 14/50 = .28$

$n = 50$

$N = 250$

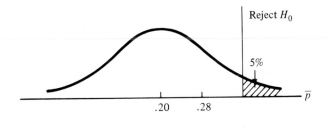

Use the FPC to compute

$\sigma_{\bar{p}} = \sqrt{\dfrac{\pi(1 - \pi)}{n}} \sqrt{\dfrac{N - n}{N - 1}}$

$= \sqrt{\dfrac{.2(.8)}{50}} \sqrt{\dfrac{250 - 50}{250 - 1}} = (.057)(.896) = .051$

$z_{\text{sample}} = \dfrac{.28 - .20}{.051} = \mathbf{1.57}$

Since z_{sample} is inside z_c, we cannot reject the null.

Conclusion: The overall proportion of our customers who will reorder is **not** more than 20%.

8. a. $H_0: \pi \leq .50$ (Howard is guessing.)

$H_1: \pi > .50$ (Howard is doing more than guessing.)

b.

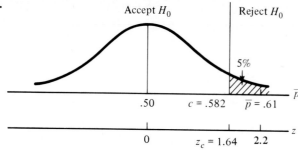

$$\sigma_{\bar{p}} = \sqrt{\frac{(.5)(.5)}{100}} = .05$$

c. $z_{\text{sample}} = \dfrac{\bar{p} - \pi}{\sigma_{\bar{p}}} = \dfrac{.61 - .50}{.05} = \textbf{2.2}$

We can reject H_0. Howard has made his case.

d. $c = .50 + 1.64(.05) = \textbf{.582}$

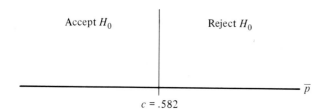

10. In Exercise 1, the test was basically

a.

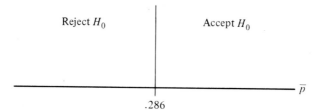

b.

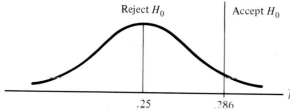

c. for the alternative distribution,

$$\sigma_{\bar{p}} = \sqrt{\frac{(.25)(.75)}{100}} = \textbf{.043}$$

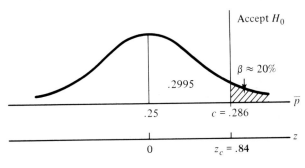

$$z_c = \frac{.286 - .25}{.043} = \mathbf{.84}$$

From the normal table, for $z = .84$, area is .2995. Therefore, β is $.5 - .2995 = \mathbf{.2005}$ or about 20%.

12. a. $H_0: \pi \le .06$ (process is in control)

$H_1: \pi > .06$ (process is out of control)

b, c.

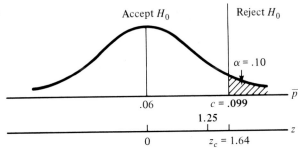

$\bar{p} = 9/100 = .09$

c. For the null distribution

$$\sigma_{\bar{p}} = \sqrt{\frac{(.06)(.94)}{100}} = .024$$

$$z_{\text{sample}} = \frac{.09 - .06}{.024} = \mathbf{1.25}$$

Conclusion: We can **not** reject the null.
Therefore, we'll conclude the process is in control.

d. On the original ($\bar{p}$) scale,

$c = .06 + 1.64(.024) = \mathbf{.099}$

Accept H_0 Reject H_0

$c = .099$

e. Type I error—believing the process is "out of control" when in fact it is "in control."
Type II error—believing the process is "in control" when in fact it is *not*.

f. Superimposing the alternative distribution on the test axis:

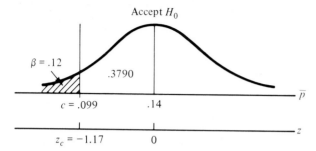

Accept H_0

$\beta = .12$

.3790

$c = .099$ .14

$z_c = -1.17$ 0

For the alternative distribution,

$$\sigma_{\bar{p}} = \sqrt{\frac{(.19)(.86)}{100}} = .035$$

$$z_c = \frac{.099 - .14}{.035} = -1.17$$

From the normal table, the area for $z = 1.17$ is .3790.
β, therefore, is $.5 - .3790 = .1210$, or approx. 12%.

15. a, b. $H_0: \mu_1 = \mu_2$ or $\mu_1 - \mu_2 = 0$

$H_1: \mu_1 \neq \mu_2$ or $\mu_1 - \mu_2 \neq 0$

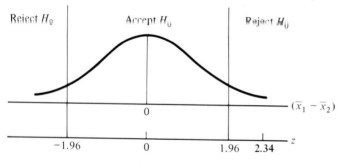

Reject H_0 Accept H_0 Reject H_0

0

-1.96 0   1.96   2.34

$$\sigma_{\bar{x}_1 - \bar{x}_2} = \sqrt{\frac{(40)^2}{200} + \frac{(32)^2}{150}} = \sqrt{8 + 6.83} = 3.85$$

c. $z_{sample} = \dfrac{(230 - 221) - 0}{3.85} = \mathbf{2.34}$

Conclusion: Reject H_0.

d.

Reject H_0 Accept H_0 Reject H_0 $c_L = 0 - 1.96(3.85) = -7.55$

$c_U = 0 + 1.96(3.85) = 7.55$

$c_L = -7.55$ $c_U = 7.55$

17. $H_0: \mu_1 \leq \mu_2$ (Acme doesn't last longer.)

$H_1: \mu_1 > \mu_2$ (Acme does last longer.)

or,

$H_0: \mu_1 - \mu_2 \leq 0$

$H_1: \mu_1 - \mu_2 > 0$

$(\bar{x}_1 - \bar{x}_2) = (12.7 - 11.8) = .9$

$\sigma_{\bar{x}_1 - \bar{x}_2} = \sqrt{\dfrac{1.4^2}{60} + \dfrac{1.8^2}{60}} = .294$

$z_{\text{sample}} = \dfrac{(12.7 - 11.8) - 0}{.294}$

$= \mathbf{3.06}$

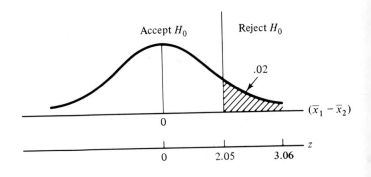

Conclusion: Reject the "no longer" null hypothesis and believe Acme's claim.

19. $H_0: \mu_1 \geq \mu_2$ or $\mu_1 - \mu_2 \geq 0$ (The new drug is not faster.)

$H_1: \mu_1 < \mu_2$ or $\mu_1 - \mu_2 < 0$ (The new drug is faster.)

where $\mu_1 =$ the average recovery time for all those who might be given the new drug

$\mu_2 =$ the average recovery time for all those who would not be given the new drug

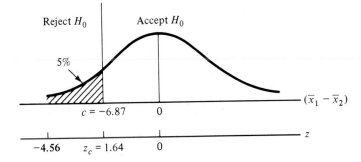

$\bar{x}_1 - \bar{x}_2 = 79.3 - 98.4 = \mathbf{-19.1}$

$\sigma_{\bar{x}_1 - \bar{x}_2} = \sqrt{\dfrac{(19.2)^2}{100} + \dfrac{(37.2)^2}{100}} = 4.19$

$z_{\text{sample}} = \dfrac{-19.1 - 0}{4.19} = \mathbf{-4.56}$

Conclusion: Reject the no faster position (i.e., reject H_0).
On the original scale, the boundary c would be $0 - 1.64(4.19) = \mathbf{-6.87}$

21. $H_0: \mu_1 = \mu_2$ or $\mu_1 - \mu_2 = 0$ (Program A's average performance is the same as Program B's.)

$H_1: \mu_1 \neq \mu_2$ or $\mu_1 - \mu_2 \neq 0$ (Program A's average performance is different from Program B's.)

Pop 1
All program
A participants

Pop 2
All program
B participants

$n_1 = 10$

$n_2 = 10$

$\bar{x}_1 = 15.2$

$\bar{x}_2 = 12.5$

$s_1 = 4.9$

$s_2 = 5.4$

$\bar{x}_1 - \bar{x}_2 = 15.2 - 12.5 = 2.7$

$$t_{\text{sample}} = \frac{(15.2 - 12.5) - 0}{\sigma_{\bar{x}_1 - \bar{x}_2}}$$

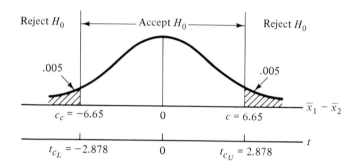

In order to produce $\sigma_{\bar{x}_1 - \bar{x}_2}$ we need to pool sample standard deviations:

$$S_{\text{pooled}} = \sqrt{\frac{(10-1)(4.9)^2 + (10-1)(5.4)^2}{10 + 10 - 2}}$$

$$= \sqrt{26.59} = \mathbf{5.16}$$

so

$$\sigma_{\bar{x}_1 - \bar{x}_2} = \sqrt{\frac{(5.16)^2}{10} + \frac{(5.16)^2}{10}} = 2.31$$

t_{sample} is then

$$\frac{(15.2 - 12.5) - 0}{2.31} = 1.17$$

Since t_{sample} is inside t_c, we cannot reject H_0.
On the original scale,

Reject H_0 Accept H_0 Reject H_0

$\bar{x}_1 - \bar{x}_2$

$c = -6.65$ $c = 6.65$

$c_L = 0 - 2.878(2.31) = -6.65$

$c_U = 0 + 2.878(2.31) = +6.65$

23. $H_0: \mu_1 = \mu_2$ or $\mu_1 - \mu_2 = 0$

$H_1: \mu_1 \neq \mu_2$ or $\mu_1 - \mu_2 \neq 0$

For the Madison sample:

$\bar{x}_1 = 519$ $n_1 = 6$

$s_1 = 76.4$ $\left(s = \sqrt{\dfrac{\sum(x - \bar{x})^2}{n - 1}} \right)$

For the Jefferson sample:

$\bar{x}_2 = 474$ $n_2 = 5$

$s_2 = 110.1$

$(\bar{x}_1 - \bar{x}_2) = 519 - 474 = \mathbf{45}$

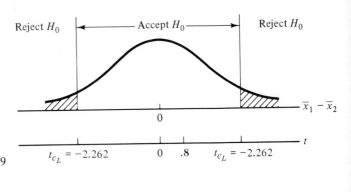

$$S_{pooled} = \sqrt{\frac{5(76.4)^2 + 4(110.1)^2}{6 + 5 - 2}} = 92.9$$

$$\sigma_{\bar{x}_1 - \bar{x}_2} = \sqrt{\frac{92.9^2}{6} + \frac{92.9^2}{5}} = 56.3$$

$$t_{c_L} = -2.262, \; t_{c_U} = +2.262$$

$$t_{sample} = \frac{(519 - 474) - 0}{56.3} = .8$$

Conclusion: We cannot reject the null hypothesis.

25. a. The test:

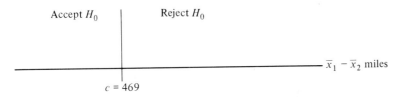

b.

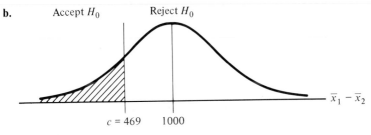

c.

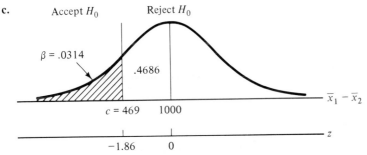

$$z_c = \frac{469 - 1000}{286} = -1.86$$

From the normal table, for $z = 1.86$, the area is .4686. Therefore, $\beta = .5 - .4686 = .0314$.

27. $H_0: \mu_1 - \mu_2 = 0$ (The average maintenance times are the same.)

$H_1: \mu_1 - \mu_2 \neq 0$ (The average maintenance times are different.)

Pop 1	Pop 2
110 Fords	130 Chevys
$\bar{x}_1 = 19$	$\bar{x}_2 = 17$
$s_1 = 6$	$s_2 = 4$
$n_1 = 36$	$n_2 = 49$
$N_1 = 110$	$N_2 = 130$

$\bar{x}_1 - \bar{x}_2 = 19 - 17 = 2$ days

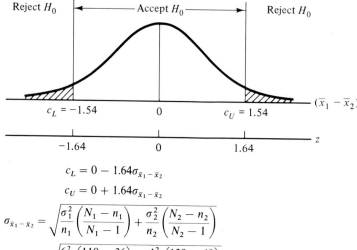

$$c_L = 0 - 1.64\sigma_{\bar{x}_1 - \bar{x}_2}$$

$$c_U = 0 + 1.64\sigma_{\bar{x}_1 - \bar{x}_2}$$

$$\sigma_{\bar{x}_1 - \bar{x}_2} = \sqrt{\frac{\sigma_1^2}{n_1}\left(\frac{N_1 - n_1}{N_1 - 1}\right) + \frac{\sigma_2^2}{n_2}\left(\frac{N_2 - n_2}{N_2 - 1}\right)}$$

$$= \sqrt{\frac{6^2}{36}\left(\frac{110 - 36}{110 - 1}\right) + \frac{4^2}{49}\left(\frac{130 - 49}{130 - 1}\right)}$$

$$= \sqrt{1(.679) + .326(.628)}$$

$$= .94$$

So

| Reject H_0 | Accept H_0 | Reject H_0 |

$c_L = -1.54$ $c_U = 1.54$; 2 ; $(\bar{x}_1 - \bar{x}_2)$

$$c_L = 0 - 1.64(.94) = -1.54$$
$$c_U = 0 + 1.64(.94) = +1.54$$

Since our one $\bar{x}_1 - \bar{x}_2 = 19 - 17 - 2$, we will reject the no difference null hypothesis.

29. a, b. $H_0: \pi_1 - \pi_2 = 0$ (The proportion of single-parent households is the same.)

$H_1: \pi_1 - \pi_2 \neq 0$ (The proportion of single-parent households is not the same.)

c. $\bar{p}_{\text{pooled}} = \dfrac{120 + 160}{2000} = .14$ **d.** $\sigma_{\bar{p}_1 - \bar{p}_2} = \sqrt{\dfrac{(.14)(.86)}{1000} + \dfrac{(.14)(.86)}{1000}} = .015$

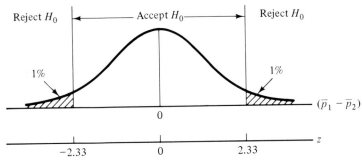

e. $z_{sample} = \dfrac{\bar{p}_1 - \bar{p}_2}{\sigma_{\bar{p}_1 - \bar{p}_2}} = \dfrac{.16 - .12}{.015} = 2.67$

f. Since z_{sample} is outside the 2.33 boundary, we should reject the "no difference" null hypothesis.

g. $c_L = 0 - 2.33(.015) = -.035$

$c_U = 0 + 2.33(.015) = \quad .035$

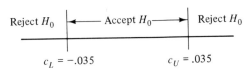

31. a. $H_0: \pi_1 - \pi_2 \leq 0$ (Turbo's on-time rate is no better.)

$H_1: \pi_1 - \pi_2 > 0$ (Turbo's on-time rate is better.)

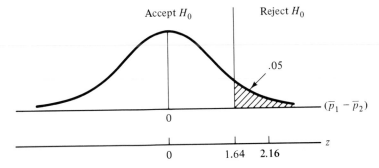

$\bar{p}_1 = .717$

$\bar{p}_2 = .583$

$\bar{p}_{pooled} = \dfrac{86 + 70}{240} = .65$

$\sigma_{\bar{x}_1 - \bar{x}_2} = \sqrt{\dfrac{(.65)(.35)}{120} + \dfrac{(.65)(.35)}{120}} = .062$

$z_{sample} = \dfrac{.717 - .583}{.062} = \mathbf{2.16}$

Since z_{sample} is outside the 1.64 cutoff, we can **reject** the "no better" null hypothesis.

b. Population 1: All Turbo deliveries
Population 2: All Ittle/Waite deliveries

33. $H_0: \pi_1 - \pi_2 = 0$ (The East German and West German populations show the same support for reunification.)

$H_1: \pi_1 - \pi_2 \neq 0$ (The East German and West German populations don't show the same support for unification.)

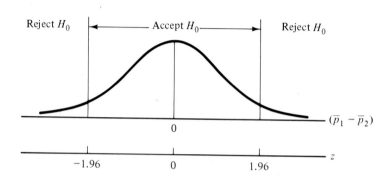

$\bar{p}_1 = .63$

$\bar{p}_2 = .55$

$$\bar{p}_{pooled} = \frac{1000(.63) + 1000(.55)}{2000} = .59$$

$$\sigma_{\bar{p}_1 - \bar{p}_2} = \sqrt{\frac{(.59)(.41)}{1000} + \frac{(.59)(.41)}{1000}} = .022$$

$$z_{sample} = \frac{(.63 - .55) - 0}{.022} = 3.64$$

Conclusion: Reject the no difference null hypothesis.

35. $H_0: \pi_1 - \pi_2 = 0$ (The defective rates are equal.)

$H_1: \pi_1 - \pi_2 \neq 0$ (The defective rates are not equal.)

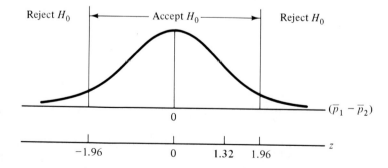

$$\bar{p}_1 = \frac{23}{120} = .192$$

$$\bar{p}_2 = \frac{18}{130} = .138$$

$$\bar{p}_{pooled} = \frac{23 + 18}{250} = \textbf{.164}$$

$$\sigma_{\bar{p}_1 - \bar{p}_2} = \sqrt{\frac{(.164)(.836)}{120}\left(\frac{450 - 120}{450 - 1}\right) + \frac{(.164)(.836)}{130}\left(\frac{560 - 130}{560 - 1}\right)}$$

$$= .041$$

$$z_{sample} = \frac{(.192 - .138) - 0}{.041} = \textbf{1.32}$$

Conclusion: We cannot reject the "no difference" null hypothesis.

CHAPTER 9

1.

x	y	xy	x^2
50	310	15,500	2,500
80	300	24,000	6,400
90	420	37,800	8,100
110	410	45,100	12,100
120	460	55,200	14,400
Totals 450	1900	177,600	43,500

$$\bar{x} = 90$$

$$\bar{y} = 380$$

$$b = \frac{5(177,600) - (450)(1900)}{5(43,500) - (450)^2}$$

$$= \frac{33,000}{15,000} = \textbf{2.2}$$

$$a = 380 - (2.2)90 = \textbf{182}$$

2.

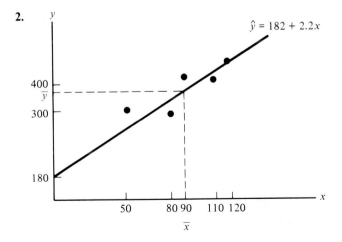

$\hat{y} = 182 + 2.2x$

3.

x	y	$\hat{y}$	$y - \hat{y}$	$(y - \hat{y})^2$
50	310	292	18	324
80	300	358	−58	3364
90	420	380	40	1600
110	410	424	−14	196
120	460	446	14	196
			0	**5680**

$\hat{y} = 182 + 2.2x$

S.E. of E. $= \sqrt{\dfrac{5680}{5}} = \mathbf{33.7}$

4.

				Tot var		Exp var
x	y	$\hat{y}$	$(y - \bar{y})$	$(y - \bar{y})^2$	$(\hat{y} - \bar{y})$	$(\hat{y} - \bar{y})^2$
50	310	292	−70	4,900	−88	7,744
80	300	358	−80	6,400	−22	484
90	420	380	40	1,600	0	0
110	410	424	30	900	44	1,936
120	460	446	80	6,400	66	4,356
	1900		0	**20,200**	0	**14,520**

$\bar{y} = \dfrac{1900}{5} = 380$

$\hat{y} = 182 + 2.2x \qquad r^2 = \dfrac{14,520}{20,200} = \mathbf{.719}$

5. $r = \sqrt{.719} = \mathbf{+.848}$

6.

			Unexp	Exp	Total
x	y	$\hat{y}$	$(y - \hat{y})^2$	$(\hat{y} - \bar{y})^2$	$(y - \bar{y})^2$
50	310	292	324	7,744	4,900
80	300	358	3,364	484	6,400
90	420	380	1,600	0	1,600
110	410	424	196	1,936	900
120	460	446	196	4,356	6,400
			5,680	14,520	20,200

Tot = Exp + Unexp

20,200 = 14,520 + 5,680

7. a. $s_{y.x} = \sqrt{\dfrac{\sum(y - \hat{y})^2}{n - 2}} = \sqrt{\dfrac{5680}{5 - 2}} = \mathbf{43.5}$

 b. $a \pm t s_a$

 $182 \pm 3.182(s_a)$

 where $s_a = s_{y.x}\sqrt{\dfrac{\sum x^2}{n\sum(x - \bar{x})^2}}$

 $= 43.5\sqrt{\dfrac{\sum x^2}{n\sum(x - \bar{x})^2}}$

The table shows the x^2 and $(x - \bar{x})^2$ terms:

x	y	x^2	$(x - \bar{x})^2$
50	310	2,500	1600
80	300	6,400	100
90	420	8,100	0
110	410	12,100	400
120	460	14,400	900
450		43,500	3000

$$\bar{x} = \frac{450}{5} = 90$$

so $\quad s_a = 43.5 \sqrt{\dfrac{43,500}{5(3,000)}}$

$$= 43.5(1.7) = 74.1$$

The 95% interval:

$$182 \pm 3.182(74.1) \quad \text{or} \quad \mathbf{182 \pm 235.8}$$

c. $b \pm t s_b$

$2.2 \pm 3.182 s_b$

where $s_b = \dfrac{s_{y.x}}{\sqrt{\sum(x - \bar{x})^2}}$

$$= \frac{43.5}{\sqrt{3000}} = \mathbf{.794}$$

so the 95% interval is

$2.2 \pm 3.182(.794)$

$\mathbf{2.2 \pm 2.53}$

d. $H_0: \beta = 0$ \quad (no linear relationship)

$H_1: \beta \neq 0$ \quad (a linear relationship)

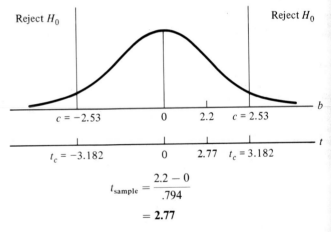

$$t_{sample} = \frac{2.2 - 0}{.794}$$

$$= \mathbf{2.77}$$

We cannot reject the "no relationship" null.

8. From problem 1,

$$\hat{y} = 182 + 2.2x \qquad s_{y.x} = 43.5 \qquad n = 5$$

x	y	x^2	$(x - \bar{x})^2$
50	310	2,500	1,600
80	300	6,400	100
90	420	8,100	0
110	410	12,100	400
120	460	14,400	900
$\bar{x} =$ 90		43,500	3,000

a. For $x = x^* = 100$

$\hat{y} = 182 + 2.2(100) = \textbf{402}$

Interval for $E(y_{100})$:

$$402 \pm 3.182\left(43.5\sqrt{\frac{1}{5} + \frac{(100 - 90)^2}{3000}}\right)$$

$= 402 \pm 3.182(43.5)(.483) = \textbf{402} \pm \textbf{66.9}$

b. For $x = x^* = 200$

$\hat{y} = 182 + 2.2(200) = \textbf{622}$

Interval for $E(y_{200})$:

$$622 \pm 3.182\left(43.5\sqrt{\frac{1}{5} + \frac{(200 - 90)^2}{3000}}\right)$$

$= 622 \pm 3.182(43.5)(2.06) = \textbf{622} \pm \textbf{285.1}$

9. From problem 1,

$\hat{y} = 182 + 2.2x \qquad s_{y.x} = 43.5 \qquad n = 5$

$\sum(x - \bar{x})^2 = 3000$

a. For $x = x^* = 100$

$\hat{y} = 182 + 2.2(100) = 402$

The interval for y_{100} is

$$\hat{y} \pm t\left(s_{y.x}\sqrt{1 + \frac{1}{n} + \frac{(x^* - \bar{x})^2}{\sum(x - \bar{x})^2}}\right)$$

$$= 402 \pm 3.182\left(43.5\sqrt{1 + \frac{1}{5} + \frac{(100 - 90)^2}{3000}}\right)$$

$= 402 \pm 3.182(43.5)(1.11)$

$= \textbf{402} \pm \textbf{153.6}$

b. For $x = x^* = 200$

$\hat{y} = 182 + 2.2(200) = 622$

The interval for y_{200} is

$$622 \pm 3.182\left(43.5\sqrt{1 + \frac{1}{5} + \frac{(200 - 90)^2}{3000}}\right)$$

$= 622 \pm 3.182(43.5)(2.29)$

$= \textbf{622} \pm \textbf{317}$

10.

x	y	xy	x^2
11	150	1650	121
13	140	1820	169
15	100	1500	225
17	110	1870	289
Totals 56	500	6840	804

$\bar{x} = 14$

$\bar{y} = 125 \qquad b = \dfrac{4(6840) - (56)(500)}{4(804) - (56)^2}$

$= \dfrac{640}{80} = -8$

$a = 125 - (-8)(14) = \textbf{237}$

16. a. $s_{y \cdot x} = \sqrt{\dfrac{420}{4-2}} = \mathbf{14.5}$

b.

x	y	x^2	$(x - \bar{x})^2$
11	150	121	9
13	140	169	1
15	100	225	1
17	110	289	9
56		804	20

$\bar{x} = \dfrac{56}{4} = 14$

$s_a = s_{y \cdot x} \sqrt{\dfrac{\sum x^2}{n \sum (x - \bar{x})^2}}$

$= 14.5 \sqrt{\dfrac{804}{4(20)}} = \mathbf{45.9}$

The 95% interval is

$a \pm t s_a$

$= 237 \pm 4.303(45.9)$

$= \mathbf{237 \pm 197.5}$

c. $s_b = \dfrac{s_{y \cdot x}}{\sqrt{\sum (x - \bar{x})^2}} = \dfrac{14.5}{\sqrt{20}} = \mathbf{3.24}$

The interval is

$b \pm t s_b$

$= -8 \pm 4.303(3.24)$

$= \mathbf{-8 \pm 13.9}$

d. $H_0 : \beta = 0$ (no linear relationship)

$H_1 : \beta \neq 0$ (linear relationship)

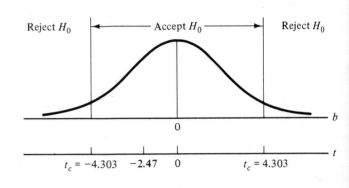

$t_{\text{sample}} = \dfrac{-8 - 0}{3.24} = \mathbf{-2.47}$

Cannot reject the "no relationship" null

18. From problem 2,

$\hat{y} = 237 - 8x$ $n = 4$ $\bar{x} = 14$ $\sum (x - \bar{x})^2 = 20$ $s_{y \cdot x} = 14.5$

a. For $x = x^* = 5$, •

$\hat{y} = 237 - 8(5) = 197$

The y_5 interval is

$\hat{y} \pm t \left(s_{y \cdot x} \sqrt{1 + \dfrac{1}{n} + \dfrac{(x^* - \bar{x})^2}{\sum (x - \bar{x})^2}} \right)$

$= 197 \pm 4.303(14.5) \left(\sqrt{1 + \dfrac{1}{4} + \dfrac{(5 - 14)^2}{20}} \right)$

$= 197 \pm 4.303(14.5)(2.3)$

$= \mathbf{197 \pm 143.5}$

b. For $x = x^* = 20$

$\hat{y} = 237 - 8(20) = 77$

The y_{20} interval is

$77 \pm 4.303(14.5) \left(\sqrt{1 + \dfrac{1}{4} + \dfrac{(20 - 14)^2}{20}} \right)$

$= 77 \pm 4.303(14.5)(1.75)$

$= \mathbf{77 \pm 109.2}$

28. a. $s_{y \cdot x} = \sqrt{\dfrac{\text{Unexp}}{n - 2}} = \sqrt{\dfrac{\sum (y - \hat{y})^2}{n - 2}} = \sqrt{\dfrac{8000}{10 - 2}} = \mathbf{31.6}$

b. α should be in the interval

$$a \pm ts_a = 10 \pm 2.306s_a$$

Here $\quad s_a = s_{y.x}\sqrt{\dfrac{\sum x^2}{n\sum(x-\bar{x})^2}}$

$$= 31.6\sqrt{\dfrac{7000}{10(20{,}000)}}$$

$$= 5.91$$

so the interval is

$$10 \pm 2.306(5.91)$$

$$\mathbf{10 \pm 13.6}$$

c. $b \pm ts_b = 2 \pm 2.306s_b$

Here $\quad s_b = \dfrac{s_{y.x}}{\sqrt{\sum(x-\bar{x})^2}} = \dfrac{31.6}{\sqrt{20{,}000}} = .22$

so

$$2 \pm 2.306(.22)$$

$$\mathbf{2 \pm .51}$$

d. $H_0: \beta = 0$
$H_1: \beta \neq 0$

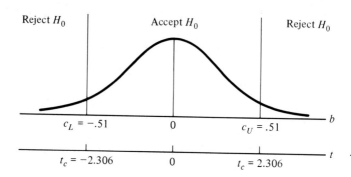

$s_b = .22$

$$t_{\text{sample}} = \dfrac{2\;\;0}{.22} = \mathbf{9.09}$$

Since $t_{\text{sample}} > t_c$, reject the "no relationship" null hypothesis.

31. a.

x	y	xy	x^2
10	180	1800	100
8	150	1200	64
12	230	2760	144
6	140	840	36
36	700	6600	344

$\bar{x} = 9 \qquad \bar{y} = 175$

so $\hat{y} = \mathbf{40 + 15x}$

$$b = \dfrac{n\sum xy - \sum x \sum y}{n\sum x^2 - (\sum x)^2}$$

$$= \dfrac{4(6600) - 36(700)}{4(344) - (36)^2}$$

$$= \dfrac{1200}{80} = 15$$

$$a = \bar{y} - b\bar{x}$$

$$a = 175 - 15(9) = 40$$

b.

x	y	$\hat{y}$	Total $(y-\bar{y})^2$	Exp $(\hat{y}-\bar{y})^2$	Unexp $(y-\hat{y})^2$
10	180	190	25	225	100
8	150	160	625	225	100
12	230	220	3025	2025	100
6	140	130	1225	2025	100
			4900	4500	400

$\bar{y} = 175$

$$r^2 = \frac{\text{Exp}}{\text{Tot}} = \frac{4500}{4900} = \mathbf{.918}$$

c. $s_{y \cdot x} = \sqrt{\dfrac{\text{Unexp}}{n-2}} = \sqrt{\dfrac{400}{2}} = \mathbf{14.14}$

d.

x	y	$(x - \bar{x})^2$
10	180	1
8	150	1
12	230	9
6	140	9
		20

$s_b = \dfrac{s_{y \cdot x}}{\sqrt{\sum(x - \bar{x})^2}} = \dfrac{14.14}{\sqrt{20}} = \mathbf{3.16}$

$\bar{x} = \dfrac{36}{4} = 9$

$H_0: \beta = 0$ (no linear relationship)

$H_1: \beta \neq 0$ (linear relationship)

Reject H_0 Accept H_0 Reject H_0

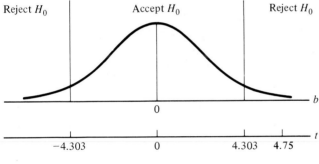

-4.303 0   4.303   4.75

$$t_{\text{sample}} = \frac{b - 0}{s_b} = \frac{15}{3.16} = \mathbf{4.75}$$

We can **reject** the "no relationship" null.

e. For $x = x^* = 11$, $\hat{y} = 40 + 15(11) = 205$.
The $E(y_{11})$ interval is

$$205 \pm 4.303(14.14)\left(\sqrt{\frac{1}{4} + \frac{(11 - 9)^2}{20}}\right)$$

$= 205 \pm 4.303(14.14)(.67)$

$= \mathbf{205 \pm 40.8}$

f. For $x = x^* = 11$, $\hat{y} = 40 + 15(11) = 205$
The y_{11} interval is

$$205 \pm 4.303(14.14)\left(\sqrt{1 + \frac{1}{4} + \frac{(11 - 9)^2}{20}}\right)$$

$= 205 \pm 4.303(14.14)(1.2)$

$= \mathbf{205 \pm 73}$

33. a. Standard error is the standard deviation of the "slope" sampling distribution, sometimes called the standard error of the regression coefficient (b).

b. Regression SS = Explained variability
Residual SS = Unexplained variability
Total SS = Total variability

c. $s_{y.x} = \sqrt{\dfrac{\text{Residual SS}}{n-2}} = \sqrt{\dfrac{34000}{25-2}} = \mathbf{38.4}$

d. $H_0: \beta = 0$ (no linear relationship)
$H_1: \beta \neq 0$ (linear relationship)

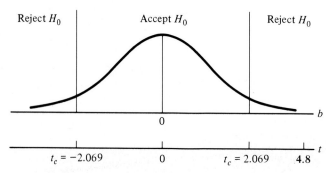

Reject H_0 Accept H_0 Reject H_0

$t_c = -2.069$ 0 $t_c = 2.069$ 4.8

$t_{\text{sample}} = \dfrac{6-0}{1.25} = \mathbf{4.8.}$ **Reject H_0**

e. $6 \pm 2.069(1.25)$ or $\mathbf{6 \pm 2.59}$

INDEX